# Integrating Ecohydraulics in River Restoration

# Integrating Ecohydraulics in River Restoration: Advances in Science and Applications

Special Issue Editors

**José Maria Santos**
**Isabel Boavida**

MDPI • Basel • Beijing • Wuhan • Barcelona • Belgrade • Manchester • Tokyo • Cluj • Tianjin

*Special Issue Editors*
José Maria Santos
University of Lisbon
Portugal

Isabel Boavida
University of Lisbon
Portugal

*Editorial Office*
MDPI
St. Alban-Anlage 66
4052 Basel, Switzerland

This is a reprint of articles from the Special Issue published online in the open access journal *Sustainability* (ISSN 2071-1050) (available at: https://www.mdpi.com/journal/sustainability/special_issues/River_Restoration).

For citation purposes, cite each article independently as indicated on the article page online and as indicated below:

LastName, A.A.; LastName, B.B.; LastName, C.C. Article Title. *Journal Name* **Year**, *Article Number*, Page Range.

**ISBN 978-3-03928-328-6 (Pbk)**
**ISBN 978-3-03928-329-3 (PDF)**

Cover image courtesy of Isabel Boavida.

# Contents

# About the Special Issue Editors

**José Maria Santos** (Assistant Professor) is an Assistant Professor from the School of Agriculture, University of Lisbon, Portugal. He received his Ph.D. (2004) on the topic 'River Regulation Effects on Fish Assemblages and the Role of fish Passes'. He worked as a senior researcher from 2018 to 2019 and from 2008 to 2017 as an assistant researcher of the Forest Research Centre (CEF) of the University of Lisbon. His fields of expertise include ecohydraulics, water resources, river hydraulics and restoration, and habitat modeling. He has co-authored more than 50 peer-reviewed papers in international journals indexed by WoS and Scopus, and more than 25 articles in international conference proceedings. He also works on the editorial boards of two international scientific journals covering water resources and ecohydraulics subjects. He is president of the Committee of the Ecosystems and Water Quality from the Portuguese Association of Water Resources and a member of the International Association of Hydro-Environment Engineering and Research (IAHR).

**Isabel Boavida** (Senior Researcher) is a researcher at CERIS (Civil Engineering Research and Innovation for Sustainability), Instituto Superior Técnico, University of Lisbon. With a background in environmental engineering and hydraulics, she is passionate about understanding fish behavior in rivers due to flow alterations. Her research topics include hydrodynamic and habitat modeling, ecohydrology, ecological flows, river restoration, and sustainable hydropower. Isabel has been involved in research, teaching, and consultancy work in environmental engineering and river management. She has co-authored more than 20 peer-reviewed papers in international journals and more than 30 articles in international conference proceedings. During the last decade, her research focus has been on hydropower impacts regarding fish. She is thrilled to understand the effects of hydropeaking in freshwater fish and propose actions to mitigate those impacts. In this context, she has done an internship at SINTEF Norway, coordinated the EcoPeak project, and participated in the FIThydro project. Isabel is a member of the Committee of the Ecosystems and Water Quality from the Portuguese Association of Water Resources and a member of the Animal Welfare Body of Instituto Superior Técnico.

# Preface to "Integrating Ecohydraulics in River Restoration"

Rivers have been intensively degraded due to increasing anthropogenic impacts from a growing population in a continuously developing world. Conflict demands on freshwater resources, exacerbated by climate change, present a difficult dilemma for scientists and managers: until when and how much can a river (and its natural flow regime) be altered, while still maintaining processes and functions, and guaranteeing sustainable aquatic populations?

Accordingly, most rivers are suffering from pressures as a result of increased dam and weir construction, habitat degradation, flow regulation, water pollution/abstraction, and the spread of invasive species. In addition, it is expected that global warming will further stimulate conflicts in water use, leading to disturbances in river ecosystems.

Science-based knowledge regarding solutions (e.g., environmental flows, dam removal, improvement of fish passes, adoption of fish-friendly hydropower solutions, riparian vegetation management) to counteract the effects of river degradation, and melding principles of aquatic ecology and engineering hydraulics, are thus urgently needed to guide present and future river restoration actions.

This Special Issue gathers a coherent set of studies from different geographic contexts, on fundamental and applied research regarding the integration of ecohydraulics in river restoration, ranging from field studies to laboratory experiments that can be applied to real-world challenges. It contains 13 original papers covering ecohydraulic issues such as river restoration technologies, sustainable hydropower, fish passage designs and operational criteria, and habitat modeling. All papers were reviewed by international experts in ecology, hydraulics, aquatic biology, engineering, geomorphology, and hydrology.

It is, therefore, our pleasure to share these studies with the scientific community, engineers and technicians, private owners, and public authorities, in the hope that the present edition will provide a basis to improve knowledge on river restoration and management and reduce arguments between different interests and opinions.

As Guest Editors, we would like to express our broad gratefulness to MDPI, who agreed to publish this issue, and to the editorial team of *Sustainability*, for their kindness and professional support; thanks are also due to all the reviewers, for improving original manuscripts and to all the authors, for providing their papers with professionalism and scientific rigor.

The papers herein well represent the wide applicability of ecohydraulics in river restoration and serve as a basis to improve current knowledge and management and to reduce arguments between different interests and opinions.

**José Maria Santos, Isabel Boavida**
*Special Issue Editors*

*Article*

# Ecohydraulic Modelling to Support Fish Habitat Restoration Measures

**Ana Adeva-Bustos [1,*], Knut Alfredsen [1], Hans-Petter Fjeldstad [2] and Kenneth Ottosson [3]**

[1] Department of Civil and Engineering, Norwegian University of Science and Technology, 7031 Trondheim, Norway; knut.alfredsen@ntnu.no

[2] SINTEF Energy Research, 7034 Trondheim, Norway; Hans-Petter.Fjeldstad@sintef.no

[3] Hushållningssällskapet, 861 33 Timrå, Sweden; kenneth.ottosson@hushallningssallskapet.se

* Correspondence: ana.adeva.bustos@ntnu.no

Received: 14 December 2018; Accepted: 3 March 2019; Published: 12 March 2019

**Abstract:** Despite that hydromorphological restoration projects have been implemented since the 1940s, the key to improve the effectiveness of future restoration measures remains a challenge. This is in part related to the lack of adequate aims and objectives together with our limitations in understanding the effects on the physical habitat and ecosystems from interventions. This study shows the potential of using remote sensing techniques combined with hydraulic modelling to evaluate the success of physical restoration measures using habitat suitability as a quantifiable objective. Airborne light detection and ranging (LiDAR) was used to build a high-resolution two-dimensional model for Ljungan River, Sweden, using HEC-RAS 5.0. Two types of instream restoration measures were simulated according to the physical measures carried out in the river to improve salmonid habitat: (a) stones and rocks were moved from the bank sides to the main channel, and (b) a concrete wall was broken to open two channels to connect a side channel with the main river. Results showed that the hydraulic model could potentially be used to simulate the hydraulic conditions before and after instream modifications were implemented. A general improvement was found for the potential suitable habitat based on depth, velocity and shear stress values after the instream measures.

**Keywords:** instream; restoration; HEC-RAS 2D; LiDAR; cost-effectiveness; fish habitat

---

## 1. Introduction

Management of restoration action in regulated rivers might be motivated by different drivers. In countries located in North Europe and North America, where the Atlantic salmon (*Salmo salar* L.) plays an important role for both its high economic and conservation value, it is often found that the status of Atlantic salmon will have an important role in guidance of management decisions [1]. Several measures can be applied to maintain and improve Atlantic salmon populations, such as flow related measures (minimum flows, changes in operational strategies), biological measures (re-stocking) and instream measures (habitat modifications) among others. However, particularly in regulated rivers and because Atlantic salmon has a wide range of habitat requirements depending on their life stage [2], implementing effective restoration measures is still a challenge. Most of the habitat modifications measures will depend on the discharge released from the hydropower system to be effective. The difficulty increases in specific seasons when water allocation lead to a conflict between Atlantic salmon requirements and energy demand. In recent years, models that integrate hydrological, hydrodynamic and habitat has shown to be the most appropriate to evaluate habitat suitability for aquatic organisms, since they include physical variables such as depth, velocity, substrate and shelter [3,4].

These models can also help to overcome some of the most common gaps in river restoration management, such as evaluating the outcome of restoration and mitigation measures before their implementation. Benchmarks and the use of endpoints that define project goals are valuable indicators to measure the success of an action, since they are realistic and can be quantified [5]. Hydraulic parameters and their interaction with physical habitat have been used for several years as benchmarks to measure instream restoration for fish habitat. For example, the weighted usable area (WUA) is a well-established method that has been widely used in combination with habitat modelling [6] to predict and quantify physical habitat requirements per unit area. However, physical habitat simulation (PHABSIM) [7] only uses a one-dimensional (1D) routine to calculated water surface elevations, and velocities for each cross-section [8]. Several studies support the use of two-dimensional (2D) models in order to better capture spatial changes in fish habitat parameters such as depth and water velocity with a finer (cell) resolution. Crowder and Diplas [9] used a 2D model to capture changes in depth and velocity after the introduction of boulders and cobbles on a river reach. Lacey J and Millar [8] used a 2D model and combined this with WUA calculations to predict the effect of instream large woody debris and a rock groyne habitat. Boavida et al. [10] used a 2D model to assed the effect in WUA from introducing different instream structures (islands, lateral bays, and deflectors). The accuracy of results from these models benefit from high resolution bathymetry data. Recent studies have shown that of light detection and ranging (LiDAR) bathymetry can be used as a suitable tool for mapping rivers with a high density over large areas [11]. Airborne LiDAR bathymetry (ALB) data also capture elevation points for the entire foreland, including riparian areas, vegetation, ice and snow, which opens the possibility to be used in a wide range of studies [12]. ALB is a fast method for collecting data with high density (>20 points/$m^2$), with an accuracy under water of approximately 5 cm [12], covering rivers of 15–20 km in a few hours, and reaching up to 10 m depth [13]. Whereas conventional methods for mapping bathymetry can provide accurate measurements, they can have limitations due to restricted accessibility, safety precautions and time required [13,14] to fully cover the interested areas. In the other hand, ALB data requires post-processing, including filtering and removal noise and false echoes, water surface detection and correction for the refraction [15]. ALB surveys are affected by environmental conditions such as floods, rain and snow and by water turbidity, since dissolved and suspended organic material affect negatively the river bottom reflection [16]. Despite these drawbacks, ALB it is still considered more cost-effective than conventional methods due to the coverage of data obtained per unit of effort [12,17]. Therefore, the use of ALB data in fish habitat quality models can support a cost-effective design of mitigation and restoration measures, in terms of amount of suitable area created per unit of cost spent and prioritize them based on their performance.

In Sweden, during the last three decades, several river restorations have been carried out, most of them comprised of instream habitat modification measures [18] related to restore river channels that earlier were modified to transport timber. Timber floating was an important activity from ca. 1850 until 1970, and to facilitate the transportation of logs to the coastal mills the channel morphology was simplified by removing boulders and large woody debris from the channel to the river banks. In addition, secondary channels and meander bends were cut off by the construction of stone and wood levees [19]. The removal of larger stones and other obstacles and elimination of eddies and side channels has led to a loss of structural complexity and simplified flow patterns [20], which has had a profound negative impact on stream-dwelling fish and invertebrates [21] as habitat niches were removed and primary production was limited [22]. Johansson [23] found that channelization affected both fish abundance as well as species richness and composition. Findings have shown a general decreased in fish species that depend on flowing water for food, shelter, spawning and movement between different habitats. Today, 98% of the Swedish salmon rivers are affected by the modification from timber floating channelization, hydropower development and agricultural areas [18]. The loss of habitat is considered one of the major threats to fish biodiversity [24] and Sweden has around 20,000 km of rivers affected by timber channelization [22]. Based on a literature review, Nilsson et al. [25] provide a summary of the effects from implementing instream measures to restore rivers that were used for timber floating. They suggested the following main variables for geomorphology and hydraulic responses: increased channel area, increased water depth and reduction of velocities, and for ecological responses: increased habitat complexity for riparian and aquatic organisms.

In this study, hydraulic parameters and their interactions with physical habitats were used as a benchmark to evaluate the impacts from instream measures carried out in Ljungan river in Sweden. In the past, Ljungan was heavily modified for timber floating. In addition, Ljungan is extensively regulated for hydropower production. Even though, salmon and sea trout reproduce in a 19 km reach from the river mouth to the most downstream hydropower plant at Viforsen [26]. In 2015, a stakeholders group was established to improve the communication between the different interest groups in the river, including power producers, non-governmental organisations (NGOs), and the local county. Today, the stakeholders group has carried out several instream restoration measures. They have concentrated their efforts on restoring the hydromorphology to the state it had before the timber floating modifications. Therefore, the term instream restoration is used to refer to the instream modification carried out in stream habitat to recreate the physical habitat conditions that characterized the stream habitat before channelization.

This study aims to demonstrate that the use of modelling techniques supported by remote sensing data is a valuable method to plan and evaluate the success of instream restoration and mitigations measures. In order to fulfil this, the following objectives were pursued: (a) to create a 2D hydraulic model for both the situation before and after the instream modifications that adequately simulated the physical parameters (depth, velocity and shear stress), (b) to evaluate the physical parameters obtained from the hydraulic model in term of potential suitable areas for salmon and (c) combine the cost of the instream modifications with their effectiveness (in terms of potential suitable area created) to calculate the cost-effectiveness of the measures. The method presented aims to show that modelling tools with support from modern data surveying could help to decide and prioritize where to place and how to design instream measures. Calculating the cost-effectiveness for the measures is done with the purpose to share knowledge and experiences and promote this type of methodologies. Future analyses combining the biological data from monitoring and further physical measured values to contrast modelling data with measured data will validate and reinforce the potential of this method to help stakeholders, managers and decision makers to reduce the uncertainty during the planning process.

## 2. Materials and Methods

### 2.1. Study Area

The Ljungan River originates on the Norwegian border and runs through the middle part of Sweden before it reaches the Gulf of Bothnia. Its total length is 399 km with a catchment area of 12,851.1 km$^2$ and a total of 15 power plants, where Viforsen power plant is the most downstream in the system (Figure 1). The instream restoration measures carried out by the stakeholders were located at three different locations: Grenforsen (Gren), Allstaforsen (Allsta) and Nolbystrommen (Nolby). The three locations were selected by the stakeholders group judged by their potential to improve the salmonid habitat quality after restoration measures were in place (Figure 1).

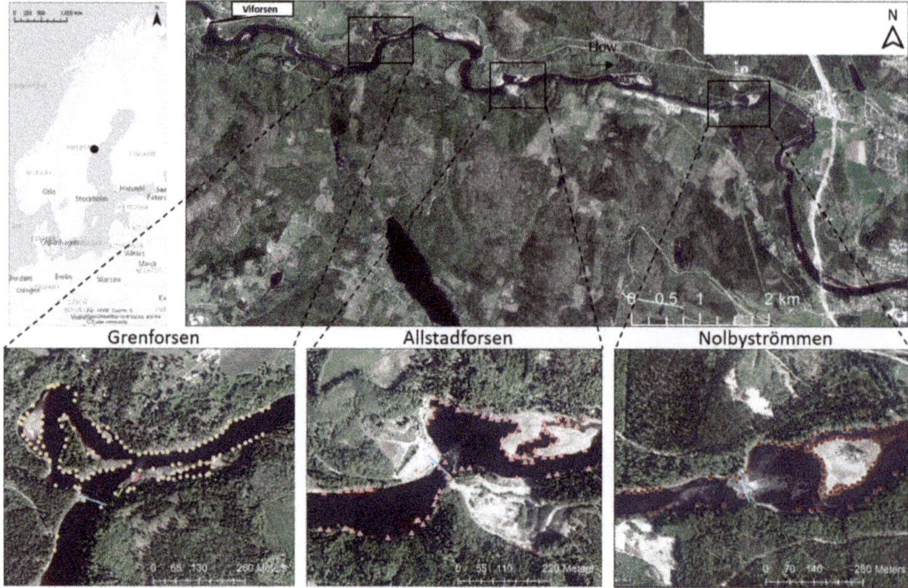



**Figure 1.** Ljungan River and the three locations in which restoration measures have been carried out. Coloured lines represent cross sections explained in Section 2.3 (Figure 2). Points, triangles and squares in Grennforsen, Allstaforsen and Nolbystromen, respectively, represent the measured water edge and are used in Section 3.1 (Figure 3) to verify the hydraulic model.

## 2.2. Terrain Modification

Bathymetry data were collected during an airborne LiDAR bathymetry (ALB) survey. It was conducted on 2 September 2015 by the company airborne hydro mapping (AHM), Austria, with the RIEGL VQ-880 G green laser camera [27] and lasted for 2–3 h to survey approximately 19 km. The total amount of ground points captured was 1,518,500, and it was delivered as cross sections with 5-m average distance. These ground points were already filtered by AHM who removed the raw data noise originated from the laser being scattered by birds, clouds, dust and other particles. The filtering process involved both automatic and manual filtering (see [15] for more details). In addition, vegetation was also removed from the point cloud by AHM in the pre-processing step. The survey was carried out with a measured flow of 58.9 $m^3$ $s^{-1}$ with an accuracy of 0.07 m for planar coordinates, and 0.03–0.04 m for mean vertical accuracy obtained from comparing LiDAR elevation points with manual measurements [13]. The maximum average depth reached was 2.8 m restricted by the dark bottom and organic material in the water. Therefore, additional manual data (14.190 points) were collected from the river bed and banks using a Sontek RiverSurveyor M9 acoustic Doppler profiler (ADCP) [28] equipped with a differential GPS system. The ADCP was mounted on a floating platform towed by a kayak and used to capture bathymetry points from Viforsen and 19 km downstream to the end of the area covered by LiDAR data. In addition, the ADCP was also used with a small rowing boat to survey additional points in Allstaforsen [13]. The ADCP surveys were carried out following a pre-specified route that was mapped based on the missing LiDAR data, however the precision to capture all the missing areas was subjective to the individual performance and to the external conditions, including security. In both cases, the GPS antenna system was used to capture the XY coordinates, whereas the ADCP was used to collect the bathymetric data with a sampling frequency of 1 Hz from the nine individual transducers which define the channel definition with an accuracy of 1% [28] and gives input to further development of a digital elevation model (see [13]). The LiDAR and ADCP points were combined into a point cloud used as the input to the empirical Bayesian kriging interpolation method using the average method for overlapping points in ArcGIS [29] to create a digital elevation model (DEM) with a resolution of 1 by 1 m. The DEM obtained was representative of the situation before the instream modifications were carried out. In order to simulate the situation after the modifications, a second DEM was created including the terrain modifications (Figure 2). These modifications were modelled by altering the DEM using ArcGIS and the raster editor (ArcMap Raster Edit) which allows changing the values of specific points in a raster. Three reaches in Ljungan river were modelled: Gren, Allsta and Nolby (Table 1). Two main measures were simulated: (a) Stones and rocks were moved from the banks to the main channel in Gren, Allsta, and Nolby and (b) a concrete wall was broken to open two channels: Gren S.Ch 1 and Gren S.Ch 2 (Figure 2).

**Table 1.** Name for the six scenarios simulated at the three locations, their status before and after modifications and the objective to fulfill after the modifications.

| Location | Sub Location | Before Modifications | After Modifications | Objective |
|---|---|---|---|---|
| Gren | Gren M.Ch | Narrow channel with high banks | Wider channel, rocks that were on the banks were placed in the middle. Gravel and cobbles were added. | Reduce water velocities, increase the wetted area and create suitable habitat for spawning. |
| | Gren S.Ch | Concrete wall was blocking water to flow in the right-side channel under low flows | Wall was opened in two channels (Gren. S.Ch 1 & Gren. S.Ch 2) so water could flow inside the right-side channel, even at low flows | Restore the right-side channel and its function as a nursery area as well as to restore connectivity. |
| | Allsta | Narrow channel with higher elevations in the banks | Wider channel, rocks that were on the banks were placed in the middle. Gravel and cobbles are added. | Reduce water velocities, increase the wetted area and create suitable habitat for spawning. |
| | Nolby | Narrow channel with higher elevations specially in the right-side bank | Wider channel, rocks that were on the right-side banks were placed in the middle. Gravel and cobbles are added. | Reduce water velocities, increase the wetted area and create suitable habitat for spawning. |

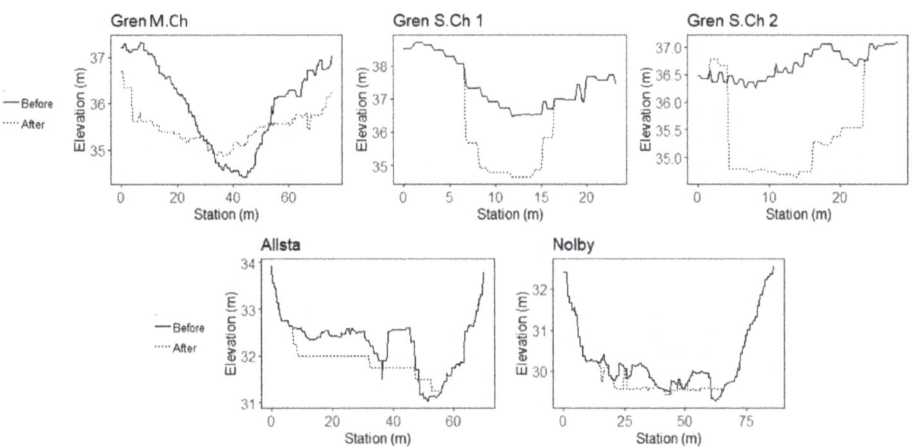


**Figure 2.** Elevation along the cross sections (colored lines in Figure 1) extracted from the digital elevation model (DEM) before (solid line) and the DEM after (dashed line) habitat adjustments in the areas were modifications took place. Figure 1 shows the cross-section lines location, colors correspond to Gren M.Ch in blue, Gren S.Ch 1 in green and Gren S.Ch 2 in pink in Grenforsen. Allsta has one in blue, and in Nolbystromen, Nolby in blue.

### 2.3. Hydraulic Modeling

A two-dimensional hydraulic model with cell size of 1 by 1 m was developed for each of the three locations using the HEC-RAS 5.0 software developed by US Army Corps of Engineers [30]. The model before the restoration measures (reference model) was calibrated for a discharge of 58.9 $m^3$ $s^{-1}$ corresponding to the discharge measured during the ALB survey. The difference between the water surface elevation simulated and the values delivered by AHM based on the LiDAR data were used for the calibration. In order to validate the situation after the modifications, a simulation was done for the observed discharge on aerial pictures (65.4 $m^3$ $s^{-1}$), and the wetted area extent from the simulation results and the water edge from the aerial picture from 2017 when the instream modifications were already in place were visually compared. The successful of the instream measures were considered under the premises that the wetted area results will be as expected under the objectives (Table 1) and in addition they will match the water covered area extent after the modifications from the aerial picture provided by Lantmäteriet (www.lantmateriet.se) with 0.25 m planar resolution.

The discharges used for the hydraulic simulations before and after instream modifications were selected based on the following criteria: discharges that are dominant during the spawning season (60 $m^3$ $s^{-1}$ and 100 $m^3$ $s^{-1}$), 138 $m^3$ $s^{-1}$ is the average flow in Ljungan and 380 $m^3$ $s^{-1}$ is the average one-day maximum discharge (Table 2). In addition, because in Gren one of the measures was designed to reconnect the side channel with the main channel also on lower flows, low discharges that could be observed particularly during summer months were also analyzed. In order to provide a detailed coverage of low discharges and due to wetted area changes in a more pronounced way at low flows changes, four discharges were selected: 20 $m^3$ $s^{-1}$, 30 $m^3$ $s^{-1}$, 35 $m^3$ $s^{-1}$, and 40 $m^3$ $s^{-1}$. These discharges were used as inputs for the upper boundary condition. In addition to the 1 by 1 m cells, break lines were included in areas were higher resolution was needed (such as along river banks, islands and side channels). Crowder and Diplas [9] showed the importance of analyzing effects at a finer scale, such as the close surrounding area after placing boulders in the river. Forcing the break lines in the mesh produced a mesh with different dimensions. Normal depth was specified for the lower boundary condition, the average channel slope at the downstream part of the reach was used as an approximation of the friction slope. For the river bed roughness, Manning's n coefficients ranged from n = 0.03 (channel with gravels and cobbles) to n = 0.15 (channel with bushes and higher resistance) [31].

**Table 2.** Parameters used for the hydraulic simulations in each reach.

| Reach | Discharge ($m^3$ $s^{-1}$) | # of Cells | Dimensions ($m^2$) | Normal Depth (m) | Manning's [1] |
|---|---|---|---|---|---|
| Gren | 20, 30, 35, 40, 60, 100, 138, 380 | 364.436 | Max: 1.92 $m^2$<br>Min: 0.01 $m^2$<br>Avg: 0.90 $m^2$ | 0.01 | 0.06 |
| Allsta | 60, 100, 138, 380 | 147.229 | Max: 1.73 $m^2$<br>Min: 0.34 $m^2$<br>Avg: 0.99 $m^2$ | 0.001 | 0.03, 0.06, 0.15 |
| Nolby | | 223.121 | Max: 1.74 $m^2$<br>Min: 0.05 $m^2$<br>Avg: 0.93 $m^2$ | 0.001 | 0.06, 0.08, 0.15 |

[1] See Appendix A, Figure A1.

### 2.4. Depth, Velocity and Shear Stress Distribution and Potential Suitable Area

Water-surface elevation, depth, velocity and shear stress values were extracted as average point values for each cell in the mesh for discharges ranging from 20 $m^3$ $s^{-1}$ to 380 $m^3$ $s^{-1}$ (Table 2) before and after the modifications. An initial comparison for the situation before and after modification for the full range of parameters (depth, velocity and shear stress) was carried out. Analyses of the potential suitable area (PSA) were carried out using literature data on preferred ranges of habitat for juvenile Atlantic salmon [2,32] and available physical habitat data (spawning areas location, substrate composition, shelter distribution) that were surveyed and mapped by Uni Research in 2014 in Ljungan [33]. Physical habitat data from field measurements were used to compare and support the data obtained from literature [2,32]. The average depths and velocities simulated from the hydraulic model were exported to GIS tools and extracted at the spawning locations [33] under the average spawning discharge conditions. The same was carried out for the nursery areas. After obtaining the simulated average depths and velocities in the studied areas, the data were compared to the ones obtained from literature. This comparison showed that the simulated values agreed with the ranges from literature, except for spawning area depths. Simulated values in Ljungan river could go up to 2 m, in contrast to the values from Armstrong and Kemp [2] and Forseth and Harby [32], which did not exceed 1.5 m. Therefore, the depth range used to identify the potential suitable area was increased accordingly. PSA was calculated as the number of square meters for depth, velocity and shear stress values that fell inside the range considered suitable (Table 3). PSA was also calculated and related to the total wetted area to obtain the percentage of PSA (PSA%). Considering the uncertainties related to habitat results from the hydraulic model and in addition the lack of detailed and observed depth, velocities and critical shear stress values in the field, the analyses of the PSA were considered separately as suggested by Scruton et al. [34]. Therefore, PSA were calculated for depth, velocity and critical shear stress individually instead of weighting and summing them into an overall PSA. Critical shear stress was included under the assumption that sediment mobility for a given particle size occurs when the bed shear stress exceeds the critical shear stress [35]. Values were selected according to the predominant substrate type in the areas [33].

**Table 3.** Values used to determine the potential suitable area based on literature data [2,32] and field data [33].

|  | **Spawning Area** | **Nursery Area** |
|---|---|---|
| Depth (m) | 0.3–2.0 | 0.05–0.9 |
| Water velocity (m s$^{-1}$) | 0.3–0.8 | 0.06–0.9 |
| Critical shear Stress (N/m$^2$) | 12.2 | 53.8 |

*2.5. Calculation of Costs Per Unit of Potential Suitable Area*

The cost of the modifications at each location were obtained from the project budget (Table 4). The total cost at each site was used to calculate the cost to create a unit of potential suitable area. This was used to compare the cost and the potential effectiveness of the modifications within sites and within the type of habitat created based on depth, velocity and shear stress values.

**Table 4.** Costs in EUR per location and action. Values were converted from Swedish Kroner to Euro using the annual average exchange rate for 2016 (0.105653917).

|  | **Excavator** | **Helicopter** | **Gravel 1–10 cm** | **Cobbles 10–100 cm** | **Coarse Cobbles 50–100 cm** | **Total** |
|---|---|---|---|---|---|---|
| Gren M.Ch | 2208 | 24,089 | 6551 | 0 | 1310 | 34,158 |
| Gren S.Ch | 2208 | 0 | 0 | 0 | 0 | 2208 |
| Allsta | 2504 | 24,089 | 6551 | 0 | 1310 | 34,454 |
| Nolby | 5404 | 12,045 | 3275 | 0 | 655 | 21,379 |
| Total | 12,324 | 60,223 | 16,376 | 0 | 3275 | 92,199 |

## 3. Results

### 3.1. Calibration & Verification

Calibration for the situation before modifications was considered good when the correlation between the observed water surface elevation and simulated water surface elevation exceeded 0.85 ($R^2 \geq 0.85$). After modification, the verification was considered good when water surface extent from the simulated results matched the situation observed from aerial pictures (Figure 3). Based on the results from calibration, the hydraulic model setup for both the before and after situation were considered adequate for simulating the effects from habitat modification on depth, velocities and shear stress.

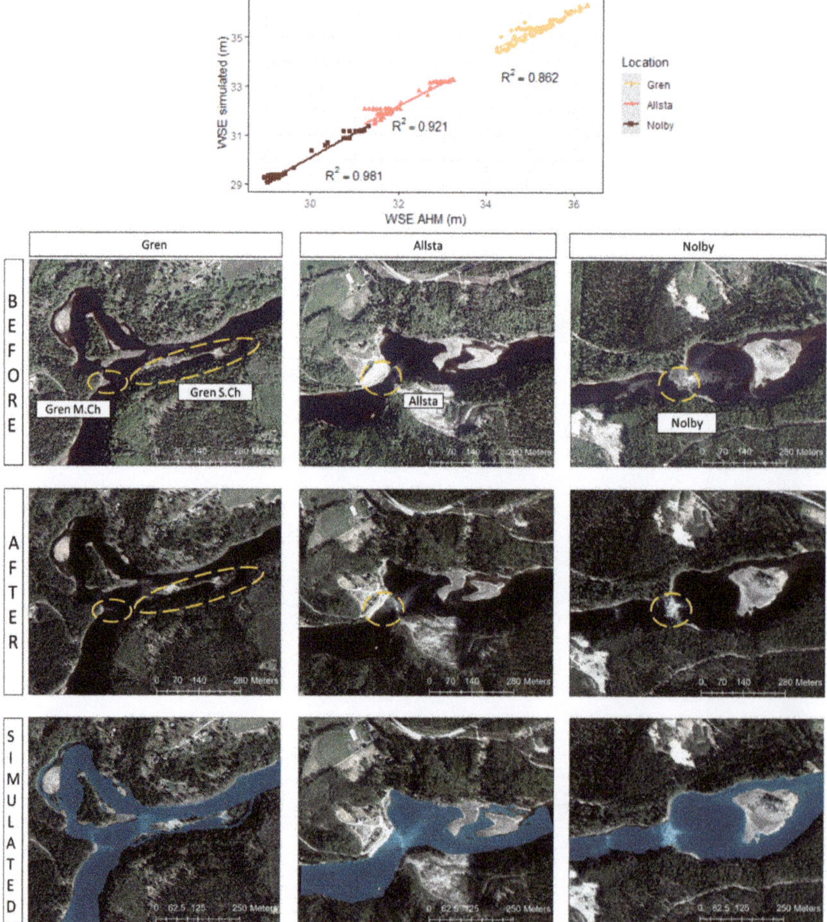


**Figure 3.** Calibration for the hydraulic model before (upper graph) showing water surface elevation simulated (WSE simulated) against water surface elevation measured by airborne hydro mapping (WSE AHM). The three panels (lower figures) show the visual verification of the hydraulic model after modifications at Gren, Allsta and Nolby. Aerial pictures from before and after modifications and simulated water surface for three locations are presented. Blue color is the water surface extent obtained from the hydraulic model and overlap by the aerial picture after modifications. Names for the sub-locations are shown, and the areas analyzed are marked with orange circles.

### 3.2. Depth, Velocity and Shear Stress Distribution

After modifications, the range of distributions for depth, velocity and shear stress was reduced. The lower values found before the modification increased after the modifications, and the higher values were reduced. This was found at Gren. M.Ch (Figure 4), Allsta (Figure 5) and Nolby (Appendix B. In Gren S.Ch (Figure 6), results showed increased values for the three parameters after modifications.

Changes in the distribution of depth and velocity, in relation to the range of potential suitable area (vertical lines) showed that the percentage of cells for discharges from 60 m$^3$ s$^{-1}$ to 138 m$^3$ s$^{-1}$ inside the specified values increased at Gren M.Ch (Figure 4), but this is not the case for the high discharge at 380 m$^3$ s$^{-1}$. The same results were found for Allsta (Figure 5) and Nolby (Appendix B). Changes in shear stress values were not that significant at any location. In Gren S.Ch (Figure 6), a general increase for the percentage of cells after instream modification was found under all discharges and for the three parameters (depth, velocity and shear stress).

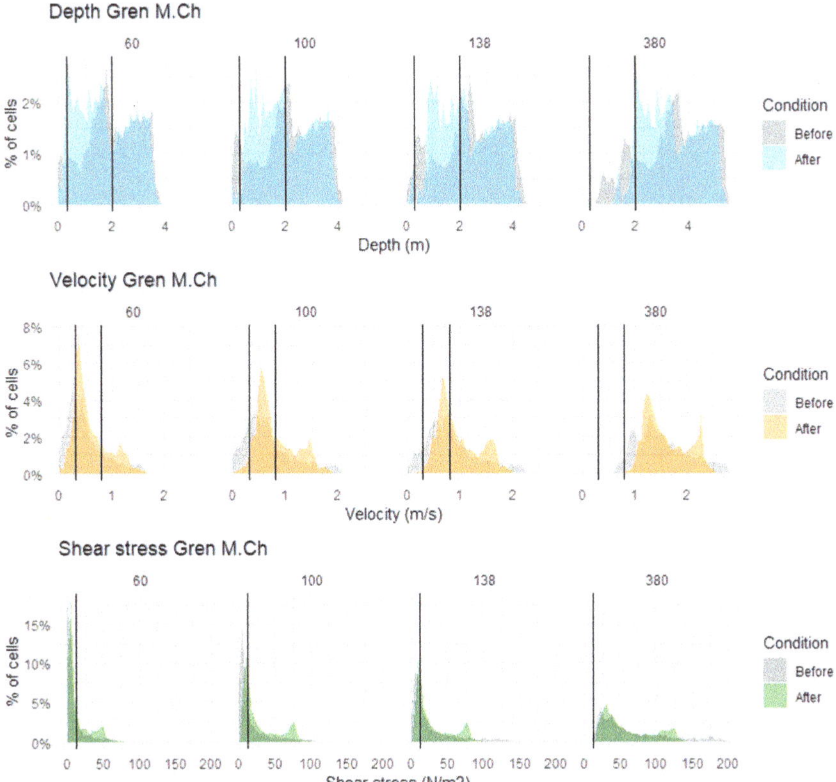


**Figure 4.** Percentage of cells for values of depth, velocity and shear stress in Gren M.Ch. for the four different simulated discharges. Vertical lines indicate the limits for the suitable range (Table 3). Darker areas appear as a result of overlapping the before and after graphs.

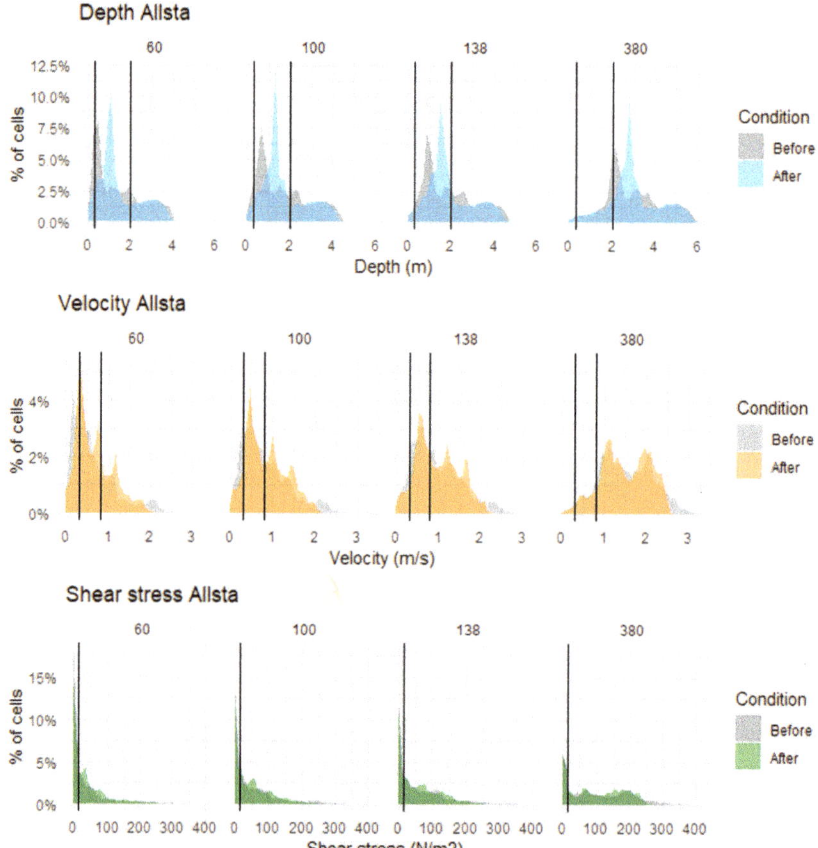


**Figure 5.** Percentage of cells for values of depth, velocity and shear stress in Allsta for the four different simulated discharges. Vertical lines indicate the limits for the suitable range (Table 3). Darker areas appear as the result of overlapping the before and after graphs.

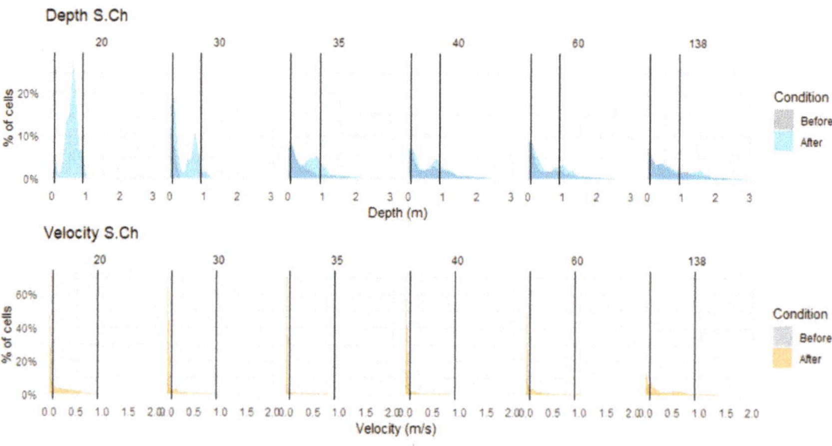


**Figure 6.** *Cont.*

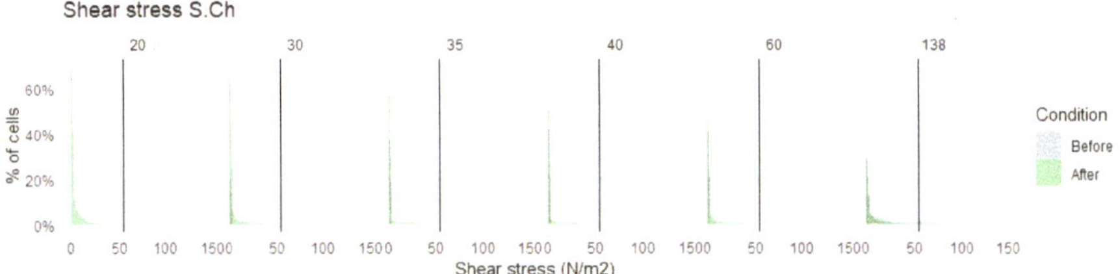



**Figure 6.** Percentage of cells for values of depth, velocity and shear stress in Gren S.Ch for the six different simulated discharges. Vertical lines indicate the limits for the suitable range (Table 3). Darker areas appear as the result of overlapping.

Wetted areas increased at all locations and with increasing discharges after the instream modifications (Figure 7). PSA% showed an increase for depth after modifications at all locations, for velocities it also improved at all locations, especially in Gren S.Ch. Shear stress PSA% values decreased after modification at all sites except in Gren S.Ch.

### 3.3. Cost Per Unit of Potential Suitable Area

The cost per unit of PSA (Figure 7) showed that the creation of PSA for depth values was the cheapest at all locations except in Grenn M.Ch, with 380 m$^3$ s$^{-1}$ as the discharge that showed the most expensive values Velocity costs showed a similar pattern as depth. Shear stress was the most expensive PSA, and Nolby exhibited the highest cost per square meter.

**Grenn M.Ch**

| Condition | Discharge (m³·s⁻¹) | WA (m²) | Depth | | | Velocity | | | Shear Stress | | |
|---|---|---|---|---|---|---|---|---|---|---|---|
| | | | PSA (m²) | PSA % | Cost (EUR/m²) | PSA (m²) | PSA % | Cost (EUR/m²) | PSA (m²) | PSA % | Cost (EUR/m²) |
| Before | 60 | 1079 | 494 | 45.78 | 0.00 | 469 | 43.47 | 0.00 | 835 | 77.39 | 0.00 |
| | 100 | 1164 | 413 | 35.48 | 0.00 | 577 | 49.57 | 0.00 | 748 | 64.26 | 0.00 |
| | 138 | 1197 | 365 | 30.49 | 0.00 | 541 | 45.20 | 0.00 | 559 | 46.70 | 0.00 |
| | 380 | 1254 | 221 | 17.62 | 0.00 | 45 | 3.59 | 0.00 | 10 | 0.80 | 0.00 |
| After | 60 | 1226 | 679 | 55.38 | 50.31 | 717 | 58.48 | 47.64 | 823 | 67.13 | 41.50 |
| | 100 | 1240 | 662 | 53.39 | 51.60 | 715 | 57.66 | 47.77 | 518 | 41.77 | 65.94 |
| | 138 | 1254 | 554 | 44.18 | 61.66 | 596 | 47.53 | 57.31 | 478 | 38.12 | 71.46 |
| | 380 | 1254 | 95 | 7.58 | 359.56 | 0 | 0.00 | 0.00 | 0 | 0.00 | 0.00 |

**Allsta**

| Condition | Discharge (m³·s⁻¹) | WA (m²) | Depth | | | Velocity | | | Shear Stress | | |
|---|---|---|---|---|---|---|---|---|---|---|---|
| | | | PSA (m²) | PSA % | Cost (EUR/m²) | PSA (m²) | PSA % | Cost (EUR/m²) | PSA (m²) | PSA % | Cost (EUR/m²) |
| Before | 60 | 9043 | 4951 | 54.75 | 0.00 | 3843 | 42.50 | 0.00 | 4532 | 50.12 | 0.00 |
| | 100 | 9514 | 5529 | 58.11 | 0.00 | 3974 | 41.77 | 0.00 | 3543 | 37.24 | 0.00 |
| | 138 | 9758 | 5516 | 56.53 | 0.00 | 3765 | 38.58 | 0.00 | 3098 | 31.75 | 0.00 |
| | 380 | 10452 | 2183 | 20.89 | 0.00 | 1061 | 10.15 | 0.00 | 2151 | 20.58 | 0.00 |
| After | 60 | 9169 | 6122 | 66.77 | 5.63 | 4286 | 46.74 | 8.04 | 3843 | 41.91 | 5.56 |
| | 100 | 9458 | 6314 | 66.76 | 5.46 | 4009 | 42.39 | 8.59 | 2960 | 31.30 | 7.22 |
| | 138 | 9682 | 6218 | 64.22 | 5.54 | 3555 | 36.72 | 9.69 | 2687 | 27.75 | 7.96 |
| | 380 | 10434 | 1523 | 14.60 | 22.62 | 970 | 9.30 | 35.52 | 2129 | 20.40 | 10.04 |

**Grenn S.Ch**

| Condition | Discharge (m³·s⁻¹) | WA (m²) | Depth | | | Velocity | | | Shear Stress | | |
|---|---|---|---|---|---|---|---|---|---|---|---|
| | | | PSA (m²) | PSA % | Cost (EUR/m²) | PSA (m²) | PSA % | Cost (EUR/m²) | PSA (m²) | PSA % | Cost (EUR/m²) |
| Before | 20 | 0 | 0 | 0.00 | 0.00 | 0 | 0.00 | 0.00 | 0 | 0.00 | 0.00 |
| | 30 | 368 | 218 | 59.24 | 0.00 | 0 | 0.00 | 0.00 | 107 | 29.08 | 0.00 |
| | 35 | 2603 | 1882 | 72.3 | 0.00 | 12 | 0.46 | 0.00 | 232 | 8.91 | 0.00 |
| | 40 | 4048 | 2310 | 57.07 | 0.00 | 40 | 0.99 | 0.00 | 697 | 17.22 | 0.00 |
| | 60 | 4772 | 2719 | 56.98 | 0.00 | 99 | 2.07 | 0.00 | 436 | 9.14 | 0.00 |
| | 138 | 7893 | 4361 | 55.25 | 0.00 | 1635 | 20.71 | 0.00 | 7617 | 96.5 | 0.00 |
| After | 20 | 792 | 751 | 94.82 | 2.94 | 272 | 34.34 | 8.12 | 450 | 56.82 | 4.91 |
| | 30 | 1726 | 1331 | 77.11 | 1.66 | 371 | 21.49 | 5.95 | 690 | 39.98 | 3.20 |
| | 35 | 4435 | 3064 | 69.09 | 0.72 | 453 | 10.21 | 4.87 | 1030 | 23.22 | 2.14 |
| | 40 | 5335 | 3221 | 60.37 | 0.69 | 869 | 16.29 | 2.54 | 2105 | 39.46 | 1.05 |
| | 60 | 6634 | 3645 | 54.94 | 0.61 | 2012 | 30.33 | 1.10 | 4772 | 71.93 | 0.46 |
| | 138 | 8930 | 4807 | 53.83 | 0.46 | 6292 | 70.46 | 0.35 | 8751 | 98 | 0.25 |

**Nolby**

| Condition | Discharge (m³·s⁻¹) | WA (m²) | Depth | | | Velocity | | | Shear Stress | | |
|---|---|---|---|---|---|---|---|---|---|---|---|
| | | | PSA (m²) | PSA % | Cost (EUR/m²) | PSA (m²) | PSA % | Cost (EUR/m²) | PSA (m²) | PSA % | Cost (EUR/m²) |
| Before | 60 | 1865 | 1565 | 83.91 | 0.00 | 1277 | 68.47 | 0.00 | 188 | 10.08 | 0.00 |
| | 100 | 1918 | 1830 | 95.41 | 0.00 | 941 | 49.06 | 0.00 | 83 | 4.33 | 0.00 |
| | 138 | 1943 | 1859 | 95.68 | 0.00 | 608 | 31.29 | 0.00 | 89 | 4.58 | 0.00 |
| | 380 | 2038 | 536 | 26.30 | 0.00 | 283 | 13.89 | 0.00 | 18 | 0.88 | 0.00 |
| After | 60 | 1915 | 1722 | 89.92 | 12.42 | 1542 | 80.52 | 13.86 | 153 | 7.99 | 139.73 |
| | 100 | 1989 | 1847 | 92.86 | 11.58 | 1180 | 59.33 | 18.12 | 32 | 1.61 | 668.10 |
| | 138 | 2001 | 1944 | 97.15 | 11.00 | 824 | 41.18 | 25.95 | 19 | 0.95 | 0.00 |
| | 380 | 2038 | 248 | 12.17 | 86.21 | 275 | 13.49 | 77.74 | 0 | 0.00 | 0.00 |

**Figure 7.** Wetted area (WA), potential suitable area (PSA) and potential suitable area percentage (PSA%) calculated as the percentage of PSA related to the WA (blue colors are for higher values and red for lower for (PSA%) and cost (Cost) per unit of potential suitable area (grey bars are used as a cost indicator). These are calculated for each location (grey rectangle), for depth, velocity and shear stress values under the discharges evaluated. When PSA was 0, costs has not been considered.

## 4. Discussion

Several concepts, tools and new techniques are used over recent decades for planning and evaluation of the feasibility of mitigation and restoration projects in rivers. The availability of high resolution bathymetry data provide the opportunity to obtain physical parameters such as depth or velocities values in a finer spatial scale distribution from 2D models compared with previous analyses carried out with 1D models, which has been reported to be more appropriate for fish habitat studies [4,9]. Despite that most of the habitat suitability studies are based on average-point depth and velocities. Pisaturo et al. [36] found that 3D modelling for habitat suitability gave significantly different results compared with 2D modelling results. This could be important if habitat suitability indexes were based on e.g., bottom flow velocity [37]. On the other hand, the 3D modelling will incur higher computational costs [38]. Based on the need for efficient computation and since we do not have the necessary fish habitat data to utilize the extra information provided by the 3D model, we decided to run a 2D simulation in this project. The 2D modelling approach has been proved to successfully simulate the water covered areas and flow patterns which is important for planning mitigation measures. The same model setup has also been shown to be efficient in evaluating the drying out areas at different flow regimes [39], which is also relevant for the study in Ljungan. This study used LiDAR bathymetry data to build a hydraulic model and analyze the depth, velocity and shear stress distributions which were used as a benchmark to evaluate the success of instream measures carried out in Ljungan river. The two types of instream measures simulated, returning stones and rocks to the river channel and opening a wall to reconnect a side channel with the main river, showed that spawning and nursery areas were potentially improved after the modifications. In addition, it also showed that Allsta was the location where the instream modification was most cost-effective. The procedure presented in this study could be used to design mitigation and restoration measures during the planning process for anticipated impacts but also in future restoration measures to improve the effectiveness aiming to improve fish habitat conditions.

### 4.1. Hydraulic Responses and PSA

The instream modifications aimed to improve the habitat for Atlantic salmon spawning showed an increase of water depth values and a reduction of flow velocity values comparing to the situation before modification. These findings were also observed by Gardeström et al. [40] who analyzed the abiotic effects from restoring a channelized river in Sweden, and found that after restoration, water velocities were significantly reduced and channels were wider. These effects can be related to the changes in the geomorphology, as explained by Nilsson and Lepori [25], after the reduction in the bank elevations and the return of stones and rocks to the channel, the channel structure will be more complex and so roughness and flow resistance will increase. This leads to a reduction in flow velocity and to an increase in the wetted area. Water depth could increase or decrease depending on the relative changes done in the channel. According to the values established from literature and field data [2,32] the suitable area for water depth and velocities increased at all sites for the discharges expected during the spawning season and also for the average discharge. The area showing values under the critical shear stress range chosen by literature [35] decreased after the measures at all sites. Despite values for shear stress PSA after the measures were reduced, still they exhibited a PSA% similar to the one for velocities, which was also observed for the velocity and shear stress distribution range. The same results were found by Bair [41], where the range of shear stress and velocities were reduced after large wood were placed in the river. The shear stress results and analyses for PSA presented in this study could be considered as conservative because of the following reasons: (a) Wilcock and McArdell [42] observed partial mobility of the bed to occur between critical shear stress value and twice the shear stress value, and full mobility was observed above twice the shear stress value, while the PSA in this study was calculated based just on values lower than the critical shear stress, (b) the critical value selected from literature was the minimum value indicated within a wider range for coarse gravel for the spawning area and the highest value for very coarse gravel for the nursery area, and (c) the shear stress obtained from the hydraulic model is an average value for each cell face instead of the river bed shear stress.

The instream modifications aimed to restore the nursery area reconnecting the side channel and showed increased values at Gren S.Ch for water depth, velocities and shear stress as discharge also increased. This is in agreement with Nilsson and Lepori [25], who described the predicted changes after the reconnection of cut-off side channels as increased channel area, increased total flow resistance and water volume. The largest changes were seen at lower flows, and this was expected since the channel was only disconnected under low flow conditions. At higher flows, water was still flowing into the channel from a small area with lower elevation some meters downstream of the concrete wall also before the modification. Based on the literature data to evaluate the increase in suitability nursery habitat [2,32], instream modification has led to an increase in the nursery habitat for water depth, velocity and shear stress. However, even if velocities showed an improvement after modifications the amount of potential suitable habitat created after modifications for the velocity is small compared with the amount of habitat for depth. Unlike most of the WUA studies, in this study the preferred values were considered by separate [34], however, this results may indicate that the potential suitable nursery area is not appropriate if velocities are not increased. It is also important to highlight that these values contain the uncertainties related to an inappropriate representation of the instream modification from the terrain modifications, and as for the other sites, further field data collection will be needed to corroborate these results.

*4.2. Expected Ecological Responses*

Several studies support that restoration measures will generally have a positive effect on fish production [43,44]. Restoration measures for spawning areas has been reported as a success by Gard [45]. They used a 2D hydraulic model and predicted an increase of WUA for salmon spawning, which was also supported by the biological monitoring data. Fjeldstad et al. [46] used 2D hydraulic modelling techniques to predict that the removal of weirs and the addition of spawning gravel would create favorable conditions for Atlantic salmon spawning, which was also corroborated with biological data. High shear stress values could cause gravel to be flushed away and consequently, salmon eggs could be scoured, however, McKean and Tonina [47] found that even at higher shear stress values a lower amount of gravel was found to be mobile at high flows. In addition, they also discussed that salmon eggs are usually buried 15–50 cm below the streambed surface, protecting the eggs to be flushed away. The reconnection of the side channel could provide suitable refuge habitat for juvenile fish [2,25]. Using a 2D hydraulic model, Koljonen et al. [48] predicted an increase in weighted usable habitat for juveniles, which was related to an observed increase of juveniles densities. However, difficult winter conditions overrode the density improvements for the next summer juveniles. Gard [45] found lower fry densities after modification than before the modifications in rearing habitats, but they suggested that their model could be used to design additional instream modifications, such as the addition of boulders and the construction of a side channel, could increase the shelter and would modify the depth and velocities values to the ones preferred by fry and juvenile salmonids. Bair [41] found an increase in the suitable habitat and shelter for salmon juveniles after large pieces of wood were added to the river bed. At the same time, shear stress values and velocity decreased. Despite these positive results, Palmer et al. [49] reported from an extensive literature review that no clear evidence for biological improvement was found after instream restorations, suggesting that a reason could be the deterioration of the habitat created after the instream measures. However, Marttila et al. [50] found that long term changes in habitat structure after restoration either remained unaltered or were reinforced through time. They were able to discard the hypothesis that the low biological improvements found after instream restoration were related to a long-term deterioration of the habitat. Therefore, based on our results for potential suitable areas and the literature findings, there are reasons to believe that future analyses of monitoring data will corroborate that the restoration measures in Ljungan improved the usable nursery and spawning areas in the river as a result of the implemented habitat adjustments.

### 4.3. Cost-Effectiveness

In terms of cost-effectiveness for the location of the instream measure, Allsta was the most cost-effective site where the most suitable areas were obtained for the money spent. In addition, creating suitable ranges of depth was found to be the most cost-effective measure for both spawning and nursery areas. Based on these results, Allsta and Gren S.C have shown to be the areas to prioritize for cost-effective restoration measures. However, it is important to notice that further instream works might be needed to improve the PSA for velocity values in the side channel. The term effectiveness in this study has been used to describe the effectiveness of instream measures to improve habitat conditions. However, benefits cannot be calculated until data from biological monitoring is available after the modifications in River Ljungan.

Re-stocking measures, which consist of the release of young salmon in order to compensate for the loss of habitat and migration corridors, has been carried out in many hydropower regulated rivers [51,52], but they are usually reported as a costly and ineffective mitigation measure because young salmon die in high numbers before maturing to spawning adults. [51,53] In Ljungan, re-stocking of salmon was carried out in the past, but in the last 15 years, salmon was re-stocked only in 2004 with 13,200 young salmon [54]. The cost per salmon release for re-stoking in Ljungan after bread from egg to two-year-old smolt is 2.21 EUR (21 SEK), and the costs for capture of spawning fish (trapped at Viforsen) is 16,904 EUR (160,000 SEK) per year. As an approximation, the cost for the release during 2004 could be calculated as 46,076 EUR. The total cost calculated in this study for all the instream modifications was 92,199 EUR, considering the cost over a 40-year time span at 2.5% p.a. amortization [55], the annuity cost is 3648 EUR/year. Considering the low effectiveness from the restocking measures, it is expected that instream measures will be a more cost-effective measure in Ljungan than the previous re-stocking programs. It is important to note that instream mitigation measures could require a follow-up maintenance [32], e.g., due to sediments deposited which could clog the interstitial spaces, affecting spawning and nursery areas. This maintenance measures will incur in an increment in the cost per unit of PSA that has not been included in this study. This however will be very dependent on the river and the sediment dynamics. Barlaup and Gabrielsen [44] did not find degradation of the restored spawning habitat from sediment depositions, the amount of sediments accumulated was washed away when the spawning fish built their redds. Still, Pulg et al. [56] found a degradation of the restored spawning areas from sediment deposition, expecting unsuitable areas for reproduction after five or six years. Follow-up mitigation measures could be done, among others, by harrowing the gravels using an excavator [32], or could be done addressing the sedimentation source which will require large scale river restoration [56]. Monitoring and follow up will be needed in Ljungan to determine the maintenance and cost evaluation of the degradation of spawning areas, based on initial cost elements and necessary frequency.

The suitable habitat data presented in this study is considered as a first estimation, and field data to calibrate and corroborate the results will be needed to complete the validity of the model. Biological data will make it possible to evaluate the effects from the instream restoration measures. Despite the lack of such data, this study has shown the capability to transform a cloud of bathymetry points into a user-friendly method and techniques that helps to get easily interpretable outcomes that can positively influence management decisions. It is important to highlight that the LiDAR survey was affected by turbidity and the low bottom reflection in some areas, which were supplemented by manual measurements. However, this was not the case in other rivers like Tokkeåi or Hallingdal where water conditions were clear and average depths registered were up to 5 and 6 m [13,57]. Despite this limitation, the use of LiDAR bathymetry data as inputs for the hydraulic model has shown the potential of using this type of technique to model large river reaches with high resolution and use the results to evaluate fish habitat suitability and to support cost-effectiveness mitigation and restoration measures. In this particular case, the method presented had been applied after the instream measures were carried out, but it could also support the design before its implementation, promoting that stakeholders and water managers could test alternative scenarios [58]. Therefore, a method that supports restoration measures in order to fulfill ecological and stakeholder outcomes, and that future efforts will benefit from the understanding gained, was defined by Palmer et al. [59] as the most effective restoration.

**Author Contributions:** K.A. and A.A.-B. conceived the idea, A.A.-B. and K.A. participated in field measurements. A.A.-B. developed the model, data analysis and wrote the manuscript, K.O. provided relevant data and information to carry out this manuscript. K.A., H.-P.F. and K.O. supervised the modelling and writing process and contributed input to analysis and presentations, and they checked and commented on the manuscript.

**Funding:** This work was supported by the Norwegian University of Science and Technology under a strategic scholarship awarded to the Centre of Environmental Design of Renewable Energy funded by the Norwegian Research Council.

**Acknowledgments:** This work was supported by Norwegian University of Science and Technology under a strategic scholarship awarded to the Centre of Environmental Design of Renewable Energy funded by the Norwegian Research Council. In addition, Statkraft AB in Sweden financially supported this study. We would like to thank Carole Rozier and Aurelie Gosset (Department of Civil and Environmental Engineering, NTNU) for their support with the modelling process and Ingrid Alne (Department of Civil and Environmental Engineering, NTNU) for collection of field measurements. Thanks to Angela Odelberg and to the rest of the members in the stakeholders group in Ljungan for their support.

**Conflicts of Interest:** The authors declare no conflicts of interests.

## Appendix A

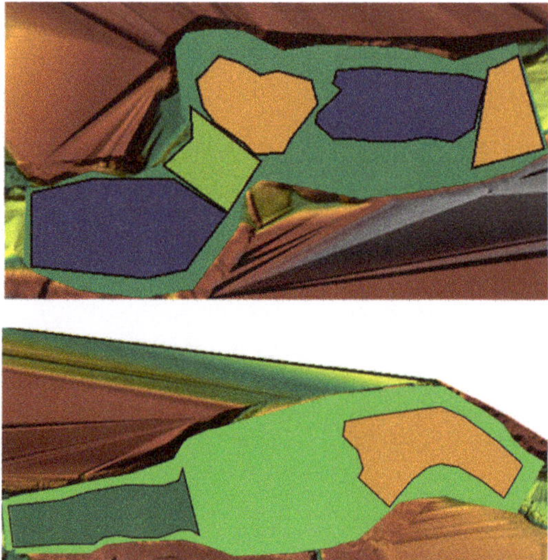

**Figure A1.** Manning areas used for the hydraulic simulations for Allsta (**upper** panel) and Nolby (**lower** panel). Blue is for areas with n = 0.03, light green n = 0.06, dark green n = 0.09, and orange n = 0.15.

## Appendix B

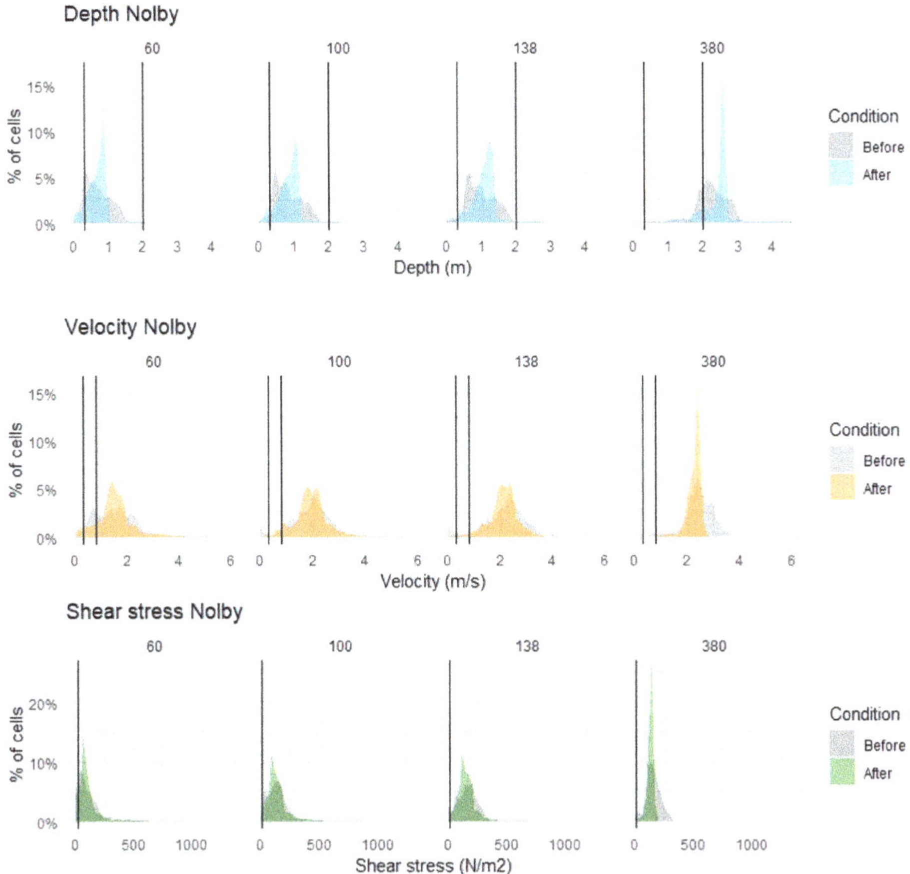


**Figure A2.** Percentage of cells under values for depth, velocity and shear stress in Nolby. Vertical lines indicate the limits for the suitable range (Table 3).

## References

1. Erwin, S.O.; Jacobson, R.B.; Elliott, C.M. Quantifying habitat benefits of channel reconfigurations on a highly regulated river system, Lower Missouri River, USA. *Ecol. Eng.* **2017**, *103*, 59–75. [CrossRef]
2. Armstrong, J.D.; Kemp, P.S.; Kennedy, G.J.A.; Ladle, M.; Milner, N.J. Habitat requirements of Atlantic salmon and brown trout in rivers and streams. *Fish. Res.* **2003**, *62*, 143–170. [CrossRef]
3. Stamou, A.; Polydera, A.; Papadonikolaki, G.; Martínez-Capel, F.; Muñoz-Mas, R.; Papadaki, C.; Zogaris, S.; Bui, M.D.; Rutschmann, P.; Dimitriou, E. Determination of environmental flows in rivers using an integrated hydrological-hydrodynamic-habitat modelling approach. *J. Environ. Manag.* **2018**, *209*, 273–285. [CrossRef] [PubMed]
4. Fabris, L.; Malcolm, I.A.; Buddendorf, W.B.; Millidine, K.J.; Tetzlaff, D.; Soulsby, C. Hydraulic modelling of the spatial and temporal variability in Atlantic salmon parr habitat availability in an upland stream. *Sci. Total Environ.* **2017**, *601–602*, 1046–1059. [CrossRef]

5. Friberg, N.; Angelopoulos, N.V.; Buijse, A.D.; Cowx, I.G.; Kail, J.; Moe, T.F.; Moir, H.; O'Hare, M.T.; Verdonschot, P.F.M.; Wolter, C. Chapter Eleven—Effective River Restoration in the 21st Century: From Trial and Error to Novel Evidence-Based Approaches. In *Advances in Ecological Research*; Dumbrell, A.J., Kordas, R.L., Woodward, G., Eds.; Academic Press: Cambridge, MA, USA, 2016; Volume 55, pp. 535–611.

6. Milhous, R.T.; Updike, M.A.; Schneider, D.M. *Physical Habitat Simulation System Reference Manual: Version II*; US Fish and Wildlife Service: Falls Church, VA, USA, 1989.

7. Waddle, T. *PHABSIM for Windows User's Manual and Exercises*; Geological Survey: Denver, CO, USA, 2001.

8. Jay Lacey, R.W.; Millar, R.G. Reach scale hydraulic assessment of instream salmonid habitat restoration 1. *JAWRA J. Am. Water Resour. Assoc.* **2004**, *40*, 1631–1644. [CrossRef]

9. Crowder, D.; Diplas, P. Using two-dimensional hydrodynamic models at scales of ecological importance. *J. Hydrol.* **2000**, *230*, 172–191. [CrossRef]

10. Boavida, I.; Santos, J.M.; Cortes, R.V.; Pinheiro, A.N.; Ferreira, M.T. Assessment of instream structures for habitat improvement for two critically endangered fish species. *Aquat. Ecol.* **2011**, *45*, 113–124. [CrossRef]

11. Mandlburger, G.; Hauer, C.; Wieser, M.; Pfeifer, N. Topo-bathymetric LiDAR for monitoring river morphodynamics and instream habitats—A case study at the Pielach River. *Remote Sens.* **2015**, *7*, 6160–6195. [CrossRef]

12. AHM. *New Possibilitiesin Bathymetricandtopographicsurvey*; A. GmbH: Innsbruck, Austria, 2015.

13. Alne, I.S. *Topo-Bathymetric LiDAR for Hydraulic Modeling—Evaluation of LiDAR Data From Two Rivers*; NTNU: Trondheim, Norway, 2016.

14. McKean, J.; Nagel, D.; Tonina, D.; Bailey, P.; Wright, C.W.; Bohn, C.; Nayegandhi, A. Remote Sensing of Channels and Riparian Zones with a Narrow-Beam Aquatic-Terrestrial LIDAR. *Remote Sens.* **2009**, *1*, 1065–1096. [CrossRef]

15. Andersen, M.S.; Gergely, Á.; Al-Hamdani, Z.; Steinbacher, F.; Larsen, L.R.; Ernstsen, V.B.J.H.; Sciences, E.S. Processing and performance of topobathymetric lidar data for geomorphometric and morphological classification in a high-energy tidal environment. *Hydrol. Earth Syst. Sci.* **2017**, *21*, 43–63. [CrossRef]

16. Mandlburger, G.; Pfennigbauer, M.; Steinbacher, F.; Pfeifer, N. Airborne Hydrographic LiDAR Mapping–Potential of a new technique for capturing shallow water bodies. In Proceedings of the 19th International Congress on Modelling and Simulation, Perth, Australia, 12–16 December 2011; pp. 12–16.

17. Klemas, V.; Pieterse, A. Using remote sensing to map and monitor water resources in arid and semiarid regions. In *Advances in Watershed Science and Assessment*; Springer: Berlin, Germany, 2015; pp. 33–60.

18. HELCOM. *Salmon and Sea Trout Populations and Rivers in the Baltic Sea—HELCOM Assessment of Salmon (Salmo salar) and Sea Trout (Salmo trutta) Populations and Habitats in Rivers Flowing to the Baltic Sea. Baltic Sea Environment Proceedings no. 126A*; H. Commission: Helsinki, Finland, 2011; p. 79.

19. Helfield, J.M.; Capon, S.J.; Nilsson, C.; Jansson, R.; Palm, D. Restoration of rivers used for timber floating: Effects on riparian plant diversity. *Ecol. Appl.* **2007**, *17*, 840–851. [CrossRef] [PubMed]

20. Petersen Jr, R.; Madsen, B.; Wilzbach, M.; Magadza, C.; Paarlberg, A.; Kullberg, A.; Cummins, K. Stream management: Emerging global similarities. *Ambio* **1987**, *16*, 166–179.

21. Muotka, T.; Paavola, R.; Haapala, A.; Novikmec, M.; Laasonen, P.J.B.C. Long-term recovery of stream habitat structure and benthic invertebrate communities from in-stream restoration. *Biol. Conserv.* **2002**, *105*, 243–253. [CrossRef]

22. Törnlund, E.; Östlund, L.J.E. Floating timber in northern Sweden: The construction of floatways and transformation of rivers. *Environ. Hist.* **2002**, *8*, 85–106. [CrossRef]

23. Johansson, U. *Stream Channelization Effects on Fish Abundance and Species Composition*; Department of Physics, Chemistry and Biology, Linköping universitet: Linköping, Sweden, 2013.

24. Dudgeon, D.; Arthington, A.H.; Gessner, M.O.; Kawabata, Z.-I.; Knowler, D.J.; Lévêque, C.; Naiman, R.J.; Prieur-Richard, A.-H.; Soto, D.; Stiassny, M.L.J.; et al. Freshwater biodiversity: Importance, threats, status and conservation challenges. *Biol. Rev.* **2006**, *81*, 163–182. [CrossRef]

25. Nilsson, C.; Lepori, F.; Malmqvist, B.; Törnlund, E.; Hjerdt, N.; Helfield, J.M.; Palm, D.; Östergren, J.; Jansson, R.; Brännäs, E.J.E. Forecasting environmental responses to restoration of rivers used as log floatways: An interdisciplinary challenge. *Ecosystems* **2005**, *8*, 779–800. [CrossRef]

26. HELCOM. *Salmon and Sea Trout Populations and Rivers in Sweden—HELCOM Assessment of Salmon (Salmo salar) and Sea Trout (Salmo trutta) Populations and Habitats in Rivers Flowing to the Baltic Sea*; Balt. Sea Environ. Proc. No. 126B; H. Commission: Helsinki, Finland, 2011; p. 110.

27. Riegl. RIEGL-VQ-880-G Data Sheet. 2014. Available online: http://www.riegl.com/uploads/tx_pxprriegldownloads/Infosheet_VQ-880-G_2016-05-23.pdf (accessed on 19 November 2018).

28. Sontek. RiverSurveyor Specifications. 2016. Available online: http://www.quantum-hydrometrie.de/RiverSurveyor-S5-M9.pdf (accessed on 19 November 2018).

29. ESRI. *ArcGIS Release 10.5*; E.S.R. Institute: Redlands, CA, USA, 2016.

30. HEC. *HEC-RAS River Analysis System Version 5.0—User Manual*; Hydrologic Engineering Center: Davis, CA, USA, 2016.

31. Chow, V.T. Open-channel hydraulics. In *Open-Channel Hydraulics*; McGraw-Hill: New York, NY, USA, 1959.

32. Forseth, T.; Harby, A.; Ugedal, O.; Pulg, U.; Fjeldstad, H.-P.; Robertsen, G.; Barlaup, B.T.; Alfredsen, K.; Sundt, H.; Saltveit, S.J. *Handbook for Environmental Design in Regulated Salmon Rivers*; NINA: Trondheim, Norway, 2014.

33. Skoglund, H.; Gabrielsen, S.-E.; Wiers, T. *Survey of Salmon Spawning and Juvenile Habitat in River Ljungan in Sweden 2014*; Uni Research Environment: Bergen, Norway, 2015. (In Norwegian)

34. Scruton, D.A.; Heggenes, J.; Valentin, S.; Harby, A.; Bakken, T.H. Field sampling design and spatial scale in habitat–hydraulic modelling: Comparison of three models. *Fish. Manag. Ecol.* **1998**, *5*, 225–240. [CrossRef]

35. Berenbrock, C.; Tranmer, A.W. *Simulation of Flow, Sediment Transport, and Sediment Mobility of the Lower Coeur d'Alene River*; U.S. Geological Survey Scientific Investigations Report 2008–5093; Geological Survey: Reston, ID, USA, 2008; p. 164.

36. Pisaturo, G.R.; Righetti, M.; Dumbser, M.; Noack, M.; Schneider, M.; Cavedon, V. The role of 3D-hydraulics in habitat modelling of hydropeaking events. *Sci. Total Environ.* **2017**, *575*, 219–230. [CrossRef]

37. Heggenes, J. Flexible summer habitat selection by wild, allopatric brown trout in lotic environments. *Trans. Am. Fish. Soc.* **2002**, *131*, 287–298. [CrossRef]

38. Niayifar, A.; Oldroyd, H.J.; Lane, S.N.; Perona, P. Modeling Macroroughness Contribution to Fish Habitat Suitability Curves. *Water Resour. Res.* **2018**, *54*, 9306–9320. [CrossRef]

39. Juárez, A.; Adeva-Bustos, A.; Alfredsen, K.; Dønnum, B.O. Performance of A Two-Dimensional Hydraulic Model for the Evaluation of Stranding Areas and Characterization of Rapid Fluctuations in Hydropeaking Rivers. *Water* **2019**, *11*, 201. [CrossRef]

40. Gardeström, J.; Holmqvist, D.; Polvi, L.E.; Nilsson, C. Demonstration Restoration Measures in Tributaries of the Vindel River Catchment. *Ecol. Soc.* **2013**, *18*. [CrossRef]

41. Bair, R. *Modeling Large Wood Impacts on Stream Hydrodynamics and Juvenile Salmon Habitat*; Oregon State University: Eugene, OR, USA, 2016.

42. Wilcock, P.R.; McArdell, B.W. Surface-based fractional transport rates: Mobilization thresholds and partial transport of a sand-gravel sediment. *Water Resour. Res.* **1993**, *29*, 1297–1312. [CrossRef]

43. Scruton, D.A.; Anderson, T.C.; King, L.W. Pamehac Brook: A case study of the restoration of a Newfoundland, Canada, river impacted by flow diversion for pulpwood transportation. *Aquat. Conserv. Mar. Freshw. Ecosyst.* **1998**, *8*, 145–157. [CrossRef]

44. Barlaup, B.T.; Gabrielsen, S.E.; Skoglund, H.; Wiers, T. Addition of spawning gravel—A means to restore spawning habitat of atlantic salmon (*Salmo salar* L.), and Anadromous and resident brown trout (*Salmo trutta* L.) in regulated rivers. *River Res. Appl.* **2008**, *24*, 543–550. [CrossRef]

45. Gard, M. Modeling changes in salmon spawning and rearing habitat associated with river channel restoration. *Int. J. River Basin Manag.* **2006**, *4*, 201–211. [CrossRef]

46. Fjeldstad, H.P.; Barlaup, B.T.; Stickler, M.; Gabrielsen, S.E.; Alfredsen, K. Removal of weirs and the influence on physical habitat for salmonids in a norwegian river. *River Res. Appl.* **2012**, *28*, 753–763. [CrossRef]

47. McKean, J.; Tonina, D. Bed stability in unconfined gravel bed mountain streams: With implications for salmon spawning viability in future climates. *J. Geophys. Res. Earth Surf.* **2013**, *118*, 1227–1240. [CrossRef]

48. Koljonen, S.; Huusko, A.; Mäki-Petäys, A.; Louhi, P.; Muotka, T. Assessing Habitat Suitability for Juvenile Atlantic Salmon in Relation to In-Stream Restoration and Discharge Variability. *Restor. Ecol.* **2013**, *21*, 344–352. [CrossRef]

49. Palmer, M.A.; Menninger, H.L.; Bernhardt, E. River restoration, habitat heterogeneity and biodiversity: A failure of theory or practice? *Freshw. Biol.* **2010**, *55*, 205–222. [CrossRef]

50. Marttila, M.; Louhi, P.; Huusko, A.; Mäki-Petäys, A.; Yrjänä, T.; Muotka, T. Long-term performance of in-stream restoration measures in boreal streams. *Ecohydrology* **2016**, *9*, 280–289. [CrossRef]

51. ICES. *Summary of ICES Advice on the Exploitation of Baltic Sea fish Stocks in 2018*; The Fisheries Secretariat Stockholm: Stockholm, Sweden, 2017.

52. Johnsen, B.O.; Arnekleiv, J.V.; Asplin, L.; Barlaup, B.T.; Næsje, T.F.; Rossenland, B.O.; Saltveit, S.J.; Tvede, A. Hydropower Development- Ecological Eeffects. In *Atlantic Salmon Ecology*; Aas, Ø., Einum, S., Klemetsen, A., Skurdal, J., Eds.; John Wiley & Sons Ltd.: Chichester, UK, 2011.

53. Saltveit, S.J. The effects of stocking Atlantic salmon, Salmo salar, in a Norwegian regulated river. *Fish. Manag. Ecol.* **2006**, *13*, 197–205. [CrossRef]

54. Palmé, A.; Wennerström, L.; Guban, P.; Ryman, N.; Laikre, L. *Compromising Baltic Salmon Genetic Diversity: Conservation Genetic Risks Associated with Compensatory Releases of Salmon in the Baltic Sea*; Havs-och vattenmyndigheten: Göteborg, Sweden, 2012.

55. Szałkiewicz, E.; Jusik, S.; Grygoruk, M.J.S. Status of and Perspectives on River Restoration in Europe: 310,000 Euros per Hectare of Restored River. *Sustainability* **2018**, *10*, 129. [CrossRef]

56. Pulg, U.; Barlaup, B.T.; Sternecker, K.; Trepl, L.; Unfer, G. Restoration of spawning habitats of brown trout (*Salmo trutta*) in a regulated chalk stream. *River Res. Appl.* **2013**, *29*, 172–182. [CrossRef]

57. Skeie, L. *Hydrauliskc Modelling of Tokkeåi-Hydraulisk Modellering av Kraftverksdrift i Tokkeåi*; NTNU: Trondheim, Norway, 2017. (In Norwegian)

58. Kennen, J.G.; Stein, E.D.; Webb, J.A. Evaluating and managing environmental water regimes in a water-scarce and uncertain future. *Freshw. Biol.* **2018**, *63*, 733–737. [CrossRef]

59. Palmer, M.A.; Bernhardt, E.S.; Allan, J.D.; Lake, P.S.; Alexander, G.; Brooks, S.; Carr, J.; Clayton, S.; Dahm, C.N.; Follstad Shah, J.; et al. Standards for ecologically successful river restoration. *J. Appl. Ecol.* **2005**, *42*, 208–217. [CrossRef]

*Article*

# Dam Removal Effects on Benthic Macroinvertebrate Dynamics: A New England Stream Case Study (Connecticut, USA)


Helen M. Poulos [1,*], Kate E. Miller [1,2], Ross Heinemann [3], Michelle L. Kraczkowski [1,4], Adam W. Whelchel [5] and Barry Chernoff [1,4,6]

[1]  College of the Environment, Wesleyan University, Middletown, CT 06459, USA; kmiller@newhaven.edu (K.E.M.); mkraczkowski@usj.edu (M.L.K.); bchernoff@wesleyan.edu (B.C.)
[2]  Biology and Environmental Science Department, University of New Haven, West Haven, CT 06510, USA
[3]  Department of Earth and Environmental Sciences, Wesleyan University, Middletown, CT 06459, USA; rheinemann@wesleyan.edu
[4]  Department of Biology, University of Saint Joseph, West Hartford, CT 06117, USA
[5]  The Nature Conservancy, New Haven, CT 06510, USA; awhelchel@tnc.org
[6]  Department of Biology, Wesleyan University, Middletown, CT 06459, USA
*  Correspondence: hpoulos@wesleyan.edu


Received: 11 April 2019; Accepted: 14 May 2019; Published: 21 May 2019


**Abstract:** Dam removal is an increasingly common stream restoration tool. Yet, removing dams from small streams also represents a major disturbance to rivers that can have varied impacts on environmental conditions and aquatic biota. We examined the effects of dam removal on the structure, function, and composition of benthic macroinvertebrate (BMI) communities in a temperate New England stream. We examined the effects of dam removal over the dam removal time-series using linear mixed effects models, autoregressive models, non-metric multidimensional scaling, and indicator and similarity analyses. The results indicated that the dam removal stimulated major shifts in BMI community structure and composition above and below the dam, and that the BMI communities are becoming more similar over time. The mixed model analysis revealed that BMI functional groups and diversity were significantly influenced by sample site and several BMI groups also experienced significant interactions between site and dam stage ($P < 0.05$), while the multivariate analyses revealed that community structure continues to differ among sites, even three years after dam removal. Our findings indicate that stream restoration through dam removal can have site-specific influences on BMI communities, that interactions among BMI taxa are important determinants of the post-dam removal community, and that the post-dam-removal BMI community continues to be in a state of reorganization.

**Keywords:** dam removal; benthic macroinvertebrates; community composition; community stability; community reorganization


## 1. Introduction

Worldwide, the damming of rivers disrupts hydrological cycles and negatively impacts the structure and function of riparian ecosystems [1,2]. The alteration and fragmentation of natural fluvial systems by dams have had myriad cascading impacts on stream ecology, geomorphology and hydrography [2–5]. This water flow obstruction by dams alters stream nutrient cycling, sedimentation, thermal regimes, and river-corridor organism mobility [6–8], all of which pose significant threats to river biodiversity and ecological processes both above and below dams [9–11].

Dam removal is an increasingly common management strategy for restoring river and stream ecosystem structure and function [4,12,13]. Thousands of dams have been removed over the last century, and they continue to be removed at an increasing rate [14]. The vast majority of dam removal efforts are concentrated in North American and Europe, however, Asia and South America also comprise regions of increasing dam removal activity [15,16]. While this widespread interest in dam removal provides a means for restoring river flow and connectivity, dam removal itself is also an ecological perturbation to stream systems that have often existed in altered hydrological conditions from decades to centuries [17]. Thus, river restoration can have varied impacts on aquatic system structure and function [13,18,19] because new flow patterns can change and redistribute substrate and create changes in habitat availability and quality for both fish and benthic macroinvertebrate (BMI) taxa [20].

Sedimentation and deposition after the dam removal event can also affect the system for years [13,21–23], impacting BMI abundance, community structure, species richness, and riparian ecosystem function [10,24,25]. These effects can be highly variable, long-lasting, and dependent on variation in dam characteristics and the rate of stabilization of physical conditions after dam removal [20,26–29]. Short-term effects (i.e., within months) can include an immediate decline in BMI density and diversity [30,31], while long-term (years) effects are often characterized by a shift in BMI community composition from lentic- to lotic-specialist taxa as stream flow increases with time-since-dam-removal [29,32].

Past studies indicate that, for invertebrates, dam removal can stimulate a significant shift above the dam site from lentic BMI taxa dominance (e.g., Chironomids, Oligochaetes, and other non-insects) to more diverse assemblages that include a mixture of taxa including riffle-specialists (i.e., "EPT" taxa: Ephemeroptera, Plecoptera, Trichoptera) [33,34]. In some cases, downstream recovery can resemble the pre-dam community [21], while in other cases, the post-dam removal BMI community composition resembles neither the pre-dam community nor that of other nearby free flowing stream reaches [10].

Due to the complexity of responses of BMI communities to dam removal, longer-term, community-scale studies that start before dam removal and continue several years afterwards are needed for identifying the diverse, long-term effects of river restoration on stream assemblages and species interactions [5,24,26,35–39]. The present study investigates the effects of dam removal on BMI community dynamics in the Eightmile River, CT, USA. A small dam that had been in place since the 1760s, was removed from the river on the Zemko property of The Nature Conservancy. Our study began in 2005, one year before drawdown (in 2006) and continued for three years following dam removal (in 2007) until the fall of 2010. Poulos, Miller, Kraczkowski, Whelchel, Heineman and Chernoff [37] and Poulos and Chernoff [36] examined the effects of the Zemko dam removal on fish assemblages and discovered that they continued to be in a state of reorganization at sites adjacent to the former dam even three years after dam removal. Our objectives for the current study build upon this knowledge by assessing how dam removal at these same three sites (above and below the dam, and at a nearby reference site) influenced BMI community structure, function, and stability. Based on the literature, we hypothesized the dam removal would (1) trigger significant, site-specific effects on BMI dynamics and function that would differ over the course of the dam removal process, and (2) that the stability of the BMI community would similarly vary in relation to site and dam stage (i.e., pre-dam removal, during drawdown, and post-dam removal).

## 2. Materials and Methods

### 2.1. Study Sites

The Eightmile River, located in Salem, CT is a largely rural and forested portion of the state (Figure 1) [40]. The river was designated a Wild and Scenic River by Congress in 2008, protecting all major branches and tributaries within the system (House Senate Report 110-94). The Connecticut Department of Environmental Protection in 1998–1999 characterized the water quality of the main stem as some of the best in the state based on the prevalence of high water quality indicator BMI taxa [41]. Sampling was conducted monthly on the same day of each month within sites during the growing season (May–October) from 2005–2010 at three main locations (Figure 1) including: the reach immediately above the dam ("ZAD") which was a pond prior to drawdown, the reach immediately below the dam ("ZBD"), and a reference site ("REF") 7.19 km downstream of the former dam (see Section 2.1.3 below for justification). Sample sites differed with regard to a number of spatial and habitat characteristics, such as basin size, canopy and substrate (see Appendix A for site characteristics and Poulos and Chernoff [36] for site-specific temporal changes in local environmental conditions). Sampling began prior to dam removal in 2005 and continued through water drawdown (which began in spring 2006), final dam removal (fall 2007), and for three years afterward.

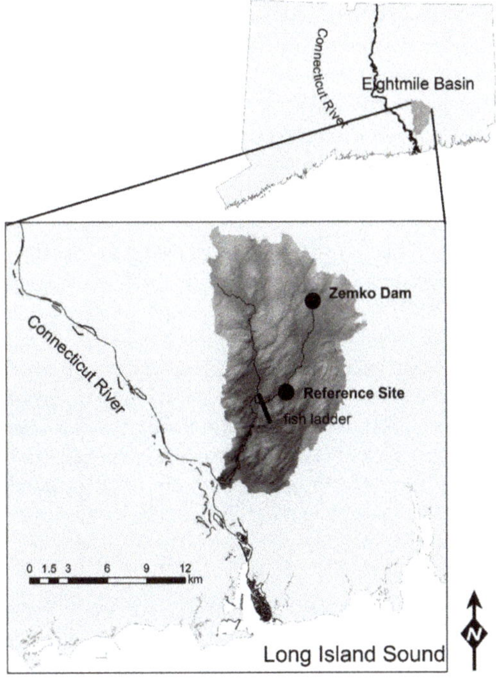


**Figure 1.** Map of the Zemko Dam and reference site on the Eightmile River, Haddam, CT.

## 2.1.1. Above the Zemko Dam (ZAD)

The East branch of the Eightmile River was modified at the Zemko site for a considerable time. A dam for milling was first constructed there in the 1720s and the resulting pond was periodically drained until the 1960s when the present-day dam was constructed. At the time of removal, the dam was a 1.5 m high, 3.7 m wide, and 24.5 m long stone- and earth-fill structure that also served as an unimproved road crossing. The dam was considered the last impasse to diadromous fish such as, *Salmo salar* (Atlantic Salmon), although a fish ladder at Bill's pond downstream may have continued to impair upstream fish and BMI movement. Details of environmental changes that occurred over the study period have been reported by Poulos and Chernoff [36].

Prior to dam removal, the substrate of the above-dam pond was largely organic. This transitioned to silt and sand as water levels decreased and the dam was removed. As the channel's flow increased over the study period, partially submerged cobble-sized rocks emerged at the upstream edge of the former pond. An approximately 14 m portion of the river at and directly above the dam was also modified post dam removal with the addition of cobble and large woody debris in order to stabilize the banks and mimic a more natural streambed structure.

### 2.1.2. Below the Dam (ZBD)

The below dam reach was largely a gravel and cobble bed with sand deposits in pools near the banks. While much of this composition remained constant, there were changes in the river-bed characteristics during the drawdown and dam removal process. We observed a build-up of sand and silt in the pool directly below the dam in addition to increased erosion and tree fall within the stream channel.

### 2.1.3. The Reference Site (REF)

An upstream reference site was not available because the dam was less than 1 km from its headwaters, so instead we chose a downstream reference site because of its relatively unimpacted river channel and bank, its location on protected land (the CT Chapter of The Nature Conservancy), and little nearby development. Its distance from the dam (7.19 km) was deemed sufficient to avoid direct dam removal impacts, and pre-study BMI sampling (2004) indicated that there was no significant difference (ANOVAs, $P > 0.05$) in numerous invertebrate metrics (total richness and abundance, EPT-, and percent dominance) from another site monitored on the free-flowing West branch of the Eightmile.

### 2.2. Benthic Macroinvertebrate Sampling

Surber sampling was chosen as the method for BMI collection based on the results of a series of experiments carried out at the reference site in 2004 by Olins [42]. An alternate method, rock bags, was used for the above dam site (ZAD) because of the water depth at this sample site (e.g., average 39 cm in 2005). At each of the three randomly selected locations within the 100 m riffle area at ZAD, a rock bag consisting of 0.008 $m^3$ of various sized rocks was placed on the substrate and after approximately one month was removed and processed in the same manner as Surber samples. At the other sites, Surber samplers (Wildco; area = 0.093 $m^2$) were used at three randomly selected locations within the designated riffle. Rocks and 3 cm deep of loose substrate were scrubbed free of organisms and organic matter, collected in the attached net (500 um), and preserved in 70% alcohol until processing. Analysis of side by side Surber and rock bag sampling indicated that there was no significant difference in BMI groups of interest between the two methods ($P > 0.05$ Olins [42]), supporting the decision to utilize rock bags at the above-dam site.

Samples were picked from a 12-square sorting tray using magnifier lights (10x) in randomized sub-samples of 25% (3 squares), and preserved in 70% alcohol. The sub-samples were then processed, and all organisms were identified to family level (using keys from Voshell and Wright [43]), which was considered sufficient for analysis of functional feeding and sensitive taxonomic groups (e.g., Tolonen et al. [44]) and to avoid differences due to rare taxa. Sample counts were limited to those families that are hydropneustic.

We chose the following groups for classification of the taxa based on the literature and use by other dam studies: (1) functional feeding groups (predator, collector-gatherer, collector-filterer, scrapers and shredders) that are utilized to learn more about community changes in response to conditions such as flow and organic matter [45,46]; (2) EPT- (minus), which included all Ephemeroptera, Plecoptera, and Trichoptera families , as they are often included in ecosystem health and water quality assessments (e.g., [47]), however, the Trichoptera family Hydropsychidae was excluded as they were ubiquitous in most samples, are more stress tolerant in Connecticut streams, and include a variety of species with differing sensitivity to water quality [48]; (3) Chironomidae (Order Diptera) abundance, as a common but variable component of the BMI community and an indicator of slow-flowing, silty, and warmer water conditions [47,49]; and (4) non-insects, consisting primarily of Olig ochaetes and Amphipods that are both highly tolerant and indicative of lentic conditions. In addition, we also examined spatiotemporal variation in individual BMI families, abundance and richness, and diversity ($H'$, $E_{var}$). $H'$ is most commonly used when evaluating species diversity [50]. $E_{var}$ is an alternate index for use in measuring species evenness where 0 is minimum evenness and 1 is maximum [51]. See Appendix C for a full list of invertebrate taxa and group composition.

### 2.3. Data Analyses

#### 2.3.1. Univariate Analyses

Prior to analysis, BMI family abundance matrices were tested for randomness with entropy analysis following Atmar and Patterson [52]. The results demonstrated that our matrices were highly significantly more ordered than 10,000 random matrices generated from a Monte Carlo process and that the data were suitable for the subsequent analyses of patterns within the matrices. We then used mixed model analyses in the nlme package of R [53], which is a common method for evaluating the effects of dam removal on stream assemblages to test for differences in the abundance and diversity of BMI groups at each sample site in relation to dam removal with fixed and random effects and interaction terms. We estimated variance components within each mixed model to account for the covariance structure of the repeated measures of the sampling intervals, followed by least squares means pairwise comparisons to test for differences in BMI assemblages both among and within sites over time. Site and dam removal were treated as fixed effects and year nested within dam stage was treated as a random effect. Dam removal was classified into three stages: (1) pre-removal (2005), (2) drawdown to removal (2006–2007), and (3) post-removal (2008–2010) for this, and all subsequent analyses.

#### 2.3.2. Non-Metric Multidimensional Scaling

We used non-metric multidimensional scaling (NMDS) to analyze community-scale differences in BMI composition among sample sites and dam stages via the vegan package in R (Oksanen et al. 2013). The relative abundance of BMI families was plotted in the NMDS solution by (1) sample site over the entire study period, and (2) dam stage for each sample site using 95% confidence ellipses. We then tested for differences in BMI ordination space among sites and among dam stages within each sample site using the ordiareatest command in vegan.

#### 2.3.3. Indicator Species Analysis

We used indicator species analysis [54] and PC-Ord Software [55] to identify key indicator BMI taxa across all years for the reference, above dam, and below dam sites. The goal of this analysis was to identify taxa that occurred in a particular habitat or location with high fidelity. The method combines abundance data from a site and faithfulness of occurrence of a taxon in a particular site. It produces indicator values for each, which are subsequently tested for statistical significance by a Monte Carlo permutation. Indicator values (0–100) are simply estimated as the relative frequency of the taxon in sample sites belonging to a particular target site group.

#### 2.3.4. Spatiotemporal Variation in BMI Community Structure

We examined differences in BMI community composition among sites and over time by examining Bray-Curtis similarity matrices using a two-way analysis of similarity (ANOSIM; $\alpha = 0.05$; 999 permutations) for both dam stage (pre-removal, drawdown, and post-removal) and by year. Spatio-temporal changes in community structure was further analyzed by performing a similarity percentage analysis (SIMPER) [56]. SIMPER was used to identify major BMI families contributing to > 50% of the total dissimilarity among sites. This method computes the percentage contribution of each taxon to the total dissimilarity between pairs of sites; those with the largest contribution to dissimilarity are those that best discriminate between site communities. All similarity analyses were performed using a paleontological statistics software package (PAST v2.17) [57].

#### 2.3.5. Multivariate Autoregressive Models (MAR)

We estimated BMI community stability in relation to sample site and dam stage by examining functional feeding group interaction strengths over the time-series of dam stage using multivariate autoregressive (MAR) models. The effect of dam stage on BMI functional feeding group (FFG) community dynamics was also examined by including dam stage as a covariate in the MAR models for the two dam sites. MAR examined the interaction strengths among BMI FFGs over the time-series at each sample site. In MAR models, variates are factors expected to affect their own dynamics and those of other groups. The MAR framework is similar to a set of simultaneously solved multiple regressions of interacting taxa, while also accounting for the serial autocorrelation in time-series data (see Beisner et al. [58]) through the calculation of autoregression coefficients that depend on the correlated response of one variable to the others in the time-series of interest. Autoregression coefficients depend upon patterns of change in the data so that, if a given variable does not change, it does not influence the changes in abundance of taxa in the dataset and the autoregression coefficients are zero. In this case, the dynamics considered were changes to abundances. Dam stage was treated as a covariate because of its potential to affect FFG abundance.

MAR models were fit and stability metrics were estimated using the MAR1 package [59] of the R Statistical Language [53]. Functional feeding group abundance data for each time-step were log transformed to better approximate the non-linear relationships in the data (sensu Ives et al. [60]) and standardized to deseasoned Z-scores prior to analysis. Data were deseasoned to facilitate model comparisons among BMIs and to remove seasonal trends in the data by dampening seasonally varying population fluctuations. Since sampling occurred during the growing season (i.e., from June to October) each year, we followed recommendations by Hampton et al. [61] and Ives, Dennis, Cottingham and Carpenter [60] by specifying a MAR model that skipped estimations between non-consecutive data points by not filling gaps in the data greater than 1 month in duration (using the fill.gap command in MAR1).

The MAR output produces several stability metrics of community resilience and reactivity that we used to examine community-level responses to dam removal. These metrics are based on the **B** species interaction matrix, as defined by Ives, Dennis, Cottingham and Carpenter [60] who developed three measures of resilience (which they termed stability), and two measures of reactivity within the MAR framework. Four of the five metrics depend upon the eigenstructure of the species interaction matrix, **B**. The first, $det(B)^{2/p}$ where det is the determinant and p is the number of taxa, measures how species interactions amplify any environmental variance (dam stage and site) in relation to the stationary distribution. The second measure, $max(\lambda_B)$, is the maximum eigenvalue of **B**. This measures the rate of return of the mean from the perturbed or transitional distribution to the stationary distribution; the largest eigenvalue corresponds to the slowest dimension of change [60]. The third measure of resilience, $max(\lambda_{\otimes})$, is the maximum eigenvalue of the Kronecker product of **B** and measures the rate of return of the variance from the perturbed or transitional distribution to the stationary distribution. The smaller the values of the three resilience measures the greater resilience that system has to perturbations.

A reactive system is one that frequently moves farther away from the stationary distribution [60] as reactivity increases stability and resilience decrease. The first measure of reactivity is $-\frac{tr(\Sigma)}{tr(V_\infty)}$, where tr is the trace, $\Sigma$ is the environmental covariance matrix and $V_\infty$ is the covariance matrix of the stationary distribution (a function of **B**); less negative values (i.e., those closer to zero) are more reactive due to the species interactions amplifying environmental variance in $V_\infty$. The second measure of reactivity depends only on **B** and is $max(\lambda_{B'B}) - 1$, where **B'** is the transpose of **B**; this measure depends upon the entire eigenstructure and is sensitive to the smaller eigenvalues; the larger the asymmetry of the eigenstructure the larger the value of the metric and the higher the reactivity of the system.

## 3. Results

### 3.1. Impacts of Dam Removal on BMI Taxonomic Groups

More than 70 BMI families were identified across all sites over the study period from 59 samples, from which we identified 11,240 individual organisms. The mixed model analyses identified significant site-level differences for virtually all of the BMI metrics in the study (Table 1, Figure 2, Appendix B). Site and dam stage interactions were significant for total BMI abundance, non-insect abundance, taxon richness, and diversity ($E_{var}$) ($P < 0.05$). Total BMI abundance was similar among sites prior to dam removal, but it diverged significantly among sites during drawdown and after dam removal ($P < 0.05$). BMI abundance was significantly higher at ZAD than at the two other sites during drawdown, and all three sites differed significantly from one another during the post-dam removal dam stage. Differences in non-insect BMI abundance over the time-series was due to significantly higher ($P < 0.05$) non-insect abundance at ZAD during dam removal. Taxon richness was similar among sites prior to dam removal, and higher at ZAD during dam removal. Post removal, taxon richness was significantly higher at REF, intermediate at ZAD, and lowest at ZBD. $E_{var}$ was consistently lower at ZAD over all sample years relative to the other two sample sites. $E_{var}$ fluctuated and differed between ZBD and REF during dam removal, but they did not differ significantly three years after dam removal.

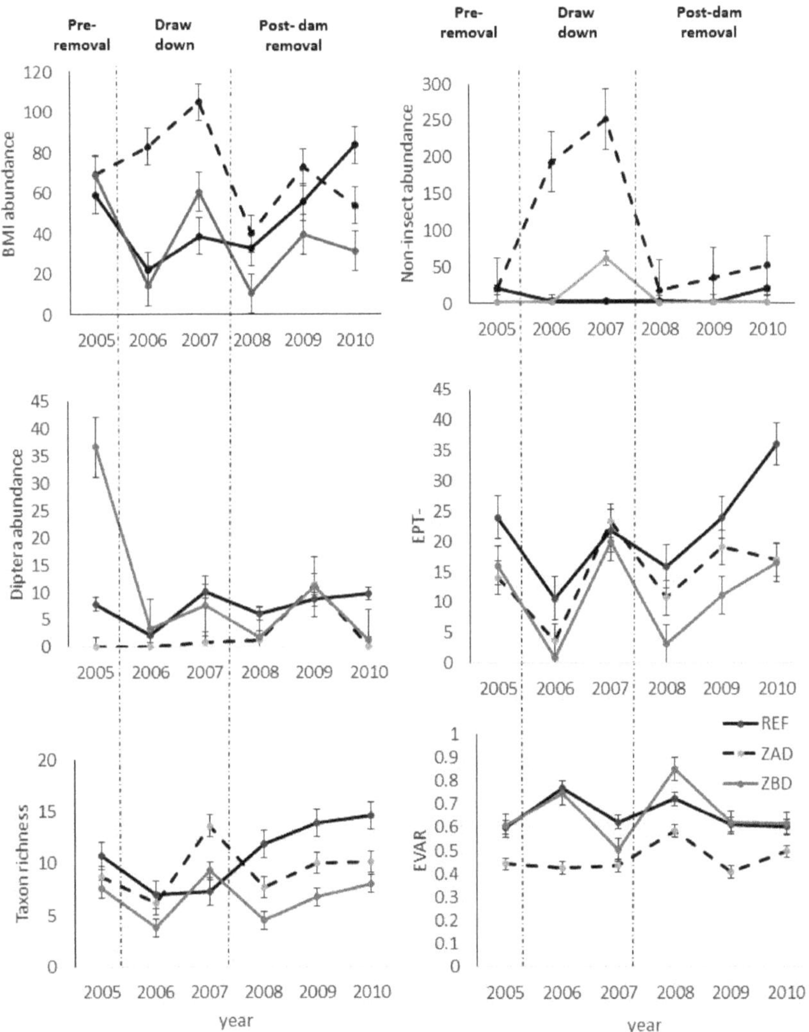


**Figure 2.** Mean (± S.E.) BMI abundances and diversity statistics for sub-samples at each site by year. Dam stages are indicated with dotted lines. Significant differences among dam stages are reported in Table 1.

**Table 1.** F-values from the modified before after control impact mixed model analyses of BMI variables in relation to the three study sites, the three stages of dam removal, and the interaction between site and dam stage. Degrees of freedom for BMI variables are 2 (site, stage) and 4 (dam*stage); for functional feeding group analysis degrees of freedom are 2 (site), and 1 (stage, interaction). Significant terms are designated by asterisks: * indicates $P < 0.05$, and ** indicates $P < 0.01$. See Appendix C for BMI abbreviations and descriptions.

| *BMI Variable* | *Interaction* | | |
| --- | --- | --- | --- |
| | **Site** | **Dam Stage** | **Site*Dam Stage** |
| *BMI abundance* | 23.34 ** | 0.8 | 3.1 * |
| *Taxon richness* | 6.4 ** | 1.4 | 3.7 * |
| *H'* | 7.2** | 2.7 | 2.5 |
| *EVAR* | 10.0 ** | 0.3 | 2.5 * |
| *EPT- abundance* | 13.0 * | 1.1 | 0.1 |
| *Diptera abundance* | 9.5 | 1.1 | 3.3 ** |
| *Non-insects* | 6.3 ** | 3.5 * | 4.1 ** |
| *Predators* | 8.3 ** | 0.2 | 2.3 |
| *Collector-gatherers* | 23.1 ** | 0.3 | 8.9 ** |
| *Collector-filterers* | 6.1 ** | 2.0 | 0.6 |
| *Scrapers* | 20.8 ** | 2.8 | 47.3 |
| *Shredders* | 46.2 ** | 6.8 | 13.2 |

### 3.2. Functional Feeding Group Dynamics

Functional feeding group abundances differed significantly by site for all FFGs in the study ($P < 0.01$) (Table 1, Figure 3, Appendices B and C). However, the interaction between site and dam stage was only significant for collector-gatherer abundance ($P = 0.031$). Significant increases in predator abundance occurred at the two dam sites during drawdown ($P < 0.05$). Collector-gatherer abundance declined significantly at ZAD, primarily due to Oligochaetes (Annelida) and Amphipods, which comprised 77% of the community before drawdown/dam removal and declined to 32% at the end of the study. Shredder abundance was consistently higher at the REF site relative to the two dam sites for all years after 2006. Collector-filterer and scraper abundances were variable among the sites. There was high variability in scraper abundance across years (Figure 3). Although there were significant site effects ($P < 0.01$, Table 1, Figure 3), the post-dam removal rebound of scrapers obscured interaction effects with stage; and scrapers also became much more abundant at REF post dam removal.

### 3.3. Impacts of Site and Dam Stage on BMI Community Composition and Structure

The NMDS results and the two-way ANOSIM analysis revealed that the three dam sites remained largely distinct over the study period. BMI families differed significantly among dam stages ($R = 0.091$, $P = 0.0374$) and sites ($R = 0.60$, $P = 0.0001$). Although the 95% confidence ellipses did not overlap among sites in the NMDS solution (Figure 4A), BMI family composition overlapped within sites among the three dam stages of pre-removal, drawdown, and post-removal (Figure 4B). The one exception was an overlap between the above and below dam sites during drawdown. Communities of all three sites shifted closer together in BMI family space post-removal.

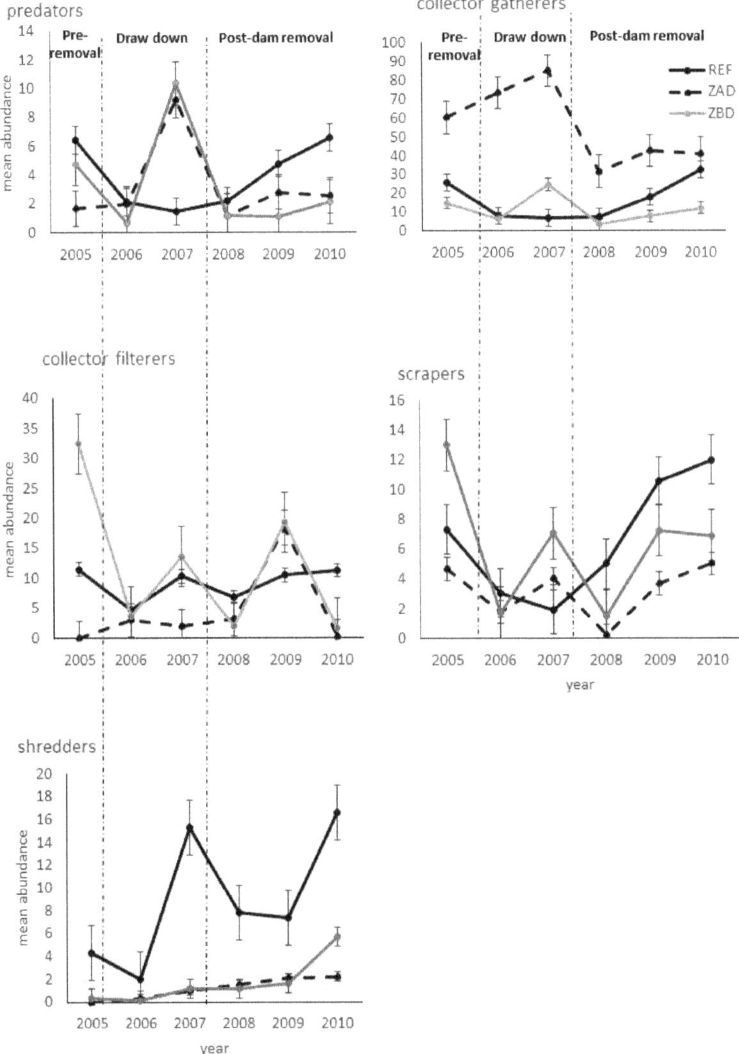



**Figure 3.** Mean (± S.E.) abundances of BMI functional feeding groups displayed by site and year. Dam stages are indicated with dotted lines. Significant differences among sites and dam stage are reported in Table 1.

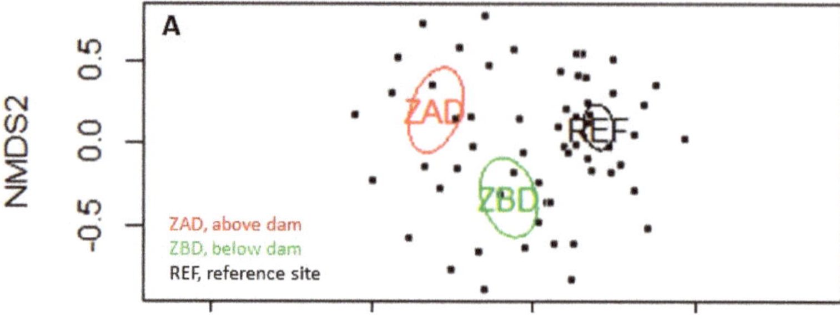


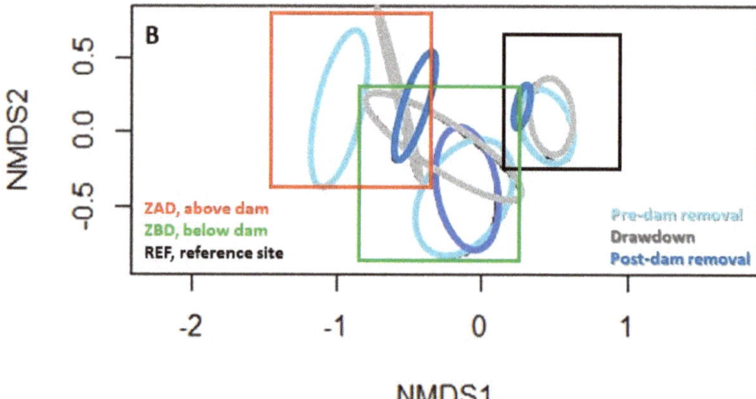


**Figure 4.** Non-metric multidimensional scaling (NMDS) of BMI families in the Eightmile River, CT, USA over the study period showing (**A**) BMI relative abundance by family (black dots) plotted with 95% confidence ellipses for each sample site over all sampling intervals, and (**B**) 95% confidence ellipses for each dam stage within each sample site (boxes). Sample sites differed significantly in BMI family space according to permutation tests ($P < 0.05$), although each site did not vary significantly in BMI family space by dam stage. The final NMDS solution was a 2-dimensional solution with a stress value of 0.22.

The SIMPER analysis highlighted that differences in BMI family composition among sites over the study were principally the result of changes in the two dam sites relative to the reference site over the course of the study (Appendix D). Such differences were explained by the abundance of more tolerant, lentic specialist taxa, including non-insects such as oligochaetes and amphipods, as well as insect taxa such as Chironomids, above the dam relative to the two other sites, and to some extent, higher Trichoptera abundance at the reference site. In addition, below the dam the post-removal stage experienced decreases in Hydropsychidae and Chironomids, which are generally considered abundant and stress-tolerant taxa.

The indicator species analysis corroborated these results and highlighted the distinctness of the reference site relative to the two dam sites. REF contained nine indicator families, six of which were EPT- taxa (Table 2). These included less stress-tolerant taxonomic groups including Trichoptera (i.e., Brachicentripodidae, Limnophilidae and Glossostomadidae), as well as other sensitive BMI groups like Plecoptera. In comparison ZBD had only two indicator taxa (both insects), one a predator that occurred in low numbers (0–5 per sample), and a scraper, while ZAD hosted seventeen indicator species from a variety of functional feeding groups. Nearly half of them were non-insect taxa, and all but two were stress-tolerant. These indicator species were prevalent at ZAD over the entire study period, regardless of dam stage, and many of them were associated with lentic stream conditions.

**Table 2.** Maximum indicator values (IV) for significant indicator taxa ($P < 0.05$) for each sample site. Taxon acronyms are explained in Appendix C.

| Taxon | IV | P | FFG | Other Groups |
|---|---|---|---|---|
| | | REF | | |
| TcLm | 43.6 | 0.0032 | SH | EPT- |
| EpEp | 47.2 | 0.0028 | CG | EPT- |
| GaHy | 46.5 | 0.0004 | SC | Non insect |
| ClPs | 62.4 | 0.0002 | SC | |
| DtTp | 61.8 | 0.0002 | SH | |
| EpIs | 51.9 | 0.0002 | CG | ETP- |
| PlPr | 85 | 0.0002 | P | EPT- |
| TcBr | 84.5 | 0.0002 | SH | EPT- |
| TcGl | 61.6 | 0.0002 | SC | EPT- |
| | | ZAD | | |
| BvSp | 20.3 | 0.0404 | CF | Non insect |
| MgSi | 15.8 | 0.0400 | P | |
| GaPh | 22.2 | 0.0130 | CG | Non insect |
| OdAs | 26.8 | 0.0110 | P | |
| PhTb | 28.7 | 0.0054 | P | Non insect |
| DtCh | 52.3 | 0.0050 | CG | Diptera |
| EpLp | 40.9 | 0.0042 | CG | EPT- |
| GaPl | 33.6 | 0.0040 | SC | Non insect |
| OdLb | 29.1 | 0.0026 | P | |
| ArHy | 34.5 | 0.0020 | P | Non insect |
| DtCr | 41.7 | 0.0012 | P | |
| OdCn | 36 | 0.0010 | P | |
| AmSc | 90 | 0.0002 | CG | Non insect |
| AnOl | 89 | 0.0002 | CG | Non insect |
| EpCn | 48.3 | 0.0002 | SG | EPT- |
| IsSw | 43.8 | 0.0002 | CG | Non insect |
| TcLt | 68.8 | 0.0002 | | EPT- |
| | | ZBD | | |
| MgCr | 39.6 | 0.0480 | P | |
| ClEl | 49.8 | 0.0144 | SC | |

*3.4. BMI Community Stability*

The MAR results identified significant interactions between the various BMI FFGs during the study. Model fits ($R^2$) were all >0.5 (Table 3). All functional feeding groups experienced positive interactive growth rates over the time-series at the REF site, as shown by the positive diagonal elements of Figure 5A. Between groups, predator growth rates were negatively affected by scraper and collector-gatherer abundance. Collector-filterers positively influenced collector-gatherers. Scrapers positively influenced abundance of both collector-gatherers and predators.

**Table 3.** Resilience and reactivity metrics from the MAR analysis. Metrics are derived from Ives, Dennis, Cottingham and Carpenter [60] as described in the methods section.

| Attribute | Metric | Reference | | Below Dam | | Above Dam | |
|---|---|---|---|---|---|---|---|
| | | Best-Fit | Bootstrap | Best-Fit | Bootstrap | Best-Fit | Bootstrap |
| Resilience | $\det(\mathbf{B})^{2/p}$ | 0 | 0 | 0.3 | 0 | 0.22 | 0.13 |
| Resilience | $\max(\lambda_\mathbf{B})$ | 0.88 | 0.89 | 0.94 | 0.93 | 1 | 0.94 |
| Resilience | $\max(\lambda_{\mathbf{B}\otimes\mathbf{B}})$ | 0.77 | 0.79 | 0.89 | 0.87 | 1 | 0.88 |
| Reactivity | $-tr(\Sigma)/tr(\mathbf{V}_\infty)$ | −0.34 | −0.39 | −0.08 | −0.18 | −0.01 | −0.21 |
| Reactivity | $\max(\lambda_{\mathbf{B}'\mathbf{B}}) - 1$ | 1.46 | 0.52 | 6.73 | 2.99 | 4.44 | 2.66 |

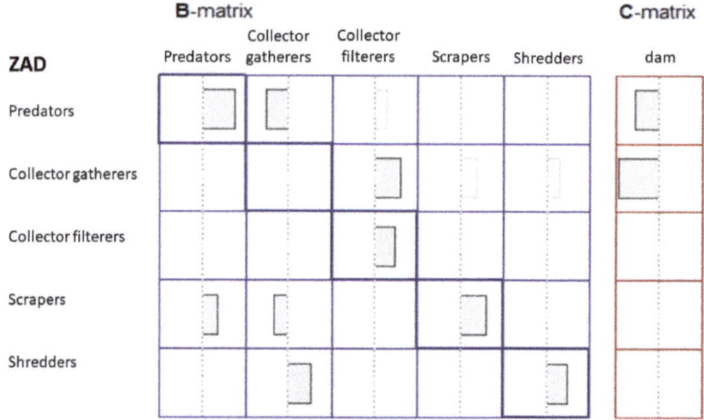


**Figure 5.** *Cont.*

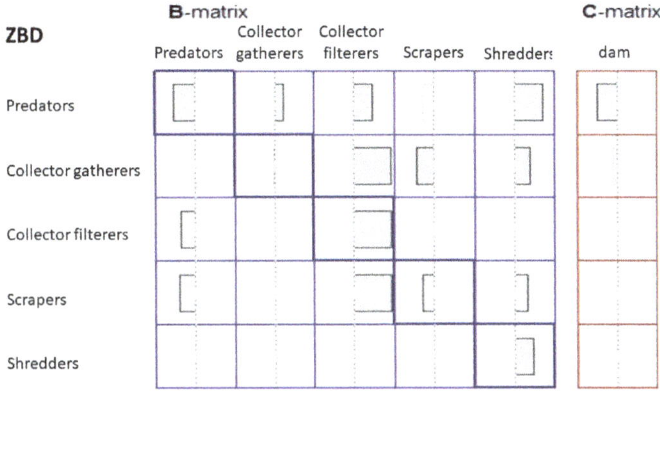



(C)

**Figure 5.** MAR model interaction strengths of the deseasoned, z-score-transformed BMI functional feeding group community on the Eightmile River at (**A**) the reference site (REF), (**B**) the above dam site (ZAD), and (**C**) the below dam site (ZBD). Bars extending to the right and left of the dotted lines represent positive and negative interactions, respectively. Interactions in the best-fit model that were excluded by bootstrapping are plotted as lighter, hatched bars. Density-dependent changes in BMI functional feeding groups are displayed along the diagonal of the B-matrix. The C-matrix in the two dam sites evaluates the effects of dam stage as a covariate influencing functional feeding group interactions.

The interactive growth rate effects were more complex at the two dam sites, and dam stage was a significant influence on predator growth rates at both sites and on collector-gatherer growth rates at ZAD. At ZAD, all taxa except collector-gatherers displayed positive growth rates (i.e., the positive diagonal element in Figure 5B). Predator growth rates positively influenced scraper abundance. Collector-gatherer growth rates positively influenced shredders but had a negative effect on predators and scrapers. Collector-filterer abundances positively affected collector-filterers. Dam removal had significant negative impacts on predator and collector-gatherer growth rates.

At ZBD, (Figure 5C), collector-filterers and shredders displayed positive growth with increasing abundance, collector-gatherers showed no significant interaction, and predators and scrapers experienced negative density-dependent growth over time. Predator abundance negatively influenced the growth rates of both collector-filterers and scrapers, while scraper abundance negatively impacted collector-gatherer growth rates. Collector-filterers positively affected growth rates of all but shredders. Only predator growth rates were negatively influenced by dam removal.

The MAR resilience metrics indicated that the BMI community of the REF site was the most resilient and least reactive site (Table 3). Most resilient means that environmental perturbations due to dam removal changed the community the least and that the rate of return of community structure was the fastest (Ives et al. 2003). The reactivity measures ranked the dam sites in opposite fashion. The first reactivity metric indicated that ZAD was more reactive. ZBD was the most reactive from the perspective of the second measure because ZBD had the most asymmetric eigenstructure.

## 4. Discussion

The effects of dam removal on BMI community structure, function, and stability are complex and spatiotemporally-variable [20,26–28]. The removal of the Zemko Dam constituted a major change to the stream system from a previously impounded state that had likely persisted for centuries prior to river restoration. Dam removal remains a difficult task, and our results and other similar studies reveal the complex and interacting effects of dam legacy, stream restoration, and BMI community reorganization that occurs in response to the restoration of river flow [10,21–23,25,62].

### 4.1. Changes in Functional Feeding Group in BMI Communities in Response to Dam Removal

The site-specific BMI community responses implied that the environmental changes caused by the dam removal itself, rather than new environmental conditions such as flow or river connectivity, were responsible for the observed changes over the study period (e.g., see Poulos and Chernoff [36]). Variation in BMI community composition is characteristic of even healthy, unperturbed stream habitats [63,64], and was exemplified in our study by the temporal variation in BMI community composition at the reference site. Species-level replacement (i.e., "turnover") may have occurred undetected, given the lack of species-specific BMI identification in this study. This may explain why there were no significant differences in diversity or richness for variables like EPT-, which are generally utilized as indicators of water quality, and thus, would be expected to change in response to river restoration [47].

While the dam removal represented a major disturbance to in-channel river traits, many of the other environmental characteristics of the sites (e.g., canopy, surrounding land cover) remained constant throughout the study period. Given the magnitude of these site differences, it is perhaps not surprising that there was only one BMI functional metric that changed with dam stage across sites, non-insect abundance, which spiked at ZAD during drawdown. Such a pattern does not typically appear in studies of other dam removals, but the greater BMI abundance at ZAD at this time period could be a product of site productivity, e.g., organic matter inputs from upstream and a lack of canopy, which is a pattern observed elsewhere [20,34,65].

Thus, we did not observe a clear shift in dominance from lentic to lotic BMI taxa in response to the dam removal. The above dam site was functionally unique, containing 17 indicator taxa, and was characterized primarily by tolerant and lentic taxa in contrast to the more varied BMI community composition at the other two sites over the study period. This site-specific dam removal impacts on the BMI indicator taxa, diversity, and FFG abundances were expected, as evidenced by the continued dominance of many tolerant, lentic-specialist taxa at the above dam site throughout the study period.

The consistently-different BMI community structure at the above and below dam sites is consistent with other prior dam removal research. The greater BMI abundance above the dam persisted across dam stages, which was, in part, due to a few especially abundant taxa (particularly during drawdown). Others have documented higher abundance of collector-gatherers above vs. below impoundments (e.g., [66]). While our sample averages would seem to support this (e.g., pre-removal average at ZAD of 60.0 per sample vs. 14.5 at ZBD), sample variability did not produce statistical differences. These differences persisted post-removal at ZAD and ZBD but were only significant during drawdown. It is worth considering whether or not there was increased patchiness and sample variability downstream from the former dam, in part due to disturbance effects that, while differing with dam stage, were likely greater at the downstream sample site [67,68].

One of the most pronounced effects was a decline in collector-filterers at ZBD. Since they feed on fine particulate organic matter [43], they may have been negatively-impacted by post-dam removal sedimentation supplied by the removal of the upstream impoundment [22,45]. Changes in macroinvertebrate assemblages in response to temporal changes in organic matter are well-documented [69–72] and demonstrate the importance of debris flow following dam removal as a regulator of stream community composition.

### 4.2. Dam Removal Effects on BMI Community Structure

Our results support the idea of a lag in recovery and increased or persistent impairment of the below dam community, as documented by others (e.g., [21,28]). Although the community composition of the three samples sites was becoming more similar over the study period, the three sample sites remained distinct in terms of BMI composition, which suggests that the BMI community continued to be in a state of reorganization, even 3 years post-dam removal. Similar continued community reorganization effects have also been well-documented in other studies (e.g., [73,74]). For example, Ahearn and Dahlgren [74] found downstream dam removal effects on BMI communities up to tens of kilometers away. This helps to explain the persistent lag of "recovery" in the below dam community and its relatively greater downstream effects, which may persist beyond the timespan of usual short-term ecological monitoring programs that evaluate stream recovery following restoration.

*4.3. Community Stability and Dam Removal*

Stream resilience includes the concepts of rates of return as well as inertia [36]. REF was the most resilient and least reactive of sites, while ZBD displayed intermediate resilience and greatest reactivity (according to one metric), and ZAD was the least resilient sample site (Table 3). The latter is not surprising because the channel width, substrate, and flow changed most dramatically. Toward the end of the study period, faster flow, a narrower and deeper channel, and even some patches of rocky substrate were established. Both dam sites were much more reactive than the reference site. The more dynamic fluctuation of the ZAD BMI community relative to the communities at ZBD and REF in response to dam removal is not surprising since this site experienced the largest changes (i.e., it went from a pool to a stream) over the study period. These results corroborate other dam removal research on BMI dynamics [21,28,75] that also identified dramatic changes in invertebrate community composition at above the dam sites. These metrics were also calculated for the fish community in a previous study [37], which also showed that the fish communities at the two dam sites were less resilient and more reactive than REF. Interestingly, both the fish and BMI data indicated that ZBD was the most reactive for the metric $\mathbf{max}\left(\lambda_{\mathbf{B'B}}\right) - 1$. In both cases, this was due to the presence of a dominant eigenvector within the species interaction matrix, $\mathbf{B}$, that resulted from a pattern of turnover.

There were no clear interannual trends across sites, but that does not remove the possibility of other environmental variables as influences on the results of the present study. For example, floods may have similar effects on substrate movement [13,21,62,76,77] and drought events can influence invertebrate movement and abundance (e.g., [68,78]). In addition to sporadic high-water events, some of which the sampling schedule naturally avoided, the study period included drought conditions of various duration and severity in three years.

The use of the far downstream site (REF) as reference was validated by metrics which included the site's greater diversity and abundance of taxa associated with higher water quality (EPT-, which also represented the majority of indicator taxa for the site), and shredders (consistent with the quality of organic matter there); similar results have been documented elsewhere [25]. Finer scale analysis could support what appears to be a trend of increasing shredder abundance at all three sites, which is often the case for undammed streams [71]. While the REF site was presumed to be beyond the reach of dam removal effects, it is interesting to note that species richness was higher there post-removal (average of 13.5 taxa/sample) than during pre-removal or drawdown (versus 9.9 and 7.1 taxa/sample, respectively). It could also be that the increased richness at REF was responsible for the decreased dissimilarity between REF and the two dam sites.

## 5. Conclusions

The goals of dam removal projects include increased connectivity, restoration of natural flow, and re-establishment of biological communities and river system function indicative of free-flowing rivers. While the goal of river restoration is clear, our results also demonstrate that the dam removal process comprises a high-magnitude change to a stream system that has experienced chronic and sustained flow alteration regime for more than 200 years. Our results indicate that community reorganization following dam removal is site-specific, and that while the above and below dam sites are part of the same stream system, dramatically different changes can occur at each site in response to dam removal. Our results are consistent with the growing body of research on the ecological effects of dam removal in that it demonstrates that stream BMI community composition and interactions shift quickly in response to changes in water quantity and quality, and that the greater the alteration of a site prior to, during, and after a dam removal event, the more reactive and less resilient the stream community.

The results from this study signal the need for implementing more widespread, comprehensive, and long-term biological monitoring of dam removals to generate better decision support tools for managers to carry out successful stream restoration. While some of the trends we observed were consistent with other prior dam removal research, our results also indicated that each stream community will respond uniquely to dam removal, and that site conditions during and following stream restoration can have significant effects on the post-dam removal biotic community composition and taxa interactions. Thus, simply restoring the hydrological regime will not necessarily bring back free-flowing stream BMI taxa within three years following a dam removal event, especially when the recovery process relies on complex interactions among taxa, local species pools, and environmental conditions. Few studies have investigated the interactions among BMI functional feeding groups in a river restoration context such as the one MAR analysis of BMI community stability. More multivariate analyses using complex statistical models and other data mining techniques on individual and multiple studies can provide great insights into general trends in stream ecosystem dynamics following dam removal worldwide. Our results also suggest that the BMI community continues to recover and change over time and that community recovery following dam removal may take decades or even centuries, especially since the original dam structure at the Zemko site had been in place for over 200 years. While costly, long-term biological monitoring of stream communities well after the dam removal itself will provide many answers about the process and mechanisms of recovery in response to dam removal.

**Author Contributions:** B.C. designed the experiment. A.W.W. orchestrated the dam removal. B.C., K.E.M., M.L.K., and R.H. conducted the BMI surveys. H.M.P. analyzed the data. All authors contributed to the writing and syntheses.

**Funding:** This project was completed using combined funds from Menakka and Essel Bailey, the Schumann Foundation, The Nature Conservancy, or project grants from Wesleyan University to BC or the College of the Environment. Student support and internships were provided by the Mellon Foundation, the Hughes Foundation, and Schumann Foundations.

**Acknowledgments:** The authors thank Valerie Marinelli and Susan Lastrina for administrative and logistical support and Sarah Donelan for field assistance.

**Conflicts of Interest:** The authors declare no conflict of interest.

## Appendix A

**Table A1.** Location, basin, and habitat characteristics of the three study sites (ZAD-above dam, ZBD-below dam, and REF-reference).

| Location, Basin and Habitat Characteristic Information on Study Sites | | | | | | | | |
| --- | --- | --- | --- | --- | --- | --- | --- | --- |
| Site | Basin Area (km$^2$) | Distance from Source (km) | Mean Wet Width (m) [I] | Dominant Substrate [II] | Canopy | % Riffles [III] | Water Slope (%) of Reach | N | W |
| ZAD | 16.92 | 0.74 | 5.25 | 43% silt, 35% sand | 0% | 0 | 0.14 | 41°29′41″ | 72°16′59″ |
| ZBD | 17.11 | 0.96 | 5.33 | 38% gravel, 30% sand | 67% | 25 [IV] | 0.15 | 41°29′34″ | 72°16′58″ |
| REF | 54.39 | 7.93 | 7.83 | 57% cobble, 22% gravel | 50% | 65 | 0.98 | 41°26′31″ | 72°18′22″ |

[I] = 2009–2010. [II] = mean % in 2009 of two greatest size categories; silt=fine, suspended, sand <0.25 cm, gravel = 0.25–5.1 cm. [III] = in study fishing reach and riffle. [IV] = Estimated.

## Appendix B

**Table A2.** Post-hoc results. Abbreviations: lsmean stands for least squares mean.

| Site | Stage | lsmean | SE | df | Lower CL | Upper CL | Post-hoc |
|---|---|---|---|---|---|---|---|
| **Site by Stage Interaction Effects** | | | | | | | |
| | | | *Abundance* | | | | |
| REF | drawdown | 29.05 | 12.30 | 70.29 | −6.04 | 64.14 | ab |
| REF | post | 58.75 | 13.58 | 67.78 | 19.97 | 97.53 | abc |
| REF | pre | 45.06 | 11.77 | 47.75 | 10.96 | 79.16 | abc |
| ZAD | drawdown | 101.55 | 14.23 | 90.14 | 61.22 | 141.87 | c |
| ZAD | post | 77.09 | 12.18 | 45.12 | 41.72 | 112.47 | bc |
| ZAD | pre | 82.13 | 35.96 | 174.10 | −18.58 | 182.84 | abc |
| ZBD | drawdown | 38.76 | 12.92 | 76.04 | 1.98 | 75.53 | ab |
| ZBD | post | 27.53 | 13.79 | 71.19 | −11.79 | 66.86 | a |
| ZBD | pre | 72.31 | 21.28 | 127.51 | 12.45 | 132.17 | abc |
| | | | *Richness* | | | | |
| REF | drawdown | 7.08 | 1.12 | 10.07 | 3.16 | 10.99 | ab |
| REF | pre | 9.86 | 1.09 | 8.91 | 5.94 | 13.79 | abc |
| REF | post | 13.49 | 1.06 | 17.51 | 10.16 | 16.81 | c |
| ZAD | pre | 8.12 | 2.40 | 95.30 | 1.33 | 14.91 | abc |
| ZAD | post | 9.39 | 0.98 | 13.06 | 6.15 | 12.64 | ab |
| ZAD | drawdown | 10.14 | 1.20 | 13.35 | 6.18 | 14.10 | bc |
| ZBD | post | 6.52 | 1.07 | 18.25 | 3.18 | 9.86 | ab |
| ZBD | drawdown | 6.70 | 1.15 | 11.04 | 2.78 | 10.62 | a |
| ZBD | pre | 7.16 | 1.60 | 33.07 | 2.43 | 11.90 | abc |
| | | | *EPT-* | | | | |
| REF | drawdown | 16.20 | 5.43 | 13.38 | −1.66 | 34.06 | a |
| REF | pre | 18.98 | 5.12 | 10.53 | 1.26 | 36.69 | a |
| REF | post | 25.85 | 5.40 | 28.02 | 9.68 | 42.02 | a |
| ZAD | pre | 11.54 | 13.63 | 147.58 | −26.71 | 49.79 | a |
| ZAD | drawdown | 14.16 | 6.01 | 19.77 | −4.49 | 32.81 | a |
| ZAD | post | 22.12 | 4.83 | 17.71 | 6.93 | 37.32 | a |
| ZBD | post | 10.60 | 5.48 | 29.70 | −5.73 | 26.94 | a |
| ZBD | drawdown | 11.08 | 5.62 | 15.21 | −7.00 | 29.16 | a |
| ZBD | pre | 13.54 | 8.63 | 46.83 | −11.47 | 38.55 | a |
| | | | *EVAR* | | | | |
| REF | drawdown | 0.70 | 0.05 | 7.60 | 0.51 | 0.89 | cd |
| REF | post | 0.64 | 0.05 | 18.85 | 0.47 | 0.80 | bd |
| REF | pre | 0.66 | 0.05 | 7.41 | 0.47 | 0.86 | abcd |
| ZAD | drawdown | 0.45 | 0.06 | 10.83 | 0.26 | 0.64 | ab |
| ZAD | post | 0.48 | 0.05 | 15.28 | 0.32 | 0.64 | ac |
| ZAD | pre | 0.46 | 0.11 | 101.39 | 0.14 | 0.78 | abcd |
| ZBD | drawdown | 0.62 | 0.05 | 8.58 | 0.43 | 0.81 | cd |
| ZBD | post | 0.68 | 0.05 | 19.47 | 0.51 | 0.84 | bd |
| ZBD | pre | 0.58 | 0.07 | 27.10 | 0.36 | 0.80 | abcd |
| | | | *Collector-gatherers* | | | | |
| REF | drawdown | 11.83 | 5.19 | 9.05 | −6.86 | 30.53 | a |
| REF | post | 16.40 | 5.76 | 27.05 | −0.90 | 33.70 | a |
| REF | pre | 22.67 | 5.62 | 12.27 | 3.89 | 41.46 | a |
| ZAD | post | 39.22 | 8.11 | 65.27 | 16.04 | 62.40 | a |
| ZAD | pre | 59.99 | 22.85 | 88.99 | −4.78 | 124.77 | ab |
| ZAD | drawdown | 80.10 | 8.68 | 50.35 | 55.03 | 105.17 | b |
| ZBD | post | 8.09 | 8.11 | 65.27 | −15.09 | 31.28 | a |
| ZBD | pre | 14.55 | 13.24 | 62.04 | −23.39 | 52.48 | a |
| ZBD | drawdown | 16.54 | 7.67 | 36.35 | −6.00 | 39.07 | a |

**Table A2.** *Cont.*

| | Site by Stage Interaction Effects | | | | | | |
|---|---|---|---|---|---|---|---|
| **Site** | **Stage** | **lsmean** | **SE** | **df** | **Lower CL** | **Upper CL** | **Post-hoc** |
| | | | *Collector-filterers* | | | | |
| REF | drawdown | 8.34 | 2.61 | 11.58 | −0.51 | 17.19 | a |
| REF | post | 6.98 | 2.85 | 30.90 | −1.49 | 15.45 | a |
| REF | pre | 9.23 | 2.81 | 15.20 | 0.18 | 18.27 | a |
| ZAD | drawdown | 2.25 | 4.26 | 53.34 | −10.02 | 14.53 | a |
| ZAD | post | 7.67 | 3.96 | 66.86 | −3.66 | 19.00 | a |
| ZAD | pre | −0.21 | 11.09 | 88.98 | −31.65 | 31.23 | a |
| ZBD | drawdown | 9.07 | 3.78 | 40.03 | −1.98 | 20.11 | ab |
| ZBD | post | 8.23 | 3.96 | 66.86 | −3.10 | 19.56 | ab |
| ZBD | pre | 32.24 | 6.48 | 67.04 | 13.72 | 50.75 | b |
| | | | Site Effects | | | | |
| | | | $H'$ | | | | |
| ZBD | NS | 1.48 | 0.07 | 44.25 | 1.32 | 1.64 | a |
| ZAD | NS | 1.48 | 0.09 | 109.44 | 1.26 | 1.70 | a |
| REF | NS | 1.85 | 0.05 | 22.62 | 1.72 | 1.98 | b |
| | | | *Shredders* | | | | |
| ZAD | NS | 0.82 | 2.14 | 84.77 | −4.38 | 6.03 | ab |
| ZBD | NS | 1.29 | 1.48 | 56.77 | −2.34 | 4.92 | a |
| REF | NS | 5.83 | 0.87 | 16.19 | 3.50 | 8.15 | b |

## Appendix C

**Table A3.** Functional feeding groups (FFG) included: CG=collector-gatherer, CF=collector-filterer, P=predator, SH=shredder, SC=scraper. A designation was given when the majority of species in a family shared a common feeding strategy, and none was specified when species had diverse feeding strategies (CT DEEP, 2004; US EPA 2011). Relative average abundance per sample denoted by shading: white = 1+, light gray = 0.1–1, dark gray = 0. Family designations from the beginning of the study were maintained to enable comparison across all sampling years. This sometimes required combining taxonomic groups: Oligochaeta include Nemertina, Baetidae include Ameltidae, and Polycentripodidae include Psychomyiidae. When FFG are not specified it is because the family is comprised of species with different feeding strategies.

| Benthic Macroinvertebrate Taxa List by Group and Site | | | | Relative Abundance by Site | | | | |
|---|---|---|---|---|---|---|---|---|
| **Order/Class** | **Family** | **Acronym** | **FFG** | **Non-Insect** | **EPT-** | **ZAD** | **ZBD** | **REF** |
| Amphipoda | | AmSc | CG | X | | | | |
| Hirudinea | | AnHr | P | X | | | | |
| Oligochaeta | | AnOl | CG | X | | | | |
| Acariformes | Hydracarina | ArHy | P | X | | | | |
| Bivalvia | Sphaeriidae | BvSp | CF | X | | | | |
| Coleoptera | Dytiscidae | ClDy | P | | | | | |
| Coleoptera | Elmidae | ClEl | SC | | | | | |
| Coleoptera | Grynidae | ClGy | P | | | | | |
| Coleoptera | Haliplidae | ClHa | SH | | | | | |
| Coleoptera | Hydrophilidae | ClHy | | | | | | |
| Coleoptera | Psephenidae | ClPs | SC | | | | | |
| Collembola | Isotomidae | CmIs | CG | X | | | | |
| Collembola | Sminthuridae | CmSm | CG | X | | | | |
| Conchostraca | Eulimnadia | CoEu | | X | | | | |
| Diptera | Athericidae | DtAt | P | | | | | |

**Table A3.** *Cont.*

| Benthic Macroinvertebrate Taxa List by Group and Site | | | | Relative Abundance by Site | | | | |
|---|---|---|---|---|---|---|---|---|
| Order/Class | Family | Acronym | FFG | Non-Insect | EPT- | ZAD | ZBD | REF |
| Diptera | Chironomidae | DtCh | CG | | | | | |
| Diptera | Ceratopogonidae | DtCr | P | | | | | |
| Diptera | Culicidae | DtCu | CF | | | | | |
| Diptera | Empididae | DtEm | P | | | | | |
| Diptera | Simuliidae | DtSm | CF | | | | | |
| Diptera | Tabanidae | DtTb | P | | | | | |
| Diptera | Tipulidae | DtTp | SH | | | | | |
| Ephemeroptera | Baetidae* | EpBt | CG | | X | | | |
| Ephemeroptera | Caenidae | EpCn | CG | | X | | | |
| Ephemeroptera | Ephemerellidae | EpEp | CG | | X | | | |
| Ephemeroptera | Heptagenidae | EpHp | SC | | X | | | |
| Ephemeroptera | Isonychiidae | EpIs | CG | | X | | | |
| Ephemeroptera | Leptohyphidae | EpLh | CG | X | X | | | |
| Ephemeroptera | Lepotophlebiidae | EpLp | CG | X | X | | | |
| Gastropoda | Hydrobiidae | GaHy | SC | X | | | | |
| Gastropoda | Physidae | GaPh | CG | X | | | | |
| Gastropoda | Planorbidae | GaPl | SC | X | | | | |
| Hemiptera | Corixidae | HmCo | P | X | | | | |
| Isopoda | | IsSw | CG | X | | | | |
| Lepidoptera | Pyralidae | LpPy | SH | | | | | |
| Megaloptera | Corydalidae | MgCr | P | | | | | |
| Megaloptera | Sialidae | MgSi | P | | | | | |
| Odonata | Aeshnidae | OdAs | P | | | | | |
| Odonata | Calopterygidae | OdCl | P | | | | | |
| Odonata | Coenagrionidae | OdCn | P | | | | | |
| Odonata | Gomphidae | OdGm | P | | | | | |
| Odonata | Libellulidae | OdLb | P | | | | | |
| Odonata | Lestidae | OdLS | P | | | | | |
| Platyhelminthes | | PhTb | P | X | | | | |
| Plecoptera | Chloroperlidae | PlCh | | | X | | | |
| Plecoptera | Capnidae | PlCp | SH | | X | | | |
| Plecoptera | Leuctridae | PlLc | SH | | X | | | |
| Plecoptera | Perlidae | PlPl | P | | X | | | |
| Plecoptera | Perlodidae | PlPr | P | | X | | | |
| Plecoptera | Taeniopterygidae | PlTn | SH | | X | | | |
| Trichoptera | Brachycentridae | TcBr | SH | | X | | | |
| Trichoptera | Glossosomatidae | TcGl | SC | | X | | | |
| Trichoptera | Hydroptilidae | TcHd | | | X | | | |
| Trichoptera | Helicopsychidae | TcHl | SC | | X | | | |
| Trichoptera | Hydropsychidae | TcHy | CF | | | | | |
| Trichoptera | Limnephilidae | TcLm | SH | | X | | | |
| Trichoptera | Lepidostomatidae | TcLp | SC | | X | | | |
| Trichoptera | Leptoceridae | TcLt | | | X | | | |

**Table A3.** *Cont.*

| Benthic Macroinvertebrate Taxa List by Group and Site | | | | Relative Abundance by Site | | | | |
| --- | --- | --- | --- | --- | --- | --- | --- | --- |
| Order/Class | Family | Acronym | FFG | Non-Insect | EPT- | ZAD | ZBD | REF |
| Trichoptera | Odontoceridae | TcOd | SC | | X | | | |
| Trichoptera | Philopotamidae | TcPh | CF | | X | | | |
| Trichoptera | Polycentropodidae | TcPl/Ps | | | X | | | |
| Trichoptera | Rhyacophilidae | TcRy | P | | X | | | |

## Appendix D

**Table A4.** Similarity percentage (SIMPER) and two-way analysis of similarity (ANOSIM) results by dam stage and site-by-site comparisons. BMI family acronyms are described in Appendix B.

| Aggregate % Dissimilarity | Dam Stage | BMI Family | % Dissimilarity per Taxon | Mean Abundance Group 1 | Mean Abundance Group 2 |
| --- | --- | --- | --- | --- | --- |
| | | **Reference-Above Dam** | | | |
| 81.7 | pre | DtCh | 32.1 | 11.1 | 34.0 |
| | | EpLp | 39.9 | 0.0 | 6.7 |
| | | AmSc | 46.5 | 0.0 | 5.7 |
| 88.2 | drawdown | AnOl | 26.3 | 0.1 | 32.8 |
| | | AmSc | 42.3 | 0.0 | 23.1 |
| 72.2 | post | DtCh | 20.8 | 14.0 | 21.2 |
| | | TcHy | 32.3 | 8.4 | 4.6 |
| | | AnOl | 43.3 | 2.5 | 9.1 |
| | | TcBr | 48.6 | 4.2 | 0.0 |
| | | **Above dam-Below Dam** | | | |
| 80.8 | pre | DtCh | 29.1 | 34.0 | 12.6 |
| | | TcHy | 44.7 | 0.0 | 24.9 |
| | | ClEl | 51.6 | 0.3 | 7.7 |
| 78.1 | drawdown | AnOl | 26.7 | 32.8 | 0.9 |
| | | AmSc | 43.2 | 23.1 | 1.0 |
| 77.3 | post | DtCh | 26.6 | 21.2 | 4.9 |
| | | AnOl | 40.3 | 9.1 | 0.2 |
| | | TcHy | 49.8 | 4.6 | 5.1 |
| | | **Reference-Below Dam** | | | |
| 76.1 | pre | TcHy | 22.0 | 6.0 | 24.9 |
| | | DtCh | 38.0 | 11 | 12.6 |
| | | ClEl | 47.0 | 2.0 | 7.7 |
| 76.9 | drawdown | DtCh | 18.6 | 4.2 | 10.3 |
| | | TcHy | 32.5 | 6.2 | 5.4 |
| | | TcBr | 45.4 | 7.8 | 0.1 |
| 70.4 | post | DtCh | 16.1 | 14.0 | 4.9 |
| | | TcHy | 29.1 | 8.4 | 5.1 |
| | | TcBr | 37.0 | 4.2 | 0.0 |
| | | EpHp | 43.4 | 1.3 | 3.9 |
| | | PlLc | 49.4 | 2.7 | 2.6 |

Families differed significantly among dam stages (R = 0.091, *P* = 0.0374) and sites (R = 0.60, *P* = 0.0001) according to a two-way analysis of similarity (ANOSIM).

## References

1. Eng, K.; Wolock, D.M.; Carlisle, D.M. River flow changes related to land and water management practices across the conterminous united states. *Sci. Total Environ.* **2013**, *463*, 414–422. [CrossRef]
2. Rosenberg, D.M.; McCully, P.; Pringle, C.M. Global-scale environmental effects of hydrological alterations: Introduction. *BioScience* **2000**, *50*, 746–751. [CrossRef]
3. Brandt, S.A. Classification of geomorphological effects downstream of dams. *Catena* **2000**, *40*, 375–401. [CrossRef]
4. Hart, D.D.; Poff, N.L. A special section on dam removal and river restoration. *BioScience* **2002**, *52*, 653–655. [CrossRef]
5. Poff, N.L.; Zimmerman, J.K.H. Ecological responses to altered flow regimes: A literature review to inform the science and management of environmental flows. *Freshw. Biol.* **2010**, *55*, 194–205. [CrossRef]
6. Perkin, J.S.; Gido, K.B.; Cooper, A.R.; Turner, T.F.; Osborne, M.J.; Johnson, E.R.; Mayes, K.B. Fragmentation and dewatering transform great plains stream fish communities. *Ecol. Monogr.* **2015**, *85*, 73–92. [CrossRef]
7. Petts, G.E. Long-term consequences of upstream impoundment. *Environ. Conserv.* **1980**, *7*, 325–332. [CrossRef]
8. Poff, N.L.; Allan, J.D.; Bain, M.B.; Karr, J.R.; Prestegaard, K.L.; Richter, B.D.; Sparks, R.E.; Stromberg, J.C. The natural flow regime. *BioScience* **1997**, *47*, 769–784. [CrossRef]
9. Alo, D.; Turner, T.F. Effects of habitat fragmentation on effective population size in the endangered rio grande silvery minnow. *Conserv. Biol.* **2005**, *19*, 1138–1148. [CrossRef]
10. Bushaw-Newton, K.L.; Hart, D.D.; Pizzuto, J.E.; Thomson, J.R.; Egan, J.; Ashley, J.T.; Johnson, T.E.; Horwitz, R.J.; Keeley, M.; Lawrence, J. An integrative approach towards understanding ecological responses to dam removal: The manatawny creek study. *J. Am. Water Resour. Assoc.* **2002**, *38*, 1581–1599. [CrossRef]
11. Poff, N.L.; Olden, J.D.; Merritt, D.M.; Pepin, D.M. Homogenization of regional river dynamics by dams and global biodiversity implications. *Proc. Natl. Acad. Sci. USA* **2007**, *104*, 5732–5737. [CrossRef]
12. Bernhardt, E.S.; Palmer, M.A. River restoration: The fuzzy logic of repairing reaches to reverse catchment scale degradation. *Ecol. Appl.* **2011**, *21*, 1926–1931. [CrossRef]
13. Stanley, E.H.; Doyle, M.W. Trading off: The ecological effects of dam removal. *Front. Ecol. Environ.* **2003**, *1*, 15–22. [CrossRef]
14. Bellmore, R.; Duda, J.J.; Craig, L.S.; Greene, S.L.; Torgersen, C.E.; Collins, M.J.; Vittum, K. Status and trends of dam removal research in the united states. *Wiley Interdiscip. Rev. Water* **2017**, *4*, e1164. [CrossRef]
15. Foley, M.M.; Bellmore, J.; O'Connor, J.E.; Duda, J.J.; East, A.E.; Grant, G.; Anderson, C.W.; Bountry, J.A.; Collins, M.J.; Connolly, P.J. Dam removal: Listening in. *Water Resour. Res.* **2017**, *53*, 5229–5246. [CrossRef]
16. Ding, L.; Chen, L.; Ding, C.; Tao, J. Global trends in dam removal and related research: A systematic review based on associated datasets and bibliometric analysis. *Chin. Geogr. Sci.* **2019**, *29*, 1–12. [CrossRef]
17. Gangloff, M.M. Taxonomic and ecological tradeoffs associated with small dam removals. *Aquat. Conserv. Mar. Freshw. Ecosyst.* **2013**, *23*, 475–480. [CrossRef]
18. Bednarek, A.T. Undamming rivers: A review of the ecological impacts of dam removal. *Environ. Manag.* **2001**, *27*, 803–814. [CrossRef]
19. Gillette, D.; Daniel, K.; Redd, C. Fish and benthic macroinvertebrate assemblage response to removal of a partially breached lowhead dam. *River Res. Appl.* **2016**, *32*, 1776–1789. [CrossRef]
20. Claeson, S.; Coffin, B. Physical and biological responses to an alternative removal strategy of a moderate-sized dam in washington, USA. *River Res. Appl.* **2015**, *32*, 1143–1152. [CrossRef]
21. Orr, C.H.; Kroiss, S.J.; Rogers, K.L.; Stanley, E.H. Downstream benthic responses to small dam removal in a coldwater stream. *River Res. Appl.* **2008**, *24*, 804–822. [CrossRef]
22. Pollard, A.I.; Reed, T. Benthic invertebrate assemblage change following dam removal in a wisconsin stream. *Hydrobiologia* **2004**, *513*, 51–58. [CrossRef]
23. Pizzuto, J. Effects of dam removal on river form and process. *BioScience* **2002**, *52*, 683–691. [CrossRef]
24. Mbaka, J.G.; Wanjiru Mwaniki, M. A global review of the downstream effects of small impoundments on stream habitat conditions and macroinvertebrates. *Environ. Rev.* **2015**, *23*, 257–262. [CrossRef]
25. Tiemann, J.S.; Gillette, D.P.; Wildhaber, M.L.; Edds, D.R. Effects of lowhead dams on riffle-dwelling fishes and macroinvertebrates in a midwestern river. *Trans. Am. Fish. Soc.* **2004**, *133*, 705–717. [CrossRef]

26. Hansen, J.F.; Hayes, D.B. Long germ implications of dam removal for macroinvertebrate communities in michigan and wisconsin rivers, united states. *River Res. Appl.* **2012**, *28*, 1540–1550. [CrossRef]

27. Louhi, P.; Mykrä, H.; Paavola, R.; Huusko, A.; Vehanen, T.; Mäki-Petäys, A.; Muotka, T. Twenty years of stream restoration in finland: Little response by benthic macroinvertebrate communities. *Ecol. Appl.* **2011**, *21*, 1950–1961. [CrossRef]

28. Renöfält, B.; Lejon, A.G.; Jonsson, M.; Nilsson, C. Long-term taxon-specific responses of macroinvertebrates to dam removal in a mid-sized swedish stream. *River Res. Appl.* **2013**, *29*, 1082–1089. [CrossRef]

29. Foley, M.M.; Warrick, J.A.; Ritchie, A.; Stevens, A.W.; Shafroth, P.B.; Duda, J.J.; Beirne, M.M.; Paradis, R.; Gelfenbaum, G.; McCoy, R. Coastal habitat and biological community response to dam removal on the elwha river. *Ecol. Monogr.* **2017**, *87*, 552–577. [CrossRef]

30. Itsukushima, R.; Ohtsuki, K.; Sato, T.; Kano, Y.; Takata, H.; Yoshikawa, H. Effects of sediment released from a check dam on sediment deposits and fish and macroinvertebrate communities in a small stream. *Water* **2019**, *11*, 716. [CrossRef]

31. Carlson, P.E.; Donadi, S.; Sandin, L. Responses of macroinvertebrate communities to small dam removals: Implications for bioassessment and restoration. *J. Appl. Ecol.* **2018**, *55*, 1896–1907. [CrossRef]

32. Tonitto, C.; Riha, S.J. Planning and implementing small dam removals: Lessons learned from dam removals across the eastern united states. *Sustain. Water Resour. Manag.* **2016**, *2*, 489–507. [CrossRef]

33. Bellmore, J.R.; Pess, G.R.; Duda, J.J.; O'connor, J.E.; East, A.E.; Foley, M.M.; Wilcox, A.C.; Major, J.J.; Shafroth, P.B.; Morley, S.A. Conceptualizing ecological responses to dam removal: If you remove it, what's to come? *BioScience* **2019**, *69*, 26–39. [CrossRef]

34. Morley, S.A.; Duda, J.J.; Coe, H.J.; Kloehn, K.K.; McHenry, M.L. Benthic invertebrates and periphyton in the elwha river basin: Current conditions and predicted response to dam removal. *Northwest Sci.* **2008**, *82*, 179–197. [CrossRef]

35. Kibler, K.; Tullos, D.; Kondolf, G. Learning from dam removal monitoring: Challenges to selecting experimental design and establishing significance of outcomes. *River Res. Appl.* **2011**, *27*, 967–975. [CrossRef]

36. Poulos, H.M.; Chernoff, B. Effects of dam removal on fish community interactions and stability in the eightmile river system, connecticut, USA. *Environ. Manag.* **2016**, *59*, 1–15. [CrossRef]

37. Poulos, H.M.; Miller, K.E.; Kraczkowski, M.L.; Welchel, A.W.; Heineman, R.; Chernoff, B. Fish assemblage response to a small dam removal in the eightmile river system, connecticut, USA. *Environ. Manag.* **2014**, *54*, 1090–1101. [CrossRef]

38. Maloney, K.O.; Dodd, H.; Butler, S.E.; Wahl, D.H. Changes in macroinvertebrate and fish assemblages in a medium-sized river following a breach of a low-head dam. *Freshw. Biol.* **2008**, *53*, 1055–1068. [CrossRef]

39. Shirey, P.; Brueseke, M.; Kenny, J.; Lamberti, G. Long-term fish community response to a reach-scale stream restoration. *Ecol. Soc.* **2016**, *21*, 11. [CrossRef]

40. Fosburgh, J.; Case, K.; Hearne, D. *Eigtmile Wild and Scenic Study*; National Park Service: Haddam, CT, USA, 2006.

41. Bellucci, C.J.; Becker, M.; Beauchene, M. Characteristics of macroinvertebrate and fish communities from 30 least disturbed small streams in connecticut. *Northeast. Nat.* **2011**, *18*, 411–444. [CrossRef]

42. Olins, H. *Comparative Methods for Sampling and Analyzing Benthic Macroinvertebrate Communities in the Lower Connecticut River Basin*; Wesleyan University: Middletown, CT, USA, 2005.

43. Voshell, J.R.; Wright, A.B. *A Guide to Common Freshwater Invertebrates of North America*; McDonald & Woodward Pub.: Granville, OH, USA, 2002.

44. Tolonen, K.E.; Leinonen, K.; Marttila, H.; Erkinaro, J.; Heino, J. Environmental predictability of taxonomic and functional community composition in high-latitude streams. *Freshw. Biol.* **2016**, *62*, 1–16. [CrossRef]

45. Takao, A.; Kawaguchi, Y.; Minagawa, T.; Kayaba, Y.; Morimoto, Y. The relationships between benthic macroinvertebrates and biotic and abiotic environmental characteristics downstream of the yahagi dam, central japan, and the state change caused by inflow from a tributary. *River Res. Appl.* **2008**, *24*, 580–597. [CrossRef]

46. Wallace, J.B.; Webster, J.R. The role of macroinvertebrates in stream ecosystem function. *Annu. Rev. Entomol.* **1996**, *41*, 115–139. [CrossRef]

47. Lenat, D.R.; Barbour, M.T. Using benthic macroinvertebrate community structure for rapid, cost-effective, water quality monitoring: Rapid bioassessment. In *Proceeding of the SIL Conference on Biological Monitoring of Aquatic Systems*; Lewis Press: Birżebbuġa, Malta, 1994; pp. 187–215.

48. Penrose, D.; Call, S.M. Volunteer monitoring of benthic macroinvertebrates: Regulatory biologists' perspectives. *J. N. Am. Benthol. Soc.* **1995**, *14*, 203–209. [CrossRef]

49. Davis, S.; Golladay, S.W.; Vellidis, G.; Pringle, C.M. Macroinvertebrate biomonitoring in intermittent coastal plain streams impacted by animal agriculture. *J. Environ. Qual.* **2003**, *32*, 1036–1043. [CrossRef]

50. Shannon, C.E.; Weaver, W. *The Mathematical Theory of Information*; Springer Nature: Basel, Switzerland, 1949.

51. Heip, C.; Engels, P. Comparing species diversity and evenness indices. *J. Mar. Biol. Assoc. U. K.* **1974**, *54*, 559–563. [CrossRef]

52. Atmar, W.; Patterson, B.D. The measure of order and disorder in the distribution of species in fragmented habitat. *Oecologia* **1993**, *96*, 373–382. [CrossRef]

53. R Development Core Team. *A Language and Environment for Statistical Computing*; R Foundation for Statistical Computing: Vienna, Austria, 2018.

54. Dufrene, M.; Legendre, P. Species assemblages and indicator species: The need for a flexible asymmetrical approach. *Ecol. Monogr.* **1997**, *67*, 345–366. [CrossRef]

55. McCune, B.; Mefford, M.J. *PC-ORD Multivariate Analysis of Ecological Data*; MjM Software Design: Gleneden Beach, OR, USA, 2011.

56. Clarke, K.R. Non-parametric multivariate analyses of changes in community structure. *Aust. J. Ecol.* **1993**, *18*, 117–143. [CrossRef]

57. Hammer, Ø.; Harper, D.A.; Ryan, P.D. Past: Paleontological statistics software package for education and data analysis. *Palaeontol. Electron.* **2001**, *4*, 9.

58. Beisner, B.E.; Haydon, D.T.; Cuddington, K. Alternative stable states in ecology. *Front. Ecol. Environ.* **2003**, *1*, 376–382. [CrossRef]

59. Scheef, L.P. *Multivariate Autoregressive Modeling for Analysis of Community Time-Series Data, version 1*; The University of Texas: Austin, TX, USA, 2013.

60. Ives, A.; Dennis, B.; Cottingham, K.; Carpenter, S. Estimating community stability and ecological interactions from time-series data. *Ecol. Monogr.* **2003**, *73*, 301–330. [CrossRef]

61. Hampton, S.E.; Holmes, E.E.; Scheef, L.P.; Scheuerell, M.D.; Katz, S.L.; Pendleton, D.E.; Ward, E.J. Quantifying effects of abiotic and biotic drivers on community dynamics with multivariate autoregressive (mar) models. *Ecology* **2013**, *94*, 2663–2669. [CrossRef]

62. Doyle, M.W.; Stanley, E.H.; Orr, C.H.; Selle, A.R.; Sethi, S.A.; Harbor, J.M. Stream ecosystem response to small dam removal: Lessons from the heartland. *Geomorphology* **2005**, *71*, 227–244. [CrossRef]

63. Walsh, C.J. Biological indicators of stream health using macroinvertebrate assemblage composition: A comparison of sensitivity to an urban gradient. *Mar. Freshw. Res.* **2006**, *57*, 37–47. [CrossRef]

64. Harmsworth, G.R.; Young, R.; Walker, D.; Clapcott, J.; James, T. Linkages between cultural and scientific indicators of river and stream health. *N. Z. J. Mar. Freshw. Res.* **2011**, *45*, 423–436. [CrossRef]

65. Stanley, E.H.; Luebke, M.A.; Doyle, M.W.; Marshall, D.W. Short-term changes in channel form and macroinvertebrate communities following low-head dam removal. *J. N. Am. Benthol. Soc.* **2002**, *21*, 172–187. [CrossRef]

66. Martínez, A.; Larrañaga, A.; Basaguren, A.; Pérez, J.; Mendoza-Lera, C.; Pozo, J. Stream regulation by small dams affects benthic macroinvertebrate communities: From structural changes to functional implications. *Hydrobiologia* **2013**, *711*, 31–42. [CrossRef]

67. Tszydel, M.; Grzybkowska, M.; Kruk, A. Influence of dam removal on trichopteran assemblages in the lowland Drzewiczka river, Poland. *Hydrobiologia* **2009**, *630*, 75–89. [CrossRef]

68. Lake, P. Disturbance, patchiness, and diversity in streams. *J. N. Am. Benthol. Soc.* **2000**, *19*, 573–592. [CrossRef]

69. Quist, M.; Spiegel, J. Population demographics of catostomids in large river ecosystems: Effects of discharge and temperature on recruitment dynamics and growth. *River Res. Appl.* **2012**, *28*, 1567–1586. [CrossRef]

70. Quist, M.C.; Schultz, R.D. Effects of management legacies on stream fish and aquatic benthic macroinvertebrate assemblages. *Environ. Manag.* **2014**, *54*, 449–464. [CrossRef]

71. Smock, L.A.; Metzler, G.M.; Gladden, J.E. Role of debris dams in the structure and functioning of low-gradient headwater streams. *Ecology* **1989**, *70*, 764–775. [CrossRef]

72. Stanford, J.A.; Ward, J.; Liss, W.J.; Frissell, C.A.; Williams, R.N.; Lichatowich, J.A.; Coutant, C.C. A general protocol for restoration of regulated rivers. *US Dep. Energy Publ.* **1996**, *43*. [CrossRef]

73. Kanehl, P.D.; Lyons, J.; Nelson, J.E. Changes in the habitat and fish community of the milwaukee river, wisconsin, following removal of the woolen mills dam. *N. Am. J. Fish. Manag.* **1997**, *17*, 387–400. [CrossRef]

74. Ahearn, D.S.; Dahlgren, R.A. Sediment and nutrient dynamics following a low-head dam removal at murphy creek, california. *Limnol. Oceanogr.* **2005**, *50*, 1752–1762. [CrossRef]

75. Chiu, M.-C.; Yeh, C.-H.; Sun, Y.-H.; Kuo, M.-H. Short-term effects of dam removal on macroinvertebrates in a taiwan stream. *Aquat. Ecol.* **2013**, *47*, 245–252. [CrossRef]

76. Doyle, M.W.; Stanley, E.H.; Luebke, M.A.; Harbor, J.M. In dam removal: Physical, biological, and societal considerations. In Proceedings of the American Society of Civil Engineers Joint Conference on Water Resources Engineering and Water Resources Planning and Management, Minneapolis, MN, USA, 30 July–2 August 2000.

77. Stanley, E.H.; Catalano, M.J.; Mercado-Silva, N.; Orr, C.H. Effects of dam removal on brook trout in a wisconsin stream. *River Res. Appl.* **2007**, *23*, 792–798. [CrossRef]

78. Hall, L.W., Jr.; Anderson, R.D.; Killen, W.D. Spatiotemporal trends analysis of benthic communities and physical habitat during non-severe drought and severe-drought years in a residential creek in california. *Calif. Fish Game* **2016**, *102*, 55–70.

 *sustainability*

*Article*

# Flow–Vegetation Interaction in a Living Shoreline Restoration and Potential Effect to Mangrove Recruitment

Kelly M. Kibler [1,*], Vasileios Kitsikoudis [2], Melinda Donnelly [3], David W. Spiering [2] and Linda Walters [4]

[1] Department of Civil, Environmental & Construction Engineering and National Center for Integrated Coastal Research, University of Central Florida, 4000 Central Florida Blvd., Orlando, FL 32816, USA

[2] Department of Civil, Environmental & Construction Engineering, University of Central Florida, 4000 Central Florida Blvd., Orlando, FL 32816, USA; vasileios.kitsikoudis@ucf.edu (V.K.); dwspiering@knights.ucf.edu (D.W.S.)

[3] Department of Biology, University of Central Florida, 4000 Central Florida Blvd., Orlando, FL 32816, USA; melinda.donnelly@ucf.edu

[4] Department of Biology and National Center for Integrated Coastal Research, University of Central Florida, 4000 Central Florida Blvd., Orlando, FL 32816, USA; linda.walters@ucf.edu

* Correspondence: kelly.kibler@ucf.edu; Tel.: +1-407-823-4150

Received: 23 April 2019; Accepted: 31 May 2019; Published: 10 June 2019

**Abstract:** Hydrodynamic differences among shorelines with no vegetation, reference vegetation (mature mangrove), and vegetation planted on restored shoreline (marsh grass and young mangrove) were compared based on field observations 6.5 years after living shoreline restoration. Mean current velocities and waves were more strongly attenuated in vegetation (from channel to shoreline: 80–98% velocity decrease and 35–36% wave height reduction) than in bare shoreline (36–72% velocity decrease, 7% wave height reduction, ANOVA: $p < 0.001$). Normalized turbulent kinetic energy dissipation rates were significantly higher in reference vegetation ($0.16 \pm 0.03$ m$^{-1}$) than in restored ($0.08 \pm 0.02$ m$^{-1}$) or bare shoreline ($0.02 \pm 0.01$ m$^{-1}$, $p < 0.001$). Significant differences in the current attenuation and turbulence dissipation rates for the reference and planted vegetation are attributed to the observed differences in vegetation array and morphology. Although the hydrodynamic analyses did not suggest limitations to recruitment, mangrove seedlings were not observed in restored vegetation, while four recruited seedlings/m were counted in the reference vegetation. The lack of recruitment in the restored shoreline may suggest a lag in morphological habitat suitability (slope, sediment texture, organic matter content) after restoration. Although hydrodynamics suggest that the restored site should be functionally similar to a reference condition, thresholds in habitat suitability may emerge over longer timescales.

**Keywords:** ecohydraulics; living shoreline; restoration; mangrove; flow-vegetation interaction; recruitment

---

## 1. Introduction

Shoreline ecotones provide essential ecosystem services, yet the modification of shoreline attributes that are vital to this functionality (e.g., slope, vegetation cover, material structure) is widespread [1,2], particularly in developed areas [3]. Given the growing human population in near-shore areas [4], combined with the intensification of coastal hazards [5,6], decisions regarding the management of the shoreline ecotone are critical to overall coastal sustainability. Nature-based solutions to shoreline stabilization, such as living shorelines, are intended to maintain the functionality of the ecotone [7], for instance by preserving or restoring the equilibrium slope, and therefore the natural hydrology and connectivity of intertidal and subtidal wetland habitats [8,9]. Living shorelines are often framed as an alternative or complement to traditional hard-armoring techniques that were designed to prevent erosion [10] and may be implemented as restoration in areas of active shoreline degradation [11], including planting or the placement of native species. Hybrid living shoreline approaches may also include engineered structural elements that were designed to dissipate hydrodynamic energy [7,12]. As investment in living shoreline restoration increases, understanding the influence of restoration designs to shoreline sediment transport and erosion will lead to more robust decision making and impactful investments in shoreline stability.

Hydrodynamic interactions are important controls to shoreline processes and functions, including the transport and retention of sediments and organic matter [13,14], the survival of planted vegetation [15–17], long-term vegetation recruitment [18,19], the provision of hydraulic habitat for estuarine species [20], and resilience to extreme events [9,11,21]. In restoring degrading shorelines with native species, restoration practitioners wish to induce physical–biological feedbacks, wherein the structure of placed or planted species reduces current and wave velocities, lessens shear stresses on substrates, and promotes the deposition of sediments and retention of organic matter. The resulting hydrodynamic and benthic environment ideally should encourage the expansion of the vegetative community, through propagation and natural recruitment. As the restored community expands, it should thus engineer conditions that lead to the further success of the community in a positive feedback mechanism. However, quantitative, field-based observations of nearshore hydrodynamics within a living shoreline are rarely undertaken. As such, there is a fundamental gap in understanding how estuarine flows (which may include both waves and tidal and/or riverine currents) interact with living shoreline structural elements and planted vegetation. For instance, while it is well known that an increased drag on vegetation attenuates mean flow velocities [22] and wave heights [23], instantaneous shear stresses may be elevated due to increased turbulence from flow–vegetation interaction, which can affect sediment transport. The critical role of turbulence in sediment transport enhancement through vegetation has been highlighted both in unidirectional flows [24,25] and oscillatory flows [26]. Turbulence can potentially affect sediment entrapment within vegetation and the vegetation expansion [27]. Current knowledge of such small-scale hydrodynamic processes derives from either numerical models (e.g., [28]) or laboratory study in flumes [29]. Mimicking distributions, geometries, and behaviors of natural (and restored) vegetation is challenging [30,31], which limits the potential for furthering understanding flow behavior in vegetation based on laboratory or numerical study alone. Furthermore, in the context of restoration, flow interaction within a restored shoreline will change over time as vegetation matures and the shoreline benthic morphology responds. Therefore, the hydrodynamic effects that are observed shortly after living shoreline restoration, when monitoring is most likely to occur, may be transient. The objective of this study is to quantify, several years after restoration, the morphological and hydrodynamic differences among shorelines with no vegetation, natural vegetation, and vegetation planted during living shoreline restoration, and to understand how the observed hydrodynamic differences influence sediment characteristics, carbon storage, and vegetation (mangrove) recruitment.

## 2. Materials and Methods

### 2.1. Study Site Description

Mosquito Lagoon is the northernmost of three estuarine lagoons comprising the extensive Indian River Lagoon system on Florida's east (Atlantic) coast. As a bar-built estuary located between Florida's mainland and barrier islands, Mosquito Lagoon is a wide, shallow (1–4 m water depth) water body containing numerous bars and vegetated islands within the main channel. Tidal exchange to the marine system occurs through an inlet in the northernmost portion of the lagoon, and a canal in the southern lagoon connects Mosquito Lagoon to the Indian River. Mosquito Lagoon is microtidal, with seasonal changes in water levels [32]. The climate is humid sub-tropical and the bulk of annual precipitation (mean 1250 mm yr$^{-1}$) is delivered as rain during the wet season from May through October. Salinity in Mosquito Lagoon ranges from 25 to 45 ppt [32]. The mean annual air temperature is 22 °C, and water temperatures range from 23 to 30 °C [33]. The barrier island comprising the east bank of the lagoon has a long history of human habitation dating to 500 B.C., and pre-historical shell middens [34] create an anthropogenically altered topography that stretches the length of the island.

Three 50-m to 100-m lengths of shoreline, consisting of one unrestored shoreline (Reference) and two restored sites (Restored—living shoreline restoration, Seawall—hybrid living shoreline stabilization) were selected for study from within a 4-km length of Mosquito Lagoon's east bank in Canaveral National Seashore, Florida (Figure 1). The Reference site is a reference-condition shoreline with no evidence of active shoreline degradation. Vegetation in the intertidal zone of the Reference site is characterized by a stable, mature stand of mangrove forest (*Rhizophora mangle, Avicennia germinans, Laguncularia racemosa*) (Figure 2a). Prior to living shoreline restoration in 2011, the Restored site was actively degrading (Figure 2c). The restoration consisted of submerged oyster mats [35] and zoned plantings of emergent marsh grasses (*Spartina alterniflora*) and mangrove (*R. mangle, A. germinans*) (for details, see [36]). While early years following the restoration indicated moderate oyster recruitment to the mats (maximum 200 oysters/m$^2$), a series of stochastic environmental events (e.g., brown tide algal bloom, drought, and historically low water levels) left recruited oysters vulnerable to predation and precipitated extensive mortality. At the time of data collection for this study, 6.5 years after restoration, few oysters inhabited the restoration site, and the mats had been either removed or buried by sediments and vegetation.

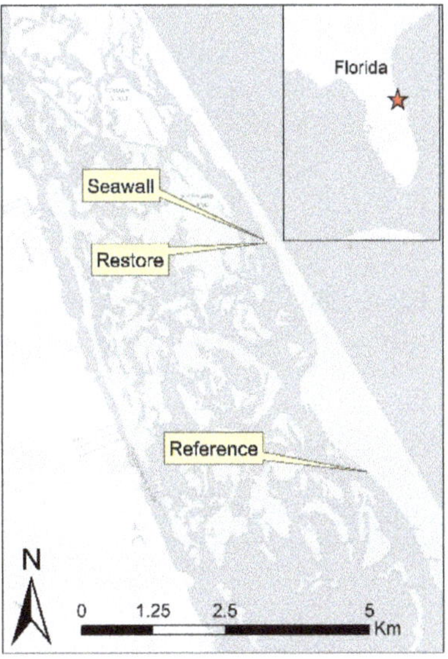


**Figure 1.** Study site locations.

**Figure 2.** Study sites: (**a**) Reference shoreline site with mature mangrove vegetation; (**b**) Seawall shoreline site at the time of hydrodynamic data observation, four years after hybrid shoreline restoration; (**c**) Restored shoreline site before restoration; (**d**) Restored shoreline site one month after restoration; (**e**) Two years after restoration; and (**f**) at the time of hydrodynamic data observation, 6.5 years after restoration.

The Seawall site is a location of active net shoreline erosion. A retaining wall had been built near the shoreline in the 1950s, and erosion had progressed to the wall footing by 2012. Bags of cement had been placed at the footing in an attempt to stabilize sediments (Figure 2b). In 2013, stabilization of the eroding shoreline with a hybrid living shoreline was undertaken. The modified living shoreline consisted of 800 mesh bags filled with oyster shells (length:width:height = 100:30:30 cm) arranged in a low sill structure (one row deep, two rows wide), and plantings of smooth cordgrass (*S. alterniflora*). At the time of this study, four years after the hybrid living shoreline restoration, the structure was in place but buried by sediment, and only sparse vegetation remained on the Seawall shoreline in front of the retaining wall.

*2.2. Field Data Collection*

Field measurements were conducted on intertidal shorelines and in subtidal channels at seasonal low and high water levels, August and November 2017, respectively, 6.5 years after living shoreline restoration and four years after hybrid restoration. Between the two sampling periods, a sizable disturbance event, Hurricane Irma, occurred in September 2017. In each site, velocity profiles (15–20 mm vertical resolution) were sampled through the full water column at 2 Hz using a 2-MHz Acoustic Doppler Current Profiler (ADCP, Nortek), stationed down-looking in the channel, approximately 10 m offshore (Figure 3). The channel velocity sensor operated in burst sampling mode with either 6600 s of continuous measurements in a 7200-s period or 512 of continuous measurements over a 600-s period. All the instruments were aligned to a common coordinate system such that $u$, $v$, $w$ components of velocity respectively corresponded to streamwise (parallel to shore), cross-shore, and vertical directions. Velocity magnitude indicates flow strength, while directionality corresponds to the direction of the tidal pulse. Positive streamwise velocities corresponded to ebb tide (from south to north), while negative streamwise velocities were observed during rising tides (from north to south). Windspeed and direction in the channel were recorded 2 m above the water surface using a Davis Wind Speed and Direction Smart Sensor (Onset, S-WCF-M003) with a 60-s interval and 1-Hz sampling rate. Water depth was recorded at shoreline and channel monitoring positions at 60-s resolution (Onset, U20L-04). Water surface deformations in the channel and on shorelines were observed at 32 Hz using sonic sensors (Ocean Sensor Systems XB Pro).

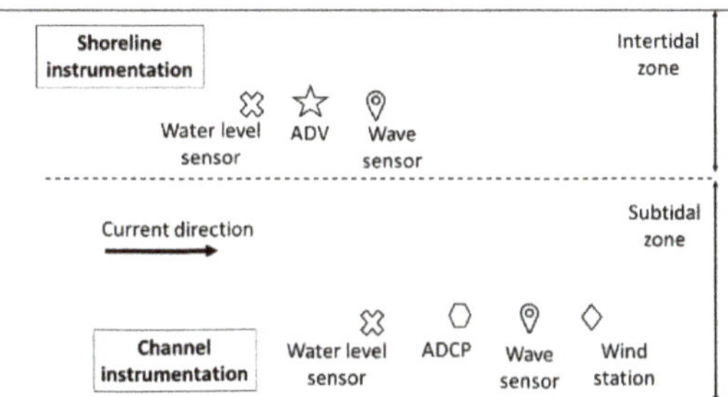


**Figure 3.** Hydrodynamic data collection schematic.

Within each shoreline, 2–3 cm vertical velocity profiles (1-mm resolution) were sampled at 100 Hz, using a down-looking Vectrino Profiler (Nortek). Shoreline velocity data were collected in "high" and "low" probe positions, with respective sampling volumes located approximately within 3.2 to 6.6 cm and −0.5 to 2.1 cm above the bed, with minimal differences among the three sites. In order to observe the near-bed boundary layer, the low sampling volume intentionally intersected the bed. Although some data cells were discarded to avoid potential bed interference [37], this configuration positioned the highest accuracy portion of the profile (5 cm from the probe, [38]) at approximately 1.0 cm above the bed. At this height, local turbulence and flow resistance are affected both by the bed roughness and by characteristics of vegetation, if present. In the high probe position, the most accurate portion of the profile was collected around 5.0 cm from the bed. At this height, the bed effect should become less prominent, and the flow pattern will be more reflective of vegetation-induced drag.

All the sites were surveyed with a CHC X91 + Real Time Kinematic (RTK) GPS survey device referencing the North American Vertical Datum of 1988 (NAVD 88). Individual $xyz$ point data were surveyed by taking the mean of six corrected samples. Transects were surveyed at 2-m intervals from the shoreline to approximately 35 m into the channel. ASCII XYZ point data were converted to an elevation surface through raster interpolation using the natural neighbor algorithm. At the Restored and Reference sites, shoreline velocity profiles were collected within vegetation (Figure 4). Vegetation within a 0.25-m$^2$ quadrat in the immediate vicinity of the probe location was characterized for stem and prop root density ($n$), species, and diameter ($d$) at 10 to 20 cm above the bed. The density of roots and stems were combined to represent the total number of individual vegetation elements in each quadrat and are referred to as "elements" in the results below. Five additional 0.25-m$^2$ quadrats were randomly selected from each shoreline, and the vegetation within was characterized similarly. Two random transects were selected in each site, beginning at the most seaward extent of vegetation and continuing until wetland vegetation was no longer observed. Species within a diameter of 10–20 cm above the bed, and cohort (approximate age) were recorded for every mangrove seedling within a 1-m wide swath along transects.

**Figure 4.** Collection of shoreline velocity profiles at (**a**) Restored shoreline site; (**b**) Seawall shoreline site; and (**c**) Reference shoreline site.

Five sediment cores (approximately 200–300 g each) were collected to a depth of 20 cm from the shorelines and channels of each site, close to the locations of hydrodynamic data collection. Both sediments and water were retained within each core sample. Individual cores were dried in the lab at 110 °C for at least 24 h. Dry sediments were gently ground, sieved at 2 mm, and 20 g samples of the <2 mm size class were processed for loss on ignition at 550 °C for 16 h. Then, cores were pooled by site for an analysis of grain size distribution using dry and wet sieving.

*2.3. Analysis of Hydrodynamic Data*

2.3.1. Post-Processing and Quality Control

On the shorelines, precise probe distances from the bed were determined by the velocity profile and no-slip condition at the bed for the low position, and with the aid of the center beam amplitude for the high position. Center beam amplitude profiles (sampling rate 2 Hz) starting 2 cm from the probe until well below the bed were analyzed. The amplitude increased considerably at the bed due to the strong return signal from the solid boundary. The bed position was selected to coincide with the minimum amplitude gradient across the recorded profile [39,40]. The ambient environment of the lagoon provided a particle-rich environment for the proper detection of Doppler shift from both velocity sensors. As a result, the signal-to-noise ratio (SNR) and correlation for most of the measurements exceeded the thresholds suggested by Nortek for obtaining reliable measurements. Most shoreline measurements had SNR >30 dB and correlation >90%. Raw data with SNR <20 dB and/or correlation <80% (less than 2% of data) were removed from the time-series and gaps were interpolated linearly. Despite the high statistics in the measurements, some outliers in the velocity time-series may still exist [41]. The time-series was despiked according to the phase-space thresholding method suggested by Goring and Nikora [42] and modified by Wahl [43]. In the channel, most of the cells exhibited very high correlation (>90%), especially in the upper flow region. Measurements with correlation <70% were discarded. Such low correlation measurements occurred relatively close to the bed, which was presumably due to strong bed acoustic interference. Wakes generated by passing boats are not considered in this analysis. Wave data influenced by boat wakes were manually removed from the data time-series. No sediment transport or bed alterations were observed during the measurements.

To calculate channel-to-shoreline hydrodynamic gradients, the instantaneous velocity time-series were decomposed into segments of 120 s, across which flow characteristics were calculated. A moving average was applied, based on 30-s successive shifts of each 120-s segment along the flow velocity time-series, across which the flow is considered to be quasi-steady. Synchronous shoreline and channel velocity time-series were paired; the final datasets for each site consisted of 71 to 158 paired segments for each site and position.

### 2.3.2. Turbulent Kinetic Energy Dissipation

The presence of irregular waves precluded the direct calculation of turbulence intensity, as the sinusoidal pattern of wave orbital velocities affected the variance of the velocity time-series, which is needed to estimate turbulence intensity. Assessment of turbulence at shorelines was based on the dissipation rate of turbulent kinetic energy (TKE), $\varepsilon$, as calculated using the structure function method [44] (Appendix A). This methodology utilizes Kolmogorov's hypotheses for turbulence cascade and links TKE dissipation to the covariance of the velocity difference between two points [45] along the vertical velocity profile. Since wave oscillations decorrelate over larger scales [46,47], comparing velocities over such short distances leads to the effective filtering of wave-induced variance. As a result, velocity differences between profile points are attributed to turbulent eddies. Herein, a second-order structure function (Equation (1)), $D(z,r)$, was formed for the estimation of TKE dissipation rate at height $z$ above the bed, with vertical velocity fluctuations $w'$ separated by distance $r$. By applying a centered difference technique, this takes the form:

$$D(z,r) = \overline{[w'(z-r/2) - w'(z+r/2)]^2} \tag{1}$$

where $w' = w - \overline{w}$, with the overbar denoting time averaging over the 120 s of the moving average. Based on Taylor's cascade theory, the structure function within the inertial subrange is also equal to [44,45]:

$$D(z,r) = C_v^2 \varepsilon(z)^{2/3} r^{2/3} \tag{2}$$

where $C_v^2$ is a constant equal to two [45]. For the calculation of dissipation $\varepsilon$, linear regression was applied to Equation (3) between $D(z,r)$ and $r^{2/3}$, in order to derive an equation of the form:

$$D(z,r) = Ar^{2/3} + N \tag{3}$$

where $A = C_v^2 \varepsilon^{2/3}$ and $N$ is error attributed to Doppler noise, which is independent of $r$ [44]. The mean dissipation rate was calculated for each 120 s of the moving average at each site and position, and was normalized by the corresponding root-mean-squared (rms) velocity, $u_{rms}$, to allow for comparison across sites and days. The rms velocity accounts for both current and wave variation [48] and is given by Equation (4):

$$u_{rms} = \sqrt{\overline{u^2}} \tag{4}$$

Data were evaluated to confirm assumptions of normal distribution and heterogeneity of variances. Current and wave attenuation and turbulent dissipation rates were compared using one-way analysis of variance (ANOVA) followed by Tukey post hoc pairwise comparisons to identify statistical significance ($\alpha = 0.05$) between shoreline types (Restored, Seawall, Reference).

## 3. Results

### 3.1. Vegetation Characteristics and Shoreline Morphologies

Shoreline morphologies and characteristics of shoreline vegetation varied among sites. The mean width of the shoreline, as defined from the water's edge at the mean low water level to the landward extent of hydrophytic vegetation, was over 25 m in the Reference site, less than 4 m in the Restored site, and below 2 m at the Seawall. Shoreline slopes in the Restored and Reference sites were similar (0.050 ± 0.001 and 0.056 ± 0.001, respectively) and considerably lower than slope of the Seawall shoreline (0.130 ± 0.005) (Table 1, Figure 5). Vegetation at the Reference site consisted of a mature stand of mangrove forest. Forest fringing the channel was inundated at all seasonal and tidal water levels, and emergent vegetation consisted primarily of *R. mangle* prop roots with a mean density of 83 elements/m$^2$, mean diameter of 28 mm, and solid volume fraction of 0.068. Many mangrove seedlings were observed in 1-m wide transects of the Reference site (4 ± 0.8 per transect meter,

Table 1), consisting of *R. mangle*, *A. germinans*, and *L. racemosa*. The majority of the Seawall site was unvegetated. Remaining vegetation consisted of one sparse patch of *S. alterniflora*. In the Restored shoreline, restoration converted a shoreline of bare sediment (Figure 2c) to a fully vegetated shoreline with up to 85% vegetation coverage (Figures 2f and 6). Six years after restoration, vegetation in the Restored shoreline consisted of *S. alterniflora*, *R. mangle*, and *A. germinans* that had been planted during the restoration, or in the case of *S. alterniflora*, had propagated vegetatively from plantings. The site's heterogeneous emergent vegetation consisted of both *S. alterniflora* stems/leaves and young *R. mangle* stems/prop roots and comprised a greater number of smaller elements (mean density of 97 elements/m$^2$ and diameter of 10 mm) as compared to the fewer, large elements in the mature Reference site. The frontal element area per unit volume ($\alpha$) and solid volume fraction ($\phi$) were both greater in the Reference site (Table 1). No mangrove seedlings were observed in transects of the Seawall or Restored site.

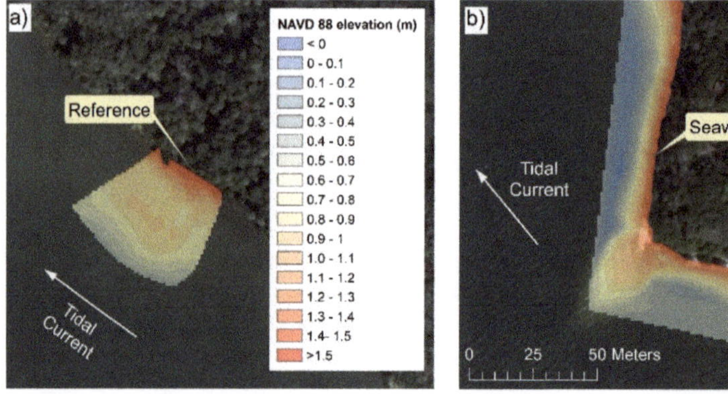


**Figure 5.** Near-shore morphologies of the (**a**) Reference, (**b**) Seawall, and Restored shoreline sites.

**Table 1.** Intertidal and nearshore subtidal slope, characteristics of vegetation across random plots and in the immediate vicinity of shoreline flow observation (probe), and mangrove seedlings per transect meter in the Restored, Seawall, and Reference sites. Site mean and standard error or coefficient of variation (CV) are reported.

| Site | Slope (m/m) Mean ± SE | | | Stem Density ($n$) (Elements/m$^2$) | | Frontal Area ($\alpha$) (m$^2$/m$^3$) | | Solid Volume Fraction ($\phi$) (m$^3$/m$^3$) | | Mangrove Seedlings Mean ± SE | | |
|---|---|---|---|---|---|---|---|---|---|---|---|---|
| | Intertidal Shoreline | Subtidal Nearshore | Overall | mean, CV | Probe | Mean, CV | PROBE | Mean, CV | Probe | Number per Transect m | Diameter (mm) | Cohort Age (years) |
| Restored | 0.050 ± 0.001 | 0.051 ± 0.001 | 0.050 ± 0.001 | 97, 0.79 | 252 | 0.94, 0.59 | 1.62 | 0.012, 1.11 | 0.009 | 0.0 ± 0.0 | 0.0 ± 0.0 | - |
| Seawall | 0.130 ± 0.005 | 0.061 ± 0.001 | 0.110 ± 0.003 | 24, 1.64 | 0 | 0.20, 1.55 | 0.00 | 0.002, 1.59 | 0.000 | 0.0 ± 0.0 | 0.0 ± 0.0 | - |
| Reference | 0.056 ± 0.001 | 0.008 ± 0.001 | 0.027 ± 0.001 | 83, 0.39 | 116 | 2.49, 0.56 | 3.12 | 0.068, 0.74 | 0.070 | 4.0 ± 0.8 | 13.0 ± 0.5 | 1 |

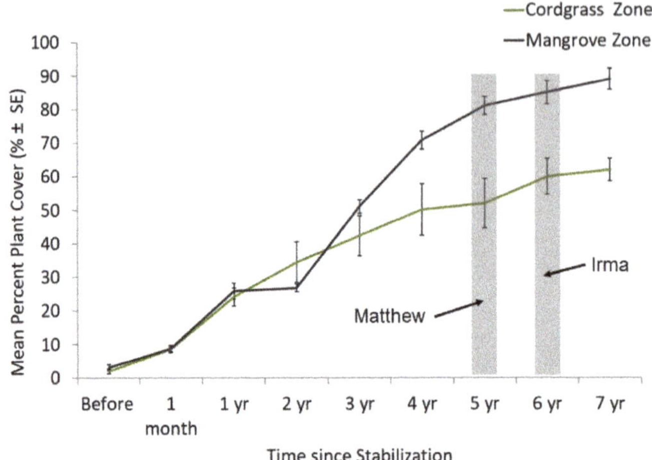


**Figure 6.** Vegetation cover at the Restored shoreline through time since restoration, with timing of extreme events (hurricanes Matthew and Irma) indicated.

### 3.2. Mean Current Attenuation from Channel to Shoreline

Depth-integrated channel velocities varied between sites and days, ranging from near zero during periods of tidal stagnation up to nearly 20 cm/s during peak tidal exchange (Figure 7). In all the shorelines and channels, streamwise velocities dominated the flow field. The shoreline velocities (1.0 and 5.0 cm above the bed) of all the sites were generally lower than the depth-integrated channel velocities, reflecting typical flow profile behavior (Figure 8). However, current attenuation rates from channel to shore differed across sites (Tables 2–4). Vegetated shorelines (Restored and Reference) attenuated currents at greater rates than the unvegetated Seawall site ($p < 0.001$ at both 1 cm and 5 cm above the bed). Additionally, currents were more strongly attenuated by mature mangrove vegetation in the Reference shoreline as compared to vegetation at the Restored site ($p < 0.001$ at both 1 cm and 5 cm above the bed). In vegetated sites, flows consistently decelerated from the channel to the shoreline (Figure 8). The mean horizontal near-bed (1.0 cm above bed) shoreline velocities in the Restored and Reference sites were respectively 84–87% and 98% lower than channel velocities. By comparison, near-bed shoreline velocities at the Seawall site were 72–74% lower than the channel velocities. Shoreline flows 5.0 cm above the bed for the most part exceeded the near-bed velocities due to the reduced bottom drag being higher within the boundary layer, but the flows were lower than the channel speeds. Similar to the near-bed results, the velocity reduction from channel to shore at the Restored and Reference sites was high (80% and 96%, respectively) relative to the Seawall (36% reduction when channel velocities exceeded 1.5 cm/s). The differences in current attenuation between sites were more pronounced 5 cm above the bed (Table 3), reflecting a lowered influence of bed interaction and stronger interaction with vegetation higher in the flow profile.

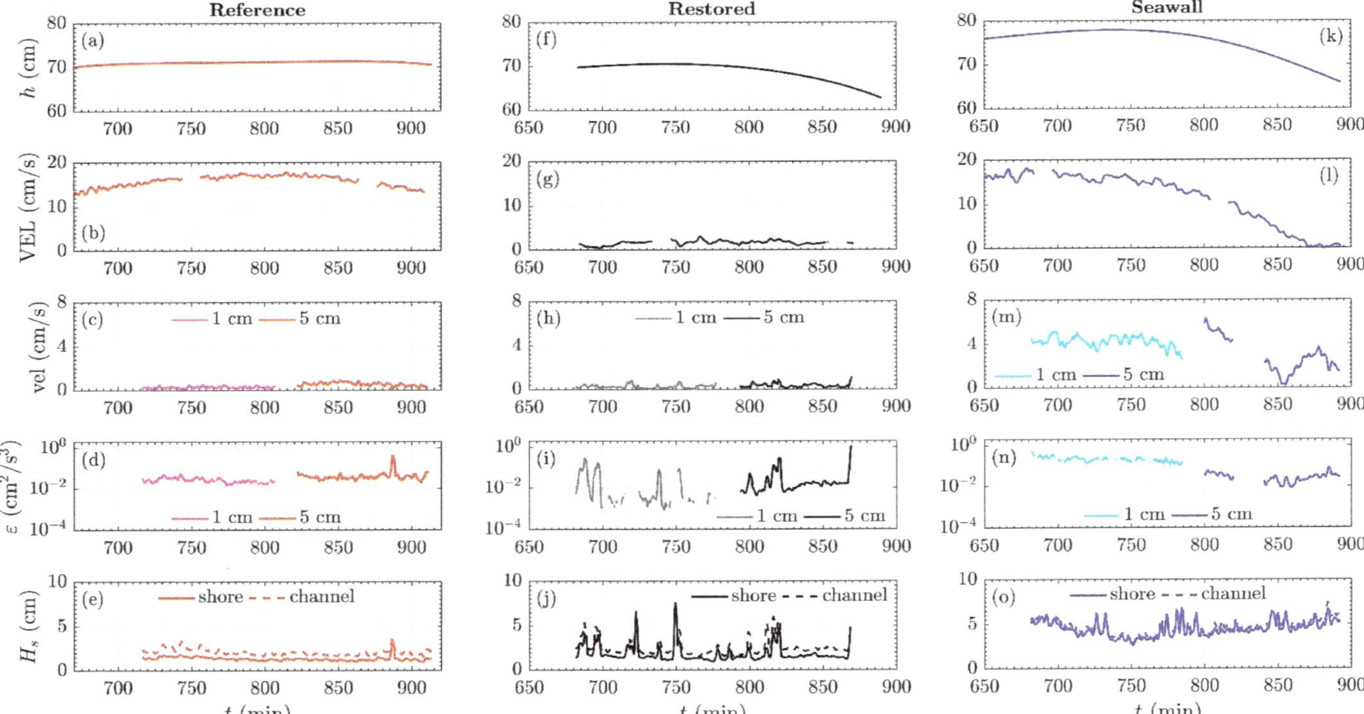



**Figure 7.** Time-series of (**a**) surface water level in the channel, $h$, (**b**) mean horizontal depth-integrated channel velocity, VEL, (**c**) mean horizontal shoreline velocity, vel, at 1.0 cm and 5.0 cm height above the bed, (**d**) turbulence dissipation rate, $\varepsilon$, and (**e**) significant wave height, $H_s$, for the Reference site, and correspondingly (**f,j**) for the Restored site and (**k–o**) for the Seawall site.

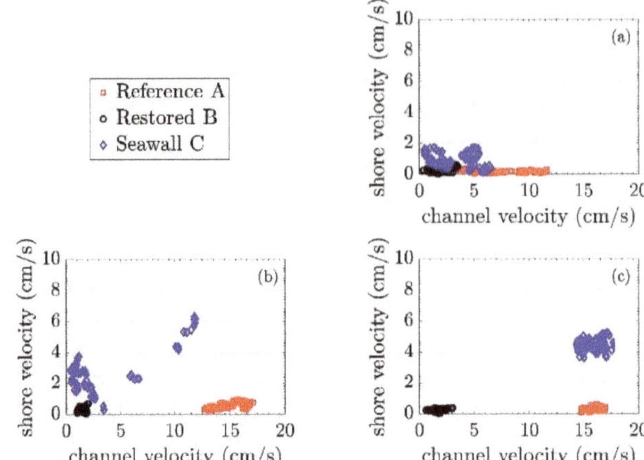


**Figure 8.** Gradients of mean horizontal velocity from channel to shoreline at the Restored, Seawall, and Reference sites: (**a**) 1.0 cm above the bed at low water levels, (**b**) 5.0 cm above the bed; and (**c**) 1.0 cm above the bed at high water levels.

**Table 2.** Mean ± SD significant wave heights and current velocities observed in shorelines and channels at the Restored, Seawall, and Reference sites.

| | Waves | | | | | | Current | | | | | |
| --- | --- | --- | --- | --- | --- | --- | --- | --- | --- | --- | --- | --- |
| | Nov 2017 Significant Wave Height (cm) | | | Aug 2017 1.0 cm Above Bed (cm/s) | | | Nov 2017 1.0 cm Above Bed (cm/s) | | | Nov 2017 5.0 cm Above Bed (cm/s) | | |
| | Channel | Shoreline | Percent Change | Channel | Shoreline | Percent Change | Channel | Shoreline | Percent Change | Channel | Shoreline | Percent Change |
| Restored | 2.4 ± 0.2 | 1.4 ± 0.1 | 36 ± 8 | 1.9 ± 0.8 | 0.3 ± 0.1 | 83.8 ± 12.0 | 1.6 ± 0.5 | 0.2 ± 0.1 | 86.9 ± 8.1 | 1.5 ± 0.3 | 0.3 ± 0.1 | 80.1 ± 9.3 |
| Seawall | 4.8 ± 0.8 | 4.5 ± 0.8 | 7 ± 6 | 3.9 ± 1.5 [+] | 0.9 ± 0.4 [+] | 74.4 ± 16.1 [+] | 16.0 ± 0.9 | 4.4 ± 0.4 | 72.3 ± 2.5 | 5.8 ± 4.1 [+] | 2.9 ± 1.8 [+] | 36.4 ± 39.3 [+] |
| Reference | 2.2 ± 0.4 | 1.4 ± 0.3 | 35 ± 6 | 7.6 ± 2.3 | 0.2 ± 0.1 | 97.6 ± 1.8 | 16.1 ± 0.5 | 0.3 ± 0.1 | 98.2 ± 0.5 | 15.2 ± 1.2 | 0.6 ± 0.2 | 96.2 ± 1.0 |

[+] when channel velocity exceeds 1.5 cm/s.

**Table 3.** Results of one-way ANOVA for wave and current attenuation rates from channel to shore, and normalized turbulent kinetic energy (TKE) dissipation at the Restored, Seawall, and Reference sites.

| | Mean Square | df | *F*-Stat | *p*-Value |
| --- | --- | --- | --- | --- |
| Wave Attenuation Rate | 9.20 * | 2 | 2694.57 | <0.001 |
| Current Attenuation 1.0 cm Above Bed | 3.32 * | 2 | 475.94 | <0.001 |
| Current Attenuation 5.0 cm Above Bed | 6.23 * | 2 | 216.89 | <0.001 |
| Normalized TKE Dissipation Rate | 0.68 * | 2 | 1660.89 | <0.001 |

* significant at the $\alpha = 0.05$ level.

**Table 4.** Pairwise comparisons of wave and current attenuation rates from channel to shore, and normalized TKE dissipation at the Restored, Seawall, and Reference sites.

| Comparison | Wave Attenuation Rate | | Current Attenuation Rate | | | | Normalized TKE Dissipation Rate | |
| --- | --- | --- | --- | --- | --- | --- | --- | --- |
| | | | 1.0 cm Above Bed | | 5.0 cm Above Bed | | 5.0 cm Above Bed | |
| | Mean ± SE Difference (%) | *p*-Value | Mean ± SE Difference (%) | *p*-Value | Mean ± SE Difference (%) | *p*-Value | Mean ± SE Difference (m$^{-1}$) | *p*-Value |
| Restored-Seawall | 29.0 ± 0.7 * | <0.001 | 12.3 ± 0.9 * | <0.001 | 43.7 ± 3.1 * | <0.001 | 0.06 ± 0.003 * | <0.001 |
| Restored-Reference | 0.5 ± 0.7 | 0.780 | −12.5 ± 0.9 * | <0.001 | −16.1 ± 2.3 * | <0.001 | −0.08 ± 0.010 * | <0.001 |
| Reference-Seawall | 28.6 ± 0.4 * | <0.001 | 24.8 ± 0.8 * | <0.001 | 59.8 ± 2.9 * | <0.001 | 0.14 ± 0.002 * | <0.001 |

* significant at the $\alpha = 0.05$ level.

When channel currents fell below 1.5 cm s$^{-1}$, behavior at the Seawall shoreline became unpredictable, and the shoreline velocities often greatly exceeded the channel depth-integrated flows. For instance, mean shoreline velocities up to 3.0 cm/s were observed when channel velocities were below 0.6 cm/s. Such erratic behavior reflects the pronounced hydrodynamic influence of waves in the unvegetated site during times of stagnant current. By comparison, within vegetation, the mean shoreline flows were consistently suppressed to below 0.90 cm/s, even when the channel flows were of relatively high magnitude (e.g., exceeding 17 cm/s in the channel near the Reference site). When similarly high channel forcings were observed in the Seawall site, the mean shoreline flows were far greater: consistently above 4 cm/s.

### 3.3. Wave Attenuation from Channel to Shoreline

Significant wave heights ($H_s$, defined as the mean of the top 1/3 wave heights, here computed across each moving average period) observed in the channels and on the shorelines at all the sites were low (<8 cm significant wave heights) and generally corresponded with prevailing wind conditions. The mean significant wave heights observed at the Seawall site (4.8 ± 0.8 cm in channel, 4.5 ± 0.8 cm on shoreline) were larger than those observed at the Reference site (2.2 ± 0.4 cm in channel, 1.4 ± 0.3 cm on shoreline) and the Restored (2.4 ± 0.2 cm in channel, 1.4 ± 0.1 cm on shoreline) site. Wave heights typically decreased from the channel to onshore (Figure 9), with stronger and more consistent attenuation observed within the vegetated sites ($p < 0.001$, Tables 3 and 4). Significant wave heights decreased from the channel to shoreline by a mean of 36 ± 8% and 35 ± 6% within the Restored and Reference sites respectively, while mean wave attenuation was 7 ± 6% at the Seawall site. Wave attenuation was similar in the Reference and Restored shorelines ($p = 0.780$). Comparing periods of similar channel wave heights across sites further confirmed that the vegetated sites attenuated waves more efficiently than the unvegetated site. When the offshore wave heights were 3–4 cm, attenuation was greatest at the Reference site (32 ± 17% reduction), followed by the Restored site (22 ± 9% reduction). By comparison, attenuation at the Seawall site was low (7 ± 4% reduction). Offshore wind waves exceeding 4 cm were not observed at the Reference or Restore sites; however, wakes from passing boats were observed to exceed 7 cm in the channel off the Restored site. When offshore wave heights ranged from 4 to 8 cm, attenuation at the Restored site (23 ± 21% reduction) was greater than at the Seawall site (7 ± 6% reduction).

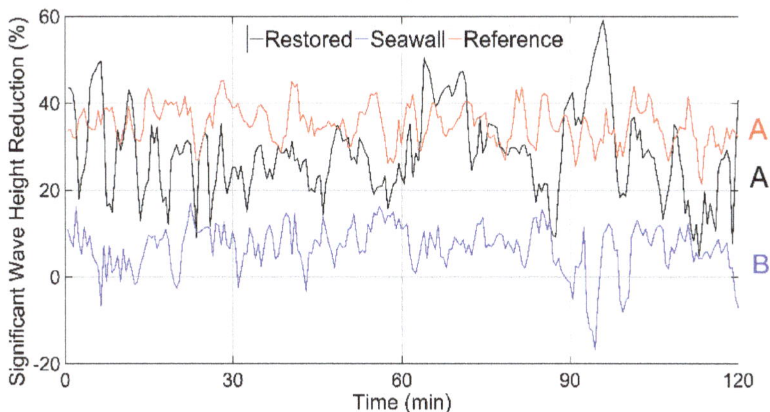


**Figure 9.** Wave attenuation rates from channel to shoreline at the Restored, Seawall, and Reference sites.

### 3.4. Comparison of Turbulent Kinetic Energy Dissipation across Sites

The linear regression that was applied in Equation (3) had a very good fit with the measurements 5.0 cm above the bed, with $R^2 > 0.90$. However, data fit closer to the bed was variable, and a cutoff threshold was established at $R^2 = 0.85$, below which measurements were discarded. The number of points utilized for the linear regression was respectively eight, 11, and 10 for the high position in the Reference, Restored, and Seawall sites, while it was consistently five for measurements at the low position for all the sites. For the low position, no velocities recorded within 4 mm of the bed were utilized, due to potential bed interference in the signal in near-bed measurements [37]. At 5 cm above the bed, the normalized TKE dissipation rate was highest at the Reference site ($0.16 \pm 0.03$ m$^{-1}$), followed by the Restored ($0.08 \pm 0.02$ m$^{-1}$) site, and the Seawall ($0.02 \pm 0.01$ m$^{-1}$) site ($p < 0.001$, Table 3, Figure 10). However, 1 cm above the bed, the three sites exhibit similar normalized dissipation rates (Reference: $0.09 \pm 0.02$ m$^{-1}$, Restored: $0.11 \pm 0.07$ m$^{-1}$, and Seawall: $0.12 \pm 0.03$ m$^{-1}$).

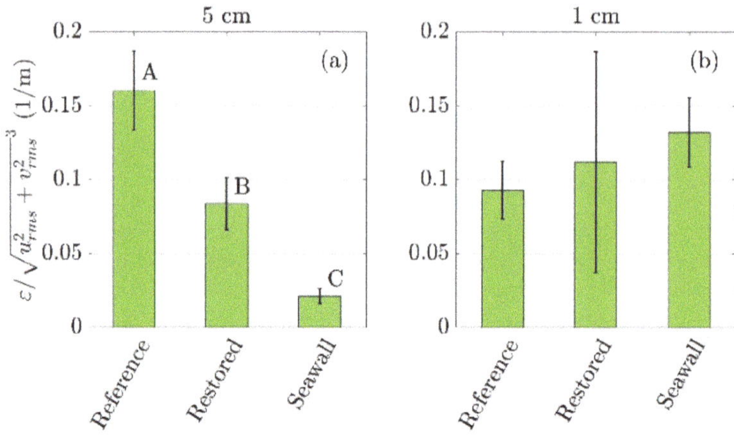


**Figure 10.** Normalized turbulent kinetic energy dissipation rate (mean ± one standard deviation) on shorelines at (**a**) 5 cm and (**b**) 1 cm height above the bed.

### 3.5. Comparison of Sediments across Sites

The bulk of sediments at all the sites consisted of coarse to fine sand. Larger particles were shell fragments, comprising up to 30% of particles by mass in the Restored and Seawall sites. Only small fractions of silt/clay were observed; these generally comprised less than 3% of the sample mass. Before Hurricane Irma, the texture of sediments in the Restored and Seawall sites were similar, and notably coarser than the Reference site sediments (Figure 11a). Surprisingly, the finest sediments were found in the channel just off the Reference shoreline, while sediments within the mature mangrove forest were slightly coarser. Changes to sediment distributions were detected in all the sites after Hurricane Irma (Figure 11b). Immediately after the hurricane, shoreline sediments at the Reference site coarsened (Figure 11b), with $d_{16}$ and $d_{50}$ increasing from 0.15 to 0.40 mm and from 0.18 to 0.60 mm, respectively. Within the mangrove vegetation $d_{84}$ remained relatively stable, but increased in the channel from 1.00 mm to 2.40 mm (Figure A1). The observed changes account for the deposition of medium to coarse sand within the Reference vegetation. A similar sediment deposition within vegetation also occurs at the Restored shoreline. Here, shoreline $d_{16}$ and $d_{50}$ remained relatively stable, but $d_{84}$ decreased notably from 6.60 to 2.20 mm due to coarse sand deposits. Meanwhile, the opposite trend in $d_{84}$ was noted in the Restored channel, coarsening after the hurricane from 4.75 to 6.30 mm, which was similar to changes observed in the Reference channel. Few changes were detected to shoreline particle distributions at the Seawall site.

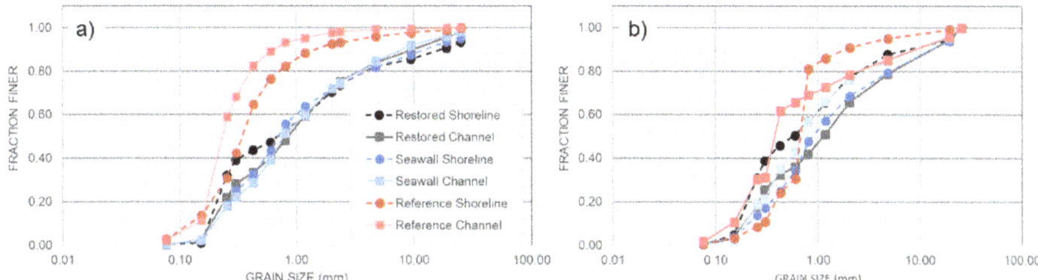

**Figure 11.** Sediment particle size distributions in shorelines and channels of the Restored, Seawall and Reference sites (**a**) during low water levels before Hurricane Irma and (**b**) two months after Hurricane Irma, during seasonal high water levels.

Organic matter (OM) content tended to be higher in the shoreline rather than channel sediments. However, at the Seawall before the hurricane, the channel and shoreline OM contents were similarly low. Sediments of vegetated shorelines contained greater percentages of carbon (respectively 4.84 ± 1.32% and 3.65 ± 1.00% in the Reference and Restored shorelines) compared to the channel and unvegetated shoreline sediments (e.g., 1.67 ± 0.21% in the Seawall shoreline) (Figure A2). Mimicking particle size results, the channel sediments off the Reference shoreline, just outside of the mature mangrove vegetation, contained high percentages of OM (3.12 ± 0.59%), which were levels that were well above the shoreline sediments in the Seawall site. Immediately after the hurricane, effects of sediment reorganization were evident. Shoreline and channel OM content became more similar at the vegetated sites, likely reflecting the large transport and deposition of sand from the channel into the vegetated shorelines during the event. However, OM content increased substantially in the Seawall shoreline, up to 3.70 ± 0.89%. During sample collection, it was observed that copious quantities of seagrass wrack had been deposited in the Seawall shoreline after the hurricane [49], which likely accounts for the increased OM content found in the sediments.

## 4. Discussion

### 4.1. Differential Hydrodynamic Effects of Bare Sediment, Restored Vegetation, and Natural Vegetation

Near-bed shoreline velocities were consistently lower than depth-integrated channel velocities at all the sites and sampling periods, which is typical flow behavior at channel margins. Similarly, wave heights at shorelines were consistently smaller than those observed in the channels, which also was expected behavior for waves across differing water depths. However, flows were consistently more strongly attenuated in the vegetated sites (Table 2, Figures 8 and 9), with the greatest attenuation rates observed in mature mangrove vegetation (Table 4). Notably, shoreline velocities within the mangrove prop roots of the Reference site remained extremely low, even when velocities in the channel become relatively strong. By comparison, at similar channel flows, velocities in the unvegetated Seawall shoreline often exceed velocities in the Reference shoreline by an order of magnitude. Comparable strong current speeds were not observed in the channel near the Restored shoreline. Therefore, while we can conclude that the planted vegetation attenuated the current at low speeds, the performance at greater currents speeds (such as those observed at the Reference site) cannot be verified. However, at larger wave heights, wave attenuation is high in planted vegetation compared to the unvegetated Seawall site.

The observed current and wave attenuation in vegetation underscores many past laboratory and field studies that reported similar conclusions regarding the effects of emergent vegetation on flows [22,50–53]. However, this is one of the first detailed comparisons of flow attenuation between reference state vegetation and restored vegetation based on field observations. The primary hydrodynamic interaction of flows within vegetation is related to an increase in flow resistance due to drag force against the vegetation, leading to slower mean flows within vegetation [54]. This effect can be clearly observed as a significantly greater current attenuation in vegetated versus unvegetated shorelines, both near-bed and higher within the flow profile at 5 cm above the bed (Tables 2 and 4). Furthermore, the greater difference in current attenuation rate at sites at 5 cm above the bed as compared to near-bed illustrates the different mechanisms of flow resistance in vegetated and non-vegetated shorelines. Near the bed, attenuation in the vegetated and unvegetated sites was more similar, as the bed boundary strongly influenced all the sites. Higher in the flow profile, the interaction with vegetation became a more significant source of flow resistance, and much stronger differences in flow attenuation were observed between the vegetated and unvegetated sites.

Prior study comparing flow attenuation through emergent marsh grasses and mangroves [55] report conclusions similar to the present study: mangrove vegetation is more effective in attenuating current than emergent marsh grasses, although herein much greater levels of flow attenuation were observed for similar bare channel current speeds. Observations reported from prior study [55] were additionally only true when the mangrove canopy was also submerged, which did not occur in this study. The presented data suggest that differences in vegetation morphologies, densities, or configuration alone are sufficient to produce the observed stronger flow attenuation through mature mangrove vegetation. Flow and wave attenuation rates at all the sites were similar across the nominal range of observed water level variations (0.2 m) within the microtidal environment. Past research has observed strong nonlinearities in flow attenuation when water level variation exceeded biological thresholds [55,56]. The results reported herein demonstrate the resilience of emergent vegetation to attenuate flows across small variations in water levels. While the wave attenuation observed at the Seawall site was low relative to the vegetated sites, attenuation rates were also consistent across the observed water levels. This observation suggests that wave attenuation at the Seawall was driven by a consistent nearshore slope and not affected by the low modular structure comprised of oyster shell bags deployed during hybrid restoration. No topographic traces of the oyster shell bag structure remained (Figure 5) after burial by a deep and uniform layer of sediments.

Turbulent Kinetic Energy Dissipation Rate

While mean flow magnitudes in vegetation may be reduced, eddies shed from the interaction of flows with vegetation components (stems, leaves, prop roots) increased the intensity of turbulent fluctuations [25,57]. Important consequences to sediment transport in vegetation derive from the combination of competing effects to mean and turbulent flows [25]. The observed difference in turbulence dissipation from the near-bed (1.0 cm from bed) as opposed to 5.0 cm above the bed allows for understanding the partitioning between bed and stem-generated turbulence (Figure 10). Observations near the bed include turbulence generated at the flow–bed boundary, and in vegetated sites, also by the flow-stem interaction. Higher within the water column, the effect of the bed boundary was smaller, and the relative contribution of the stem-generated turbulence may be inferred. We observe that the magnitudes of the turbulence dissipation rate (normalized to local flow condition) and the partitioning of turbulent dissipation between the bed and stems varied between sites. Five centimeters above the bed, normalized turbulent dissipation was significantly greater in vegetated shorelines. The difference in dissipation rates observed at the vegetated and unvegetated shorelines may reflect systematic differences in advected turbulence within obstructed versus unobstructed flows. For example, within rough canopies, the production and dissipation of turbulence may be imbalanced due to an advection of eddies shed from upstream obstructions [58]. It is likely that the greater dissipation rates observed in the vegetated sites reflect not only the local production of turbulence, but also include a greater input of advected turbulence shed from the stems immediately upstream. The turbulence dissipation rate 1.0 cm above the bed was relatively balanced for all three sites, which highlights the dominant effect of bed-generated turbulence in near-bed flow regions.

Interestingly, turbulent dissipation was significantly greater in the Reference shoreline than the Restored shoreline, which was similar to how the current attenuation was also significantly greater. The differences in current attenuation rates and normalized dissipation rates observed in the two vegetated sites likely reflect variable drag, turbulence production, and advection from the differing vegetation morphologies. Although both sites are vegetated, the vegetation characteristics are markedly different (Table 1). The mature *R. mangle* prop roots that dominate vegetation in the Reference shoreline create an array of relatively sparse, larger elements that are approximately cylindrical above typical water levels, but have varied rough morphology (through the recruitment of barnacles and oysters) in subtidal portions of stems (Figure 4c). In addition, TKE in the wake of the prop roots is likely affected by the variation in the projected frontal area with height, as the prop roots project a bell-shaped rather than rectangular frontal area (Figure 4c) [59]. Thus, the solid volume fractions represented by assuming that mangrove roots are regular cylinders with diameters similar to above the typical water level are likely to be underestimations. In addition, irregularities in the vertical variation of the frontal area of an obstacle could lead to vertical mixing in the wake, which enhances the turbulence generated from the stems [60]. At the Restored site, the young mangrove roots and *S. alterniflora* grass stems and leaves created arrays with more and thinner elements, compared to the Reference site, with varied cross-sectional morphology (Figure 4a). These array and morphological differences may partially explain the differences in current attenuation and dissipation rates observed in the Reference and Restored sites.

Analyzing the turbulence from field data influenced by both waves and current can be challenging, as it can be difficult to isolate turbulent fluctuations from periodic oscillations due to non-regular waves. The methodology applied herein to estimate the TKE dissipation rate was originally developed for deep flows in the marine environment [44]. However, more recently, it has been utilized in surf and swash zones [61,62], salt marsh vegetation [63], and flows through mangroves [46,47,64]. The actual turbulence dissipation rates observed in this study are much lower than the rates reported by Norris et al. [47,64], who observed turbulence dissipation rates up to $6.5 \cdot 10^{-4}$ m$^2$/s$^3$ in pneumatophores of black mangroves. However, both current velocity and wave heights were greater in the pneumatophores. Additionally, the structure of black and red mangrove vegetation is different. Pneumatophores are rigid stems that protrude into the flow and usually become fully submerged during high tide. The vegetation studied herein remained fully emergent across the range of flows observed. Almost all the normalized dissipation rates reported in [64] are more than an order of magnitude lower than the normalized TKE dissipation rates reported in this study, although different normalization factors were applied. Norris et al. [64] normalized dissipation by simultaneous current and orbital velocity measurements in the free-stream above the pneumatophore canopy rather than by root-mean-square velocities. Thus, it is unclear whether differences in normalized dissipation can actually be attributed to the greater turbulence observed within red mangroves and *S. alterniflora* or whether the normalization differences between the two studies preclude their direct comparison.

### 4.2. Mangrove Recruitment into Restored Shorelines

Long-term restoration success ultimately hinges upon the natural recruitment of desired species into restored areas. The retention and establishment of the first mangrove seedlings can indicate that a restored site has achieved a critical tipping point in habitat suitability. For example, Kamali and Hashim [65] observed the natural recruitment of *Avicennia marina* propagules within eight months following the installation of a wave break. In this study, 6.5 years after restoration, the growth of wetland vegetation planted at the Restored shoreline was substantial and had demonstrably altered the shoreline habitat with respect to cover and hydrodynamics. Past research indicated that propagules were supplied by dispersal from nearby mature mangrove stands and within-site reproduction, and that supply was not limited at the Restored shoreline [15]. An increased cover of vegetation following restoration, particularly of *S. alterniflora*, was expected to support mangrove recruitment by "propagule trapping" to promote retention and establishment [17,66,67]. Yet, naturally recruited mangrove seedlings were not observed in the restored vegetation. By comparison, hundreds of recruited seedlings were counted during the same year in the Reference site, where a mean of four seedlings were observed along each 1-m wide band transect meter. The recruitment rates observed in the nearby reference areas may indicate levels of recruitment that can sustain a site long-term and support observations that propagule supply was not a limiting factor at the Restored shoreline.

A detailed analysis suggests that the hydrodynamic limitations to mangrove recruitment do not explain the lack of observed recruitment at the Restored site. Shoreline hydrodynamic forces were similar in magnitude to those observed at the Reference site, where mangrove recruitment was evident. Rather, other limiting factors can perhaps explain the observed recruitment rate differences. For instance, sediment and morphological characteristics differed in the Restored and Reference shorelines. Sediments in the mature mangrove forest were finer in texture and contained more organic matter than those at the Restored shoreline, which more closely resembled sediments from the bare Seawall shoreline. The nearshore slope leading into the mangrove forest at the Reference site was very low; nearshore channel sediments were fine, and contained high organic matter content. This may suggest that the influence of the mangrove vegetation extends well beyond the actual vegetated zone. As flows approaching the vegetation are attenuated, it is possible that suspended materials are deposited upstream, over time creating a long, shallow, low-gradient approach to the shoreline. Similar morphology and fine channel sediments were not observed at the Restored site. Although vegetation cover was high, shoreline slopes were similar to the Reference shoreline, and flows were well attenuated; 6.5 years after restoration, sediments and nearshore (subtidal) slopes in the Restored site still carried evidence of the site's history. Prior to stabilization, severe erosion of the shoreline into the adjacent shell midden provided a consistent source of large shell fragments to the shoreline. The years of site degradation prior to restoration had a greater fraction of large particles, as smaller fractions were selectively winnowed from the shoreline. The nearshore slope had perhaps not yet had time to reflect the morphologic adjustments to sediment transport initiated by the addition of vegetation to the shoreline. Such lingering indicators of sediment and morphology serve to remind that restoration impact requires time. While some effects—such as vegetation growth and related changes to hydrodynamics—can be observed relatively quickly, their long-term influence to morphology may manifest over longer timescales. Thus, thresholds in habitat suitability for mangrove recruitment may take time to achieve, even when hydrodynamics suggest that a site should be functionally similar to a reference condition. Given the significant time investment that may be required to create a functional shoreline from a degraded state, intact shoreline ecotones must be recognized as the green infrastructure asset that they are, and be preserved at all cost.

*4.3. Why Did Vegetation at the Restored Site Thrive While Hybrid Restoration at the Seawall Did Not?*

Vegetation planted during restoration was able to readily establish and flourish in the Restored site, while planted vegetation was not able to persist in the Seawall site, which was stabilized using a hybrid living shoreline approach. The analysis presented herein suggests that the Restored site was successful because, from a hydrodynamic perspective, it was well-suited for the applied living shoreline restoration technique. Although the two sites are contiguous, the unique geographies of each create very different hydrodynamic environments (Figure 7), which influence the shoreline and nearshore morphology. The Seawall and Restored shorelines occupy opposite sides of a point that protrudes into the main channel; thus, their aspects differ relative to the predominant current direction (Figure 5). Furthermore, the Restored shoreline is located within a small embayment. The effect of geography to current exposure is evident. Channel currents observed at similar distances from the shoreline at the same portion of the tidal cycle on contiguous sampling days were consistently greater at the Seawall site than at the Restored site (Figure 8). In addition, the intertidal slope of the Restored shoreline was much lower than the shoreline slope in front of the Seawall, similar to the slope observed in the Reference shoreline. The observed incoming waves were larger in the Seawall channel (Figure 7). Even with similar wind exposure (all the sites had similar fetch), waves were consistently larger at the Seawall, potentially due to the greater channel slope.

The lower slope observed in the Restored shoreline, compared to the Seawall, may itself have been an effect of the planted vegetation; however, the restoration did not include manual site grading.

Alterations to the Restored shoreline slope would have manifested after restoration due to changes in sediment deposition and scour patterns. However, the similar sediment grain-size distributions found at the Seawall and Restored sites (pre-Hurricane Irma) suggest that such changes may have been nominal, even 6.5 years after restoration. It is rather more likely that the slope of the Restored shoreline was low before restoration. The combination of lower-energy hydrodynamic conditions and lower topographic gradients from the shoreline to channel likely explains why the Restored site thrived with a minimal intervention of vegetation plantings only, while the hybrid living shoreline at the Seawall was not able to persist.

The divergent trajectories of two contiguous sites highlight the importance of matching the selected restoration technique to a shoreline's unique hydrodynamic environment. The design of a living shoreline should consider the range of hydrodynamic conditions that a site may experience and the relative frequencies of different hydrodynamic events. For example, restoration should be designed for robustness during typical and frequent water levels, tidal currents, and windspeeds, as well as the higher water levels and more extreme currents and waves that a site may experience during infrequent disturbances. Hybrid shoreline designs that meld living shorelines with high-gradient armored structures (e.g., bulkheads, revetments) additionally must consider how flows may interact with the structure across different water levels. For instance, vegetation in the Restored shoreline remained robust to extreme conditions, and was not diminished by Hurricane Irma, while vegetation within the Seawall site was not able to persist through the typical site conditions. Site-specific living shoreline designs are needed to address local environmental limitations in order to improve the function and resilience of restored shorelines.

## 5. Conclusions

Despite the growing popularity of nature-based solutions to shoreline erosion, few quantitative analyses of field-based observations in restored living shorelines have been undertaken. Herein, differences in nearshore morphology, sediment, vegetation, and hydrodynamics are quantified among shorelines with no vegetation, natural vegetation, and restored vegetation several years after living shoreline restoration. High-resolution wave and velocity measurements in shorelines and channels reveal differences in mean and turbulent flow characteristics in sites with and without vegetation, as well as between vegetation types. Mean current velocities and waves are attenuated at significantly greater rates ($p < 0.001$) and normalized turbulent kinetic energy dissipation rates are significantly higher ($p < 0.001$) in vegetated shorelines than in bare shorelines. Additionally, current attenuation rates and dissipation rates are significantly greater ($p < 0.001$) in mature mangrove vegetation than in planted marsh grasses and young mangroves in restored shorelines. Differences in vegetated shorelines are explained by differences in vegetation characteristics. Mature mangrove vegetation consisted of a sparser array of larger elements with complex subsurface morphologies, while the planted marsh grass and young mangroves created a denser array of smaller elements. Although hydrodynamic analyses did not suggest that flow conditions should limit recruitment, mangrove seedlings were not observed in restored vegetation, while recruited seedlings were counted in reference vegetation. A lack of recruitment in restored shoreline may suggest a lag in morphological habitat suitability (slope, sediment texture, organic matter content) after restoration. For instance, 6.5 years after restoration, sediments on the restored shoreline remained coarser and contained less organic matter than on the reference shoreline. Although the hydrodynamics of the restored site were functionally similar to the reference conditions, other thresholds in habitat suitability may take longer to manifest. Finally, planted vegetation in one living shoreline withstood extreme conditions, while vegetation planted at a contiguous hybrid living shoreline was not able to persist. The divergent response highlights the role of a shoreline's unique hydrodynamic environment in living shoreline design. The site-specific range of hydrodynamic conditions and relative frequencies of hydrodynamic stresses should be a cornerstone of living shoreline design.

*Sustainability* **2019**, *11*, 3215

**Author Contributions:** Conceptualization, K.M.K.; methodology, K.M.K., V.K.; formal analysis, K.M.K., V.K., D.W.S., M.D.; investigation, K.M.K., V.K., D.W.S., M.D.; resources, K.M.K., M.D., L.W.; writing—original draft preparation, K.M.K., V.K.; writing—review and editing, K.M.K., M.D., L.W.; supervision, K.M.K., L.W.

**Funding:** This research was funded by the US National Science Foundation, grant number 1617374 and the University of Central Florida.

**Acknowledgments:** We would like to acknowledge Samantha Maldonado, OlaToyin Olasimbo, Sebastian Robbins, and Christopher Hagglund for sediment data preparation, and Samantha Maldonado, Leigh Durden, Mari Irving, and Kelsey Rodriguez Doran, for field assistance. Original funding for restored and hybrid sites was provided by the Indian River Lagoon National Estuary Program.

**Conflicts of Interest:** The authors declare no conflict of interest. The funders had no role in the design of the study; in the collection, analyses, or interpretation of data; in the writing of the manuscript, or in the decision to publish the results.

## Appendix A

The structure function method employed herein for the calculation of the TKE dissipation rate is valid within the inertial dissipation range. Thus, the distance separating the velocity cells, $r$, based on which the structure function is calculated, should lie within the Kolmogorov length scale, $\eta$, and the length scale of the largest eddies of the flow $L$, according to $\eta \ll r \ll L$. The Kolmogorov length scale was calculated from $\eta = \left(v^3/\varepsilon\right)^{1/4}$, where $v$ is the kinematic viscosity of water and was considered equal to $0.9 \cdot 10^{-6}$ m$^2$/s for a water temperature of 25 °C, as indicated by the Vectrino Profiler. At first, the centered difference technique for the computation of the structure function was applied to the adjacent cells from the cell of interest at height above the bed, $z$. This led to a minimum $r$ distance of 0.0019 m, which only slightly exceeded the maximum Kolmogorov length scale estimated from the induced dissipation rates at the Restored site. As a result, the minimum distance $r$ at the Restored site was taken as 0.0038 m, which corresponds to a minimum separating distance of four cells, i.e., at least two cells above and two below the cell of interest at height $z$. The range of the calculated TKE dissipation rates, after boat wakes were removed, was $1.5 \cdot 10^{-6}$–$7.5 \cdot 10^{-6}$ m$^2$/s$^3$ for the Reference site, $8.2 \cdot 10^{-8}$–$2.2 \cdot 10^{-6}$ m$^2$/s$^3$ for the Restored site, and $8.4 \cdot 10^{-7}$–$5.2 \cdot 10^{-5}$ m$^2$/s$^3$ for the Seawall site. The corresponding maximum Kolmogorov length scales were 0.0008 m for the Reference site, 0.002 m for the Restored site, and 0.001 m for the Seawall site, which were lower than the minimum separating distance $r = 0.0019$ m and $r = 0.0038$ m for the Reference and Seawall sites, and for the Restored site, respectively.

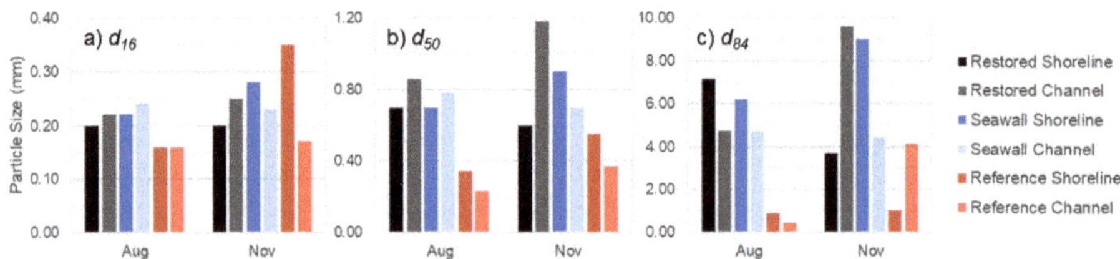


**Figure A1.** Particle size percentiles ($d_{16}$, $d_{50}$, and $d_{84}$) in Restored, Seawall, and Reference sites before and after Hurricane Irma.

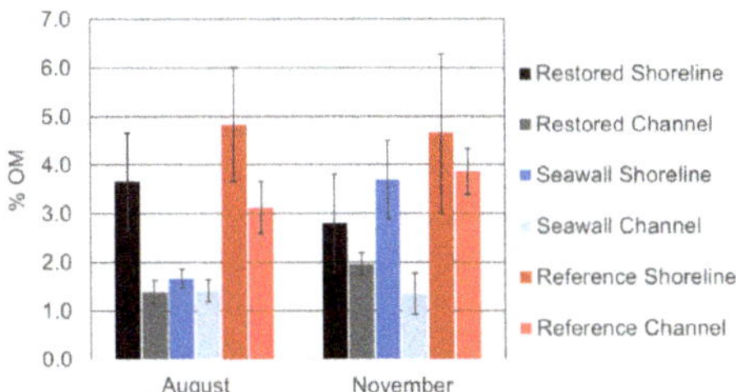


**Figure A2.** Mean carbon content ± SD in shoreline and channel sediments of the Restored, Seawall, and Reference sites, before and after Hurricane Irma.

## References


1. Lotze, H.K.; Lenihan, H.S.; Bourque, B.J.; Bradbury, R.H.; Cooke, R.G.; Kay, M.C.; Kidwell, S.M.; Kirby, M.X.; Peterson, C.H.; Jackson, J.B. Depletion, degradation, and recovery potential of estuaries and coastal seas. *Science* **2006**, *312*, 1806–1809. [CrossRef] [PubMed]

2. Gittman, R.K.; Scyphers, S.B.; Smith, C.S.; Neylan, I.P.; Grabowski, J.H. Ecological consequences of shoreline hardening: A meta-analysis. *Biosci. J.* **2016**, *66*, 763–773. [CrossRef] [PubMed]

3. Gittman, R.K.; Fodrie, F.J.; Popowich, A.M.; Keller, D.A.; Bruno, J.F.; Currin, C.A.; Piehler, M.F. Engineering away our natural defenses: An analysis of shoreline hardening in the US. *Front. Ecol. Environ.* **2015**, *13*, 301–307. [CrossRef]

4. Small, C.; Nicholls, R.J. A global analysis of human settlement in coastal zones. *J. Coast. Res.* **2003**, *19*, 584–599.

5. Hinkel, J.; Lincke, D.; Vafeidis, A.T.; Perrette, M.; Nicholls, R.J.; Tol, R.S.; Marzeion, B.; Fettweis, X.; Ionescu, C.; Levermann, A. Coastal flood damage and adaptation costs under 21st century sea-level rise. *Proc. Natl. Acad. Sci. USA* **2014**, *111*, 3292–3297. [CrossRef] [PubMed]

6. Young, I.R.; Zieger, S.; Babanin, A.V. Global trends in wind speed and wave height. *Science* **2011**, *332*, 451–455. [CrossRef] [PubMed]

7. Bilkovic, D.M.; Mitchell, M.; Mason, P.; Duhring, K. The role of living shorelines as estuarine habitat conservation strategies. *Coast. Manag.* **2016**, *44*, 161–174. [CrossRef]

8. Gedan, K.B.; Kirwan, M.L.; Wolanski, E.; Barbier, E.B.; Silliman, B.R. The present and future role of coastal wetland vegetation in protecting shorelines: Answering recent challenges to the paradigm. *Clim. Chang.* **2011**, *106*, 7–29. [CrossRef]

9. Gittman, R.K.; Popowich, A.M.; Bruno, J.F.; Peterson, C.H. Marshes with and without sills protect estuarine shorelines from erosion better than bulkheads during a Category 1 hurricane. *Ocean Coast. Manag.* **2014**, *102*, 94–102. [CrossRef]

10. Temmerman, S.; Meire, P.; Bouma, T.J.; Herman, P.M.; Ysebaert, T.; De Vriend, H.J. Ecosystem-based coastal defence in the face of global change. *Nature* **2013**, *504*, 79–83. [CrossRef]

11. Smith, C.S.; Gittman, R.K.; Neylan, I.P.; Scyphers, S.B.; Morton, J.P.; Fodrie, F.J.; Peterson, C.H. Hurricane damage along natural and hardened estuarine shorelines: Using homeowner experiences to promote nature-based coastal protection. *Mar. Policy* **2017**, *81*, 350–358. [CrossRef]

12. Swann, L. The use of living shorelines to mitigate the effects of storm events on Dauphin Island, Alabama, USA. In Proceedings of the American Fisheries Society Symposium, Ottawa, ON, Canada, 8–12 September 2008; Volume 64, p. 11.

13. Mcleod, E.; Chmura, G.L.; Bouillon, S.; Salm, R.; Björk, M.; Duarte, C.M.; Lovelock, C.E.; Schlesinger, W.H.; Silliman, B.R. A blueprint for blue carbon: Toward an improved understanding of the role of vegetated coastal habitats in sequestering $CO_2$. *Front. Ecol. Environ.* **2011**, *9*, 552–560. [CrossRef]

14. Ridge, J.T.; Rodriguez, A.B.; Fodrie, F.J. Salt Marsh and Fringing Oyster Reef Transgression in a Shallow Temperate Estuary: Implications for Restoration, Conservation and Blue Carbon. *Estuaries Coast.* **2017**, *40*, 1013–1027. [CrossRef]

15. Donnelly, M.; Shaffer, M.; Connor, S.; Sacks, P.; Walters, L. Using mangroves to stabilize coastal historic sites: Deployment success versus natural recruitment. *Hydrobiologia* **2017**, *803*, 389–401. [CrossRef]

16. Struve, J.; Falconer, R.A. Hydrodynamic and Water Quality Processes in Mangrove Regions. *J. Coast. Res.* **2001**, *27*, 65–75.

17. Lewis, R.R. Ecological engineering for successful management and restoration of mangrove forests. *Ecol. Eng.* **2005**, *24*, 403–418. [CrossRef]

18. Balke, T.; Bouma, T.; Horstman, E.; Webb, E.; Erftemeijer, P.; Herman, P. Windows of opportunity: Thresholds to mangrove seedling establishment on tidal flats. *Mar. Ecol. Prog. Ser.* **2011**, *440*, 1–9. [CrossRef]

19. Balke, T.; Swales, A.; Lovelock, C.E.; Herman, P.M.J.; Bouma, T.J. Limits to seaward expansion of mangroves: Translating physical disturbance mechanisms into seedling survival gradients. *J. Exp. Mar. Biol. Ecol.* **2015**, *467*, 16–25. [CrossRef]

20. Gittman, R.K.; Peterson, C.H.; Currin, C.A.; Joel Fodrie, F.; Piehler, M.F.; Bruno, J.F. Living shorelines can enhance the nursery role of threatened estuarine habitats. *Ecol. Appl.* **2016**, *26*, 249–263. [CrossRef]

21. Smith, C.S.; Puckett, B.; Gittman, R.K.; Peterson, C.H. Living shorelines enhanced the resilience of saltmarshes to Hurricane Matthew. *Ecol. Appl.* **2016**, *28*, 871–877. [CrossRef]

22. Nepf, H.M. Hydrodynamics of vegetated channels. *J. Hydraul. Res.* **2012**, *50*, 262–279. [CrossRef]

23. Mendez, F.J.; Losada, I.J. An empirical model to estimate the propagation of random breaking and nonbreaking waves over vegetation fields. *Coast. Eng.* **2004**, *51*, 103–118. [CrossRef]

24. Yang, J.Q.; Chung, H.; Nepf, H.M. The onset of sediment transport in vegetated channels predicted by turbulent kinetic energy. *Geophys. Res. Lett.* **2016**, *43*, 11261–11268. [CrossRef]

25. Yang, J.Q.; Nepf, H.M. A Turbulence-Based Bed-Load Transport Model for Bare and Vegetated Channels. *Geophys. Res. Lett.* **2018**, *45*, 10–428. [CrossRef]

26. Tinoco, R.O.; Coco, G. Turbulence as the Main Driver of Resuspension in Oscillatory Flow Through Vegetation. *J. Geophys. Res. Earth Surf.* **2018**, *123*, 891–904. [CrossRef]

27. Bouma, T.J.; van Duren, L.A.; Temmerman, S.; Claverie, T.; Blanco-Garcia, A.; Ysebaert, T.; Herman, P.M.J. Spatial flow and sedimentation patterns within patches of epibenthic structures: Combining field, flume and modelling experiments. *Cont. Shelf Res.* **2007**, *27*, 1020–1045. [CrossRef]

28. Chang, W.-Y.; Constantinescu, G.; Tsai, W.F. On the flow and coherent structures generated by a circular array of rigid emerged cylinders placed in an open channel with flat and deformed bed. *J. Fluid Mech.* **2017**, *831*, 1–40. [CrossRef]

29. Neary, V.S.; Constantinescu, S.G.; Bennett, S.J.; Diplas, P. Effects of vegetation on turbulence, sediment transport, and stream morphology. *J. Hydraul. Eng.* **2011**, *138*, 765–776. [CrossRef]

30. Luhar, M.; Nepf, H.M. From the blade scale to the reach scale: A characterization of aquatic vegetative drag. *Adv. Water Resour.* **2013**, *51*, 305–316. [CrossRef]

31. Horstman, E.M.; Bryan, K.R.; Mullarney, J.C.; Pilditch, C.A.; Eager, C.A. Are flow-vegetation interactions well represented by mimics? A case study of mangrove pneumatophores. *Adv. Water Resour.* **2018**, *111*, 360–371. [CrossRef]

32. Walters, L.J.; Roman, A.; Stiner, J.; Weeks, D. *Water Resource Management Plan, Canaveral National Seashore*; National Park Service: Titusville, FL, USA, 2001.

33. Down, C.; Withrow, R. *Vegetation and Other Parameters in the Brevard County Bar-Built Estuaries*; Project Report 06-73; Brevard County Health Department: Melbourne, FL, USA, 1978.

34. Hellmann, R. *Canaveral National Seashore: Archeological Overview and Assessment*; Southeast Archeological Center, National Park Service: Tallahassee, FL, USA, 2013.

35. Garvis, S.; Sacks, P.L. Walters. Assessing the Formation, Movement and Restoration of Dead Intertidal Oyster Reefs Over Time Using Remote Sensing in Canaveral National Seashore and Mosquito Lagoon, Florida. *J. Shellf. Res.* **2015**, *34*, 251–258. [CrossRef]

36. Walters, L.; Donnelly, M.; Sacks, P.; Campbell, D. Lessons learned from living shoreline stabilization in popular tourist areas: Boat wakes, volunteer support, and protecting historic structures. In *Living Shorelines: The Science and Management of Nature-Based Coastal Protection*; Bilkovic, D.M., Mitchell, M.M., la Peyre, M.K., Toft, J.D., Eds.; CRC Press: Boca Raton, FL, USA, 2017; pp. 235–248.

37. Koca, K.; Noss, C.; Anlanger, C.; Brand, A.; Lorke, A. Performance of the Vectrino Profiler at the sediment–water interface. *J. Hydraul. Res.* **2017**, *55*, 573–581. [CrossRef]

38. Thomas, R.E.; Schindfessel, L.; McLelland, S.J.; Crelle, S.; Mulder, T.D. Bias in mean velocities and noise in variances and covariances measured using a multistatic acoustic profiler: The Nortek Vectrino Profiler. *Meas. Sci. Technol.* **2017**, *28*, 075302. [CrossRef]

39. Pieterse, A.; Puleo, J.A.; McKenna, T.E.; Figlus, J. In situ measurements of shear stress, erosion and deposition in man-made tidal channels within a tidal saltmarsh. *Estuar. Coast. Shelf Sci.* **2017**, *192*, 29–41. [CrossRef]

40. Puleo, J.A.; Lanckriet, T.; Blenkinsopp, C. Bed level fluctuations in the inner surf and swash zone of a dissipative beach. *Mar. Geol.* **2014**, *349*, 99–112. [CrossRef]

41. Mori, N.; Suzuki, T.; Kakuno, S. Noise of acoustic Doppler velocimeter data in bubbly flows. *J. Eng. Mech.* **2007**, *133*, 122–125. [CrossRef]

42. Goring, D.G.; Nikora, V.I. Despiking acoustic Doppler velocimeter data. *J. Hydraul. Eng.* **2002**, *128*, 117–126. [CrossRef]
43. Wahl, T.L. Discussion of "Despiking acoustic Doppler velocimeter data". *J. Hydraul. Eng.* **2003**, *129*, 484–487. [CrossRef]
44. Wiles, P.J.; Rippeth, T.P.; Simpson, J.H.; Hendricks, P.J. A novel technique for measuring the rate of turbulent dissipation in the marine environment. *Geophys. Res. Lett.* **2006**, *33*, L21608. [CrossRef]
45. Pope, S.B. *Turbulent Flows*; Cambridge University Press: Cambridge, UK, 2000.
46. Mullarney, J.C.; Henderson, S.M.; Norris, B.K.; Bryan, K.R.; Fricke, A.T.; Sandwell, D.R.; Culling, D.P. A question of scale: How turbulence around aerial roots shapes the seabed morphology in mangrove forests of the Mekong Delta. *Oceanography* **2017**, *30*, 34–47. [CrossRef]
47. Norris, B.K.; Mullarney, J.C.; Bryan, K.R.; Henderson, S.M. Turbulence within Natural Mangrove Pneumatophore Canopies. *J. Geophys. Res. Oceans* **2019**. accepted for publication.
48. Stocking, J.B.; Rippe, J.P.; Reidenbach, M.A. Structure and dynamics of turbulent boundary layer flow over healthy and algae-covered corals. *Coral Reefs* **2016**, *35*, 1047–1059. [CrossRef]
49. Breithaupt, J.L.; Duga, E.M.; Witt, R.; Filyaw, N.; Friedland, M.J.; Donnelly; Walters, L.J.; Chambers, L.G. Carbon and nutrient fluxes from seagrass and mangrove wrack on organic and mineral sediment shorelines. *Estuar. Coast. Shelf Sci.* **2019**. in review.
50. Bouma, T.J.; Vries, M.D.; Low, E.; Kusters, L.; Herman, P.M.J.; Tanczos, I.C.; Van Regenmortel, S. Flow hydrodynamics on a mudflat and in salt marsh vegetation: Identifying general relationships for habitat characterisations. *Hydrobiologia* **2005**, *540*, 259–274. [CrossRef]
51. Tinoco, R.O.; Coco, G. Observations of the effect of emergent vegetation on sediment resuspension under unidirectional currents and waves. *Earth Surf. Dyn.* **2014**, *2*, 83–96. [CrossRef]
52. Paquier, A.E.; Haddad, J.; Lawler, S.; Ferreira, C.M. Quantification of the attenuation of storm surge components by a coastal wetland of the US Mid Atlantic. *Estuaries Coast.* **2017**, *40*, 930–946. [CrossRef]
53. Siniscalchi, F.; Nikora, V.I.; Aberle, J. Plant patch hydrodynamics in streams: Mean flow, turbulence, and drag forces. *Water Resour. Res.* **2012**, *48*. [CrossRef]
54. Nepf, H.M.; Vivoni, E.R. Flow structure in depth-limited, vegetated flow. *J. Geophys. Res. Oceans* **2000**, *105*, 28547–28557. [CrossRef]
55. Chen, Y.; Li, Y.; Cai, T.; Thompson, C.; Li, Y. A comparison of biohydrodynamic interaction within mangrove and saltmarsh boundaries. *Earth Surf. Process. Landf.* **2016**, *41*, 1967–1979. [CrossRef]
56. Koch, E.W.; Barbier, E.B.; Silliman, B.R.; Reed, D.J.; Perillo, G.M.; Hacker, S.D.; Halpern, B.S. Non-linearity in ecosystem services: Temporal and spatial variability in coastal protection. *Front. Ecol. Environ.* **2009**, *7*, 29–37. [CrossRef]
57. Yager, E.M.; Schmeeckle, M.W. The influence of vegetation on turbulence and bed load transport. *J. Geophys. Res. Earth Surf.* **2013**, *118*, 1585–1601. [CrossRef]
58. Kitsikoudis, V.; Kibler, K.M.; Walters, L.J. In-situ measurements of turbulent flow over intertidal natural and degraded oyster reefs in an estuarine lagoon. *Ecol. Eng.* **2019**. in review.
59. Maza, M.; Adler, K.; Ramos, D.; Garcia, A.M.; Nepf, H. Velocity and drag evolution from the leading edge of a model mangrove forest. *J. Geophys. Res. Oceans* **2017**, *122*, 9144–9159. [CrossRef]
60. Kitsikoudis, V.; Yagci, O.; Kirca, V.S.O.; Kellecioglu, D. Experimental investigation of channel flow through idealized isolated tree-like vegetation. *Environ. Fluid Mech.* **2016**, *16*, 1283–1308. [CrossRef]
61. Lanckriet, T.; Puleo, J.A. Near-bed turbulence dissipation measurements in the inner surf and swash zone. *J. Geophys. Res. Oceans* **2013**, *118*, 6634–6647. [CrossRef]
62. Brinkkemper, J.A.; Lanckriet, T.; Grasso, F.; Puleo, J.A.; Ruessink, B.G. Observations of turbulence within the surf and swash zone of a field-scale sandy laboratory beach. *Coast. Eng.* **2016**, *113*, 62–72. [CrossRef]
63. Pieterse, A.; Puleo, J.A.; McKenna, T.E.; Aiken, R.A. Near-bed shear stress, turbulence production and dissipation in a shallow and narrow tidal channel. *Earth Surf. Process. Landf.* **2015**, *40*, 2059–2070. [CrossRef]
64. Norris, B.K.; Mullarney, J.C.; Bryan, K.R.; Henderson, S.M. The effect of pneumatophore density on turbulence: A field study in a Sonneratia-dominated mangrove forest, Vietnam. *Cont. Shelf Res.* **2017**, *147*, 114–127. [CrossRef]
65. Kamali, B.; Hashim, R. Mangrove restoration without planting. *Ecol. Eng.* **2011**, *37*, 387–391. [CrossRef]

66. Peterson, J.M.; Bell, S.S. Tidal events and salt-marsh structure influence black mangrove (*Avicennia germinans*) recruitment across an ecotone. *Ecology* **2012**, *93*, 1648–1658. [CrossRef]
67. Donnelly, M.J.; Walters, L.J. Trapping of Rhizophora mangle by coexisting early successional species. *Estuaries Coast.* **2014**, *37*, 1562–1571. [CrossRef]

 *sustainability*

Article

# Analysing Habitat Connectivity and Home Ranges of Bigmouth Buffalo and Channel Catfish Using a Large-Scale Acoustic Receiver Network

Eva C. Enders [1],*, Colin Charles [1], Douglas A. Watkinson [1], Colin Kovachik [1], Douglas R. Leroux [1], Henry Hansen [2] and Mark A. Pegg [2]

[1] Fisheries and Oceans Canada, Freshwater Institute, 501 University Crescent, Winnipeg, MB R3T 2N6, Canada; colin.charles@dfo-mpo.gc.ca (C.C.); doug.watkinson@dfo-mpo.gc.ca (D.A.W.); colin.kovachik@dfo-mpo.gc.ca (C.K.); douglas.leroux@dfo-mpo.gc.ca (D.R.L.)

[2] School of Natural Resources, University of Nebraska-Lincoln, 3310 Holdrege Street, Lincoln, NE 68503, USA; henry.hansen@huskers.unl.edu (H.H.); mpegg2@unl.edu (M.A.P.)

*   Correspondence: eva.enders@dfo-mpo.gc.ca; Tel.: +1-204-984-4653

Received: 11 February 2019; Accepted: 26 May 2019; Published: 30 May 2019

**Abstract:** The determination if fish movement of potadromous species is impeded in a river system is often difficult, particularly when timing and extent of movements are unknown. Furthermore, evaluating river connectivity poses additional challenges. Here, we used large-scale, long-term fish movement to study and identify anthropogenic barriers to movements in the Lake Winnipeg basin including the Red, Winnipeg, and Assiniboine rivers. In the frame of the project, 80 Bigmouth Buffalo (*Ictiobus cyprinellus*) and 161 Channel Catfish (*Ictalurus punctatus*) were tagged with acoustic transmitters. Individual fish were detected with an acoustic telemetry network. Movements were subsequently analyzed using a continuous-time Markov model (CTMM). The study demonstrated large home ranges in the Lake Winnipeg basin and evidence of frequent transborder movements between Canada and the United States. The study also highlighted successful downstream fish passage at some barriers, whereas some barriers limited or completely blocked upstream movement. This biological knowledge on fish movements in the Lake Winnipeg basin highlights the need for fish passage solutions at different obstructions.

**Keywords:** fish passage; fish telemetry; river restoration; ecohydraulics; *Ictiobus cyprinellus*; *Ictalurus punctatus*

---

## 1. Introduction

River connectivity may be interrupted by dams, weirs, and culverts, resulting in fragmentation of habitat [1]. Damming of large rivers is likely the most noticeable form of river fragmentation [2] and it is often observed to lead to hydromorphological alteration of the water course and changes in the biota [3]. The fragmentation of riverine ecosystems can result in a decline of fish biodiversity [4,5]. Particularly, the blockage of the migration of anadromous (e.g., salmon) and catadromous (e.g., eel) fishes has led to population declines or even extirpation of populations [6,7]. However, there is a lack of appreciation for the movement needs of potadromous fishes and the various scales that riverine fish species may move. This makes it more challenging to demonstrate the importance of river connectivity and the dispersal of riverine fishes that are crucial for population processes such as reproduction, rearing, and feeding [8]. Several freshwater fish species undertake long distance movements if their riverine habitat corridor is not impeded and competition for feeding and spawning sites can increase as dams disconnect, isolate, and reduce the number and size of habitats [9,10]. Consequently, river restoration efforts have focused on establishing connectivity to enable longitudinal and lateral fish movement to meet the life-history requirements for these species [11]. River restoration efforts that reconnect fragmented habitats are generally successful at improving fish populations [12] and isolated habitats are quickly recolonized after the removal of barriers [13]. If the removal of a barrier is unfeasible, increasing river connectivity through the installation of effective fish passage structures can be an alternate management strategy [14–17].

Riverine fish conservation requires including various spatial scales when considering longitudinal connectivity of rivers to allow access to resource use that may be influenced by food availability, water temperature, and suitable habitats that are found in different river sections [18]. Determining the scale of freshwater fish movements and the size of their home ranges remain a research priority particularly for imperiled species [19,20]. Here, we focus on two freshwater fish species with long distance movement behavior. Bigmouth Buffalo (*Ictiobus cyprinellus*) is a filter-feeder using its very fine gill rakers to strain food items from the water [21]. Bigmouth Buffalo spawn in the spring to early summer and lay adhesive eggs on plants. Currently, little is known about the movement patterns and home ranges of Bigmouth Buffalo. Channel Catfish (*Ictalurus punctatus*) are an omnivorous, benthic fish [21]. Channel Catfish spawn during spring or summer when the water warms to an optimal temperature of 21–28 °C. A mark-recapture study using Floy tags demonstrated that Channel Catfish undergo migratory movements [22], however, the study did not allow for determination of timing or extent of these movements. Both species are of interest from a biodiversity conservation perspective in the Lake Winnipeg basin, Canada. First, the loss of access to spawning and/or the degradation of spawning habitat due to water management practices is thought to have contributed to the decline in Bigmouth Buffalo (in the Saskatchewan – Nelson River watershed; SARA 2016a, www. dfo-mpo.gc.ca/species-especes/profiles-profils/bigmouth-buffalo-grande-bouche-eng.html). Second, Channel Catfish is the only known host fish of the endangered Mapleleaf (*Quadrula quadrula*; SARA 2016b, http://dfo-mpo.gc.ca/species-especes/profiles-profils/mapleleaf-feuillederable-sk-eng.html). The mussel is in decline and appears to be limited to the Red, Assiniboine, and Roseau rivers as well as tributaries on the east side of Lake Winnipeg. Barriers result in habitat loss and fragmentation, altered flow regimes, and may increase mortality by entrainment in turbines. Consequently, knowledge on fish movement is essential to inform conservation and recovery strategies, fishery management actions, and fish passage approaches to avoid migration barriers for these fishes.

The specific objectives of this study were to: (1) describe fish movement and home range of two fish species, Bigmouth Buffalo and Channel Catfish, in the Lake Winnipeg basin, (2) determine the transitions between different regions in the Lake Winnipeg basin using continuous-time Markov models on the telemetry data, and (3) to analyze if and to what extent fish passage may be impeded by the multiple anthropogenic structures in the Lake Winnipeg system including Red, Winnipeg, and Assiniboine rivers using a large-scale acoustic receiver network.

## 2. Materials and Methods

### 2.1. Study Site

Lake Winnipeg is the largest lake in the province of Manitoba, Canada (52°7′N 97°15′W) covering 24,514 km$^2$ (Figure 1). The lake is relatively shallow, elongated, and isothermal, with a mean water depth of 12 m and spanning 416 km from north to south [23]. Lake Winnipeg's watershed measures about 982,900 km$^2$ and covers much of Alberta, Saskatchewan, Manitoba, northwestern Ontario, Minnesota, and North Dakota. The lake drains to the north into the Nelson River at an average annual rate of 2066 m$^3$·s$^{-1}$ and ultimately into Hudson Bay. Lake Winnipeg is eutrophic, receiving excessive amounts of nutrient run-off from agricultural land use. Lake Winnipeg is also one of the largest hydro reservoirs in the world and supports one of the most productive commercial and recreational fisheries for Walleye (*Sander vitreus*). Several aquatic invasive species are established in the lake including

Common Carp (*Cyprinus carpio*), Zebra Mussel (*Dreissena polymorpha*), and Spiny Water Flea (*Bythotrepes longimanus*).

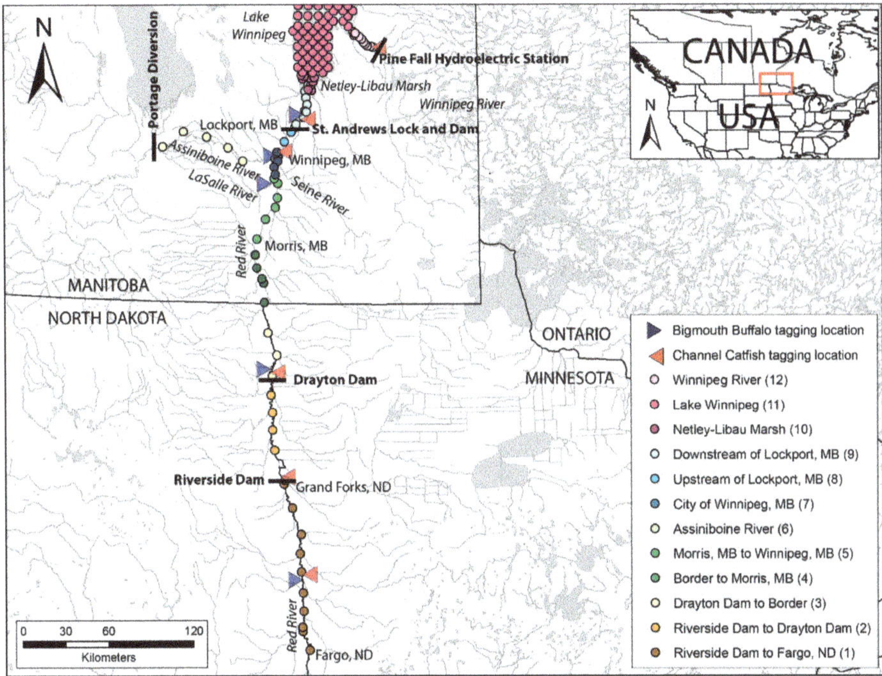


**Figure 1.** Map of the Lake Winnipeg basin including the Red, Winnipeg, and Assiniboine rivers and Netley-Libau Marsh. Tagging (▶) and receiver (●) locations are indicated. Potential barriers to fish movement are the St. Andrews Lock and Dam, Drayton Dam, and Riverside Dam on the Red River, Portage Diversion Dam on the Assiniboine River, and Pine Falls Hydroelectric Generating Station on the Winnipeg River. River sections with the Section ID in parenthesis for the movement analysis using continuous-time Markov models are indicated on the map with different color codes.

The main tributaries of Lake Winnipeg analyzed in this study include the Red River flowing into the lake from the south and the Winnipeg River from the southeast (Figure 1). The Red River originates at the confluence of the Bois de Sioux and Otter Tail rivers between the States of Minnesota and North Dakota. It is approximately 885 km long and flows northward through the Red River Valley, forming most of the border of Minnesota and North Dakota before flowing into Manitoba. It empties into Lake Winnipeg through the Netley-Libau Marsh. There are three water level control structures in the main stem of the Red River: The St. Andrews Lock and Dam, the Drayton Dam, and the Riverside Dam, that potentially pose barriers to habitat connectivity and therefore fish movement. The St. Andrews Lock and Dam is located in Lockport, Manitoba at RKM 43.6 from the mouth of the Red River (Figure 1). It was constructed in the early 20th century to facilitate commercial navigation from Lake Winnipeg to the City of Winnipeg by inundating the Lister Rapids during the navigation season. The facility is operated and maintained by Public Works and Government Services Canada and consists of a dam, a navigation lock, and a fishway. Fish can move upstream through the lock and the fishway and downstream through the lock, fishway, and the spill.

The Drayton Dam, situated at RKM 327.3 on the Red River, was constructed in 1964 to provide water supply for agricultural and municipal use. It is located approximately 3 km north of Drayton, North Dakota (Figure 1). The dam consists of a concrete weir with a spillway length of 68.5 m and a crest elevation of 3.7 m above the natural riverbed. It operates as a run-of-river control structure and has no dedicated fish passage features.

The Riverside Dam at RKM 476.5 in Grand Forks, North Dakota (Figure 1) was originally built in 1922 as a water control structure. In 2001, it was restored to a rock ramp to provide erosion control, eliminate a hydraulic roller, provide fish passage and spawning habitats, as well as whitewater boating opportunities. It consists of a rock arch rapid with a 5% slope (3% near banks). Interestingly, it was the largest full width rock ramp fishway in terms of tonnage and height in the world at time of construction [24].

The Winnipeg River flows from Lake of the Woods to Lake Winnipeg (Figure 1). There are eight hydroelectric dams on the 235 RKM long river with the most downstream facility being the Pine Falls Hydroelectric Station at Powerview, Manitoba (commissioned in 1952). Flows on the Winnipeg River are controlled through the various dams by the Lake of the Woods Control Board. None of the dams on the Winnipeg River including the Pine Falls Hydroelectric Station provide for upstream fish passage. Downstream migrants can pass through the turbines or spillways.

The Assiniboine River originates in eastern Saskatchewan. It flows east into Manitoba (Figure 1). Its junction with the Red River is in the City of Winnipeg, Manitoba. The 1070 km long meandering river is prone to spring flooding. In 1970, the Portage Diversion Dam, located at RKM 163, was completed to divert flood flows into Lake Manitoba at Portage la Prairie, Manitoba. Fish can move downstream over the spillway but there is no upstream fish passage installation at the Portage Diversion Dam.

*2.2. Fish Collection*

Depending on the fish species, different collection methods were used. Bigmouth Buffalo (n = 80) were caught by boat electrofishing at five tagging locations in the Red River (Figure 1, Table 1). Channel Catfish (n = 161) were caught by angling with barbless hook and landed with rubber nets or with hoop nets (1 m diameter, 2.5 cm bar mesh). Collection efforts for Channel Catfish were conducted at six tagging locations (Figure 1, Table 1), including the Lower Red River and Winnipeg River where high recreational fishing efforts occur for trophy Channel Catfish.

**Table 1.** Tagging location with section ID, mean length, and body mass of Bigmouth Buffalo and Channel Catfish tagged with acoustic transmitters in the frame of the Lake Winnipeg Fish Movement Study. Bigmouth Buffalo were caught by boat electrofishing and Channel Catfish by angling.

| Species | Year | Section ID | Section of Tagging Site | n | Mean (± S.D.) Length (mm) | Mean (± S.D.) Body Mass (kg) |
|---|---|---|---|---|---|---|
| Bigmouth Buffalo | 2017 | 1 | Fargo, ND to Riverside Dam | 12 | 593.25 (65.0) | 3.11 (1.0) |
| | 2017 | 3 | Drayton Dam to Border | 8 | 564.9 (48.6) | 2.72 (0.5) |
| | 2016 | 5 | Morris, MB to Winnipeg, MB | 20 | 561.2 (58.6) | 2.85 (0.9) |
| | 2016 | 7 | City of Winnipeg, MB | 20 | 621.1 (57.8) | 4.01 (1.2) |
| | 2017 | 9 | Downstream of Lockport, MB | 20 | 683.7 (83.3) | 6.23 (2.6) |
| Channel Catfish | 2017 | 1 | Fargo, ND to Riverside Dam | 9 | 671.8 (117.0) | 3.46 (2.1) |
| | 2017 | 2 | Riverside Dam to Drayton Dam | 16 | 751.4 (106.6) | 5.53 (2.5) |
| | 2017 | 3 | Drayton Dam to Border | 15 | 690.1 (93.6) | 3.65 (1.6) |
| | 2016 | 7 | City of Winnipeg, MB | 24 | 660.8 (73.7) | 3.33 (1.2) |
| | 2016 | 9 | Downstream of Lockport, MB | 67 | 751.2 (113.9) | 5.66 (2.8) |
| | 2016 | 12 | Winnipeg River | 30 | 640.0 (52.3) | 2.98 (1.1) |

*2.3. Fish Tagging*

Upon capture, fish were placed in holding tanks filled with ambient river water. Captured fish were measured and weighed immediately, only individuals with a body mass >1.2 kg were tagged, and undersized individuals (>2% tag: body weight) were released (Table 1). Acoustic telemetry transmitters (VEMCO, V16-4H, 16 mm diameter, 24 g, $6\frac{1}{2}$ years battery life, with an average transmission delay of 120 s with a pseudo random uniform interval between 80–160 s) were implanted in fish. As Channel Catfish are part of the recreational and commercial fisheries, individuals were also tagged with an external Floy tag (Floy Tag Inc., Floy T-bar anchor). Fish were placed into the Portable Electroanesthesia System (PES™, Smith-Root, Vancouver, WA, USA) to immobilize them during surgery, without use of chemical anesthetics. The PES™ was set to 100 Hz, 25% duty cycle, and 40 V. Pulsed direct current is an appropriate sedation for adult fish because it provides a surgery window of 250–350 s and fish recover quickly with minimal impact to vertebral integrity [25]. Upon sedation, fish were placed in a padded v-shaped trough. Ambient river water was continuously pumped over the gills using a recirculating flow-through pump system to maintain normal respiration during the surgical period (<5 min). A small incision was made posterior to the pectoral girdle just dorsal of the ventral midline. The acoustic transmitter was inserted posteriorly into the peritoneal cavity. Transmitter expulsion is common in Channel Catfish [26]. Consequently, for Channel Catfish a specific surgical procedure was used that tethers the transmitter around the cleithrum and/or supracleithrum near the scapula by looping a monofilament suture through the transmitter and around the bone [27]. Subsequently, the incision was closed with three to four interrupted sutures (standard surgical knots; 3-0 polydioxanone-II violet monofilament; Ethicon, Cincinnati, OH, USA). Fish were put in the recovery tank and released 10–15 min post-surgery at the tagging location. Surgical procedures were carried out in accordance with approved animal use protocols of Fisheries and Oceans Canada (FWI-ACC-2016-018, FWI-ACC-2017-001) and the University of Nebraska-Lincoln (Project ID: 1208).

### 2.4. Receiver Array

Acoustic receivers (VEMCO, VR2W and VR2Tx receivers, $n = 247$) were placed in Lake Winnipeg and the Red, Winnipeg, and Assiniboine rivers (Figure 1). Receiver spacing in the rivers varied between 5 to 30 km covering an accumulated distance of 860 RKM (Figure 1). As upstream passage at the Portage Diversion Dam and the Pine Falls Hydroelectric Station is impossible, no receivers were placed upstream of these two barriers. In Lake Winnipeg, receivers were installed on a $7 \times 7$ km grid [10]. Data from the receivers were downloaded annually (2016–2018) in the open water season, usually from June–September.

### 2.5. Data Manipulation and Analysis

All data manipulations and analyses were conducted in R [28]. The telemetry dataset for Bigmouth Buffalo and Channel Catfish consisted of >1.3 million individual detections. Subsequently, the 'dplyr' library for data manipulations was used to augment the computational efficiency of the subsequent data analyses [29]. Fish movements, home ranges, and river system connectivity between habitats were analyzed using the R package 'riverdist' [30] and the data analysis was separated by year. The R package 'riverdist' allows to read river network shape files, compute network distances as well as to as to display and calculate fish home ranges on a linear framework using telemetry data. To create the river network in R, we imported spatial coordinates (lat/long) and used the 'convUL()' function in the 'PBSmapping' package [31] to convert the lat/long coordinates into UTM and then applied the 'line2network()' function in 'riverdist' to create a river network, which was used to calculate the home range estimates. Fish movement data were used to calculate the home ranges in the Lake Winnipeg basin by species and year to describe and compare movements of Bigmouth Buffalo and Channel Catfish in the basin.

### 2.6. Continuous-Time Markov Model

Fish movement in the Lake Winnipeg basin was simulated by a continuous-time Markov model (CTMM) to recognize the continuous movement patterns that fish exhibit. Receivers were grouped using a combination of geographical proximity, geopolitical boundaries, tributaries, and/or physical barriers (see Figure 1, Table 2). A multi-state model was used to describe and quantify how fish transitioned between $k$ unique states ($S = \{1, 2, \ldots, k\}$), and transitions were only allowed between adjacent states (i.e., river sections in the Lake Winnipeg basin; Table 2). Assuming that an individual fish is able to freely change back and forth between $k$ states in continuous time, a $k \times k$ transitional intensity matrix ($Q$ matrix), was defined as follows:

$$q_{rs}(t) = \frac{\lim_{\delta t \to 0} P[S(t + \delta t) = s | S(t) = r]}{\delta t}, r \neq s$$

where $t$ is time and $q_{rs}$ is the instantaneous rate of change from the current state $r$ to the next state $s$. The rows of the $Q$ matrix sum to zero, while the diagonal entries are defined by $q_{rr} = -\sum_{s \neq r} q_{rs}$ and the off-diagonal entries can be any non-negative number [32]. Transition probabilities on both extents of the receiver network could not be predicted by the CTMM (Table 4). Subsequently, given that the Fargo, North Dakota to Riverside Dam river section was the southernmost extent of our receiver network, there is only a one way transition to an adjacent state possible. As a result, we can only predict transition from state 1 to state 2 but not transitions upstream from Fargo. Due to imperfect detections of the receivers, some fish movements in and out of a given state may have been undetected, leading to a potential of underestimating movements in the system.

**Table 2.** Transition matrix representing allowable transitions between states/sections represented in the continuous-time Markov model (CTMM). Dashes (-) represent no permissible transition while $q_{rs}$ entries represent allowable transitions between different sections. In **bold** are the transition over the barriers (i.e., St. Andrews Lock and Dam, Drayton Dam, and Riverside Dam).

| | | To State/Section | | | | | | | | | | | |
|---|---|---|---|---|---|---|---|---|---|---|---|---|---|
| | | 1 | 2 | 3 | 4 | 5 | 6 | 7 | 8 | 9 | 10 | 11 | 12 |
| | 1 | $q_{11}$ | $q_{12}$ | - | - | - | - | - | - | - | - | - | - |
| | 2 | $q_{21}$ | $q_{22}$ | $q_{23}$ | - | - | - | - | - | - | - | - | - |
| | 3 | - | $q_{32}$ | $q_{33}$ | $q_{34}$ | - | - | - | - | - | - | - | - |
| | 4 | - | - | $q_{43}$ | $q_{44}$ | $q_{45}$ | - | - | - | - | - | - | - |
| | 5 | - | - | - | $q_{54}$ | $q_{55}$ | - | $q_{57}$ | - | - | - | - | - |
| **From State/Section** | 6 | - | - | - | - | - | $q_{66}$ | $q_{67}$ | - | - | - | - | - |
| | 7 | - | - | - | - | $q_{75}$ | $q_{76}$ | $q_{77}$ | $q_{78}$ | - | - | - | - |
| | 8 | - | - | - | - | - | - | $q_{87}$ | $q_{88}$ | **$q_{89}$** | - | - | - |
| | 9 | - | - | - | - | - | - | - | $q_{98}$ | $q_{99}$ | $q_{910}$ | - | - |
| | 10 | - | - | - | - | - | - | - | - | $q_{109}$ | $q_{1010}$ | $q_{1011}$ | - |
| | 11 | - | - | - | - | - | - | - | - | - | $q_{1110}$ | $q_{1111}$ | $q_{1112}$ |
| | 12 | - | - | - | - | - | - | - | - | - | - | $q_{1211}$ | $q_{1212}$ |

**Note:** (1) Fargo, ND to Riverside Dam, (2) Riverside Dam to Drayton Dam, (3) Drayton Dam to Border, (4) Border to Morris, MB, (5) Morris, MB to Winnipeg, MB, (6) Assiniboine, (7) City of Winnipeg, MB, (8) Upstream of Lockport, MB, (9) Downstream of Lockport, MB, (10) Netley-Libau Marsh, (11) Lake Winnipeg, (12) Winnipeg River.

Additionally, the 'msm' library can calculate retention times (i.e., sojourn times or times spent in different sections) in the CTMM, where the retention times are exponentially distributed with mean $-1/q_{rr}$. Finally, a $P_{next}$ matrix was constructed that defines the probability of changing from state $r$ to state $s$ in the next transition, regardless of the time elapsed. The diagonal entries of the $P_{next}$ matrix are equal to zero.

## 3. Results

### 3.1. Home Ranges

Over the three study years, tagged Bigmouth Buffalo were consistently detected in high numbers (90–100% of tagged population) whereas Channel Catfish detection decreased by up to 40–78% of a given tagged population (Table A1). Bigmouth Buffalo consistently showed large individual home ranges varying from 4.2 to 621.9 km per year whereas individual Channel Catfish movement ranged from 3.4 to 101.3 km (Figure 2, Table 3). In all years, Bigmouth Buffalo had significantly larger home ranges than Channel Catfish (2016: mean 177.5 km vs. 32.7 km; $p < 0.001$; 2017: mean 132.6 km vs. 91.0 km; $p = 0.03$; 2018: mean 150.9 km vs. 60.0 km; $p < 0.01$, Figure 3). Fish were predominately moving in the open water season between April to October and were relatively inactive during the ice-on season from November to March (Figure 4).

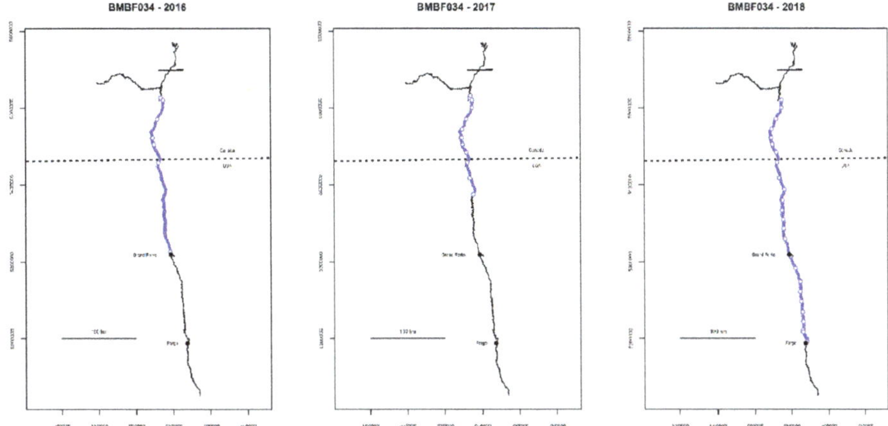

**Figure 2.** Example of the annual home range (km) of an individual Bigmouth Buffalo (BMBF034) from 2016 to 2018.

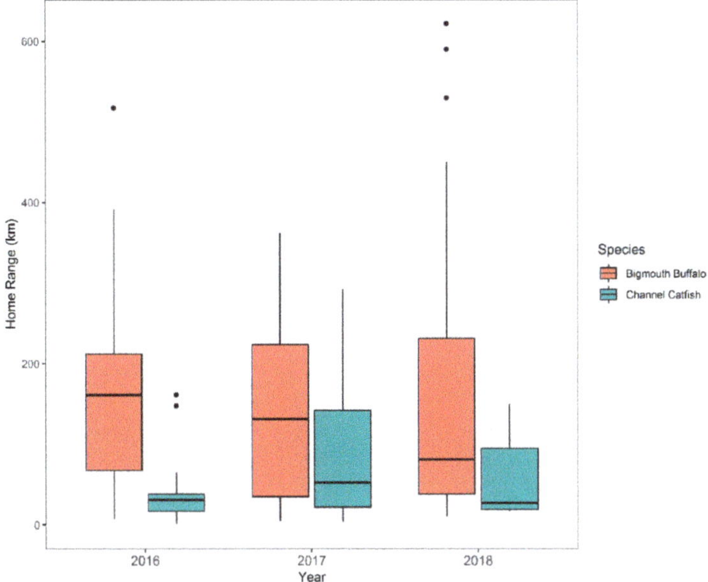

**Figure 3.** Home range (km) for Bigmouth Buffalo and Channel Catfish tagged with acoustic transmitters in various locations in the Lake Winnipeg basin.

**Table 3.** Inter-annual variations in species-specific river movements.

| Species | Year | Distance (km) | | |
|---|---|---|---|---|
| | | Minimum | Mean (± S.D.) | Maximum |
| Bigmouth Buffalo | 2016 | 7 | 177.5 (119.7) | 517.5 |
| | 2017 | 4.2 | 132.6 (105.2) | 361.9 |
| | 2018 | 10 | 150.9 (148.0) | 621.6 |
| Channel Catfish | 2016 | 0.8 | 32.7 (28.0) | 161.5 |
| | 2017 | 3.5 | 91.0 (93.8) | 292.1 |
| | 2018 | 16.8 | 60.0 (56.6) | 149.6 |

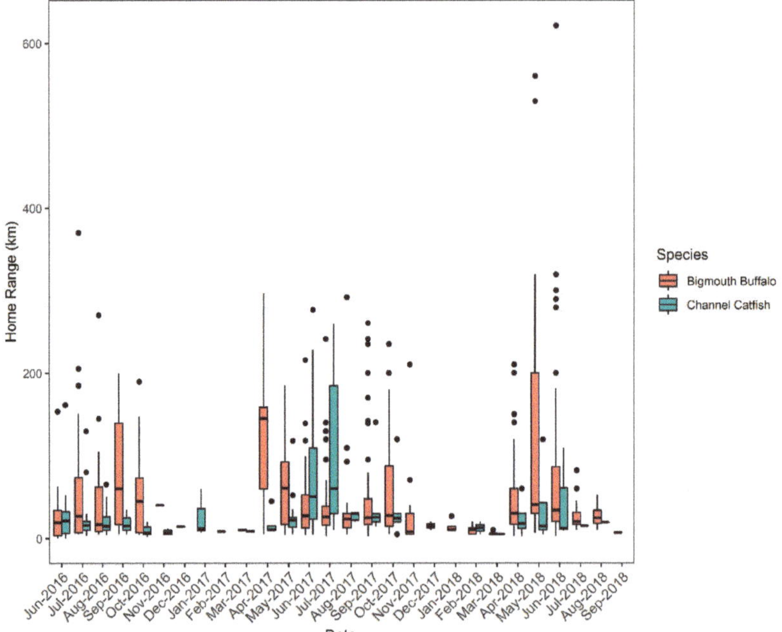

**Figure 4.** Monthly home range (km) of Bigmouth Buffalo and Channel Catfish tagged with acoustic transmitters in the Lake Winnipeg basin.

*3.2. Transitions Over Barriers in the Basin and Between River Sections*

Using the continuous-time Markov model (CTMM), we were able to predict the probability of fish transitioning to the adjacent river/lake sections (Figure 1) given the current section in which they were observed ($P_{next}$; Tables 4 and 5). For Bigmouth Buffalo in Section 2 (Riverside Dam to Drayton Dam), there was a low probability (4%) to move upstream over the rock ramp at the Riverside Dam but a 96% probability to migrate downstream over the Drayton Dam (Table 4). The probability for upstream transitions over the Drayton Dam was 46%. In regards to the St. Andrews Lock and Dam, the downstream transition probability of Bigmouth Buffalo was 3%, in comparison to an even lower upstream transition probability of 0.5%. The predicted transition probabilities between unimpeded rivers sections were relative high ranging between 39–100% (Table 2a). The CTMM did not predict an upstream transition probability for Channel Catfish at any of the three barriers on the Red River (i.e., Riverside, Drayton, and St. Andrews dams) but there was a very high probability for Channel Catfish to migrate downstream over the Drayton Dam (100%, Table 5). The predicted transition probability for Channel Catfish to migrate downstream over the St. Andrews Lock and Dam was also higher with

44% in comparison to Bigmouth Buffalo (3%). The transition probabilities for Channel Catfish did not appear to be limited during unimpeded river sections as up-and downstream transition probabilities were equal in unimpeded reaches (Table 2b).

**Table 4.** $P_{next}$ matrix to estimate the probability of Bigmouth Buffalo moving to the next upstream or downstream section in the Lake Winnipeg basin. In **bold** are the transition over the barriers (i.e., St. Andrews Lock and Dam, Drayton Dam, and Riverside Dam). *Transition probabilities on both extents of the receiver network cannot be predicted.

|  |  | To Section | | | | | | | | | | | |
|---|---|---|---|---|---|---|---|---|---|---|---|---|---|
|  |  | 1 | 2 | 3 | 4 | 5 | 6 | 7 | 8 | 9 | 10 | 11 | 12 |
| | 1 | - | **1.000*** | - | - | - | - | - | - | - | - | - | - |
| | 2 | 0.039 | - | 0.961 | - | - | - | - | - | - | - | - | - |
| | 3 | - | **0.455** | - | 0.545 | - | - | - | - | - | - | - | - |
| | 4 | - | - | 0.53 | - | 0.47 | - | - | - | - | - | - | - |
| | 5 | - | - | - | 0.352 | - | - | 0.648 | - | - | - | - | - |
| From Section | 6 | - | - | - | - | - | - | 1 | - | - | - | - | - |
| | 7 | - | - | - | - | 0.543 | 0.066 | - | 0.391 | - | - | - | - |
| | 8 | - | - | - | - | - | - | 0.966 | - | **0.034** | - | - | - |
| | 9 | - | - | - | - | - | - | - | **0.005** | - | 0.995 | - | - |
| | 10 | - | - | - | - | - | - | - | - | 0.974 | - | 0.026 | - |
| | 11 | - | - | - | - | - | - | - | - | - | **1.000*** | - | - |
| | 12 | - | - | - | - | - | - | - | - | - | - | - | - |

**Table 5.** $P_{next}$ matrix to estimate the probability of Channel Catfish moving to the next upstream or downstream section in the Lake Winnipeg basin. In **bold** are the transition over the barriers (i.e., St. Andrews Lock and Dam, Drayton Dam, and Riverside Dam). *Transition probabilities on both extents of the receiver network cannot be predicted.

|  |  | To Section | | | | | | | | | | | |
|---|---|---|---|---|---|---|---|---|---|---|---|---|---|
|  |  | 1 | 2 | 3 | 4 | 5 | 6 | 7 | 8 | 9 | 10 | 11 | 12 |
| | 1 | - | **1.000*** | - | - | - | - | - | - | - | - | - | - |
| | 2 | 0.000 | - | 1.000 | - | - | - | - | - | - | - | - | - |
| | 3 | - | **0.000** | - | 1.000 | - | - | - | - | - | - | - | - |
| | 4 | - | - | 0.000 | - | 1.000 | - | - | - | - | - | - | - |
| | 5 | - | - | - | 0.407 | - | - | 0.593 | - | - | - | - | - |
| From Section | 6 | - | - | - | - | - | - | 1.000 | - | - | - | - | - |
| | 7 | - | - | - | - | 0.598 | 0.038 | - | 0.363 | - | - | - | - |
| | 8 | - | - | - | - | - | - | 0.564 | - | **0.436** | - | - | - |
| | 9 | - | - | - | - | - | - | - | **0.000** | - | 1.000 | - | - |
| | 10 | - | - | - | - | - | - | - | - | 0.224 | - | 0.776 | - |
| | 11 | - | - | - | - | - | - | - | - | - | 0.074 | - | 0.926 |
| | 12 | - | - | - | - | - | - | - | - | - | - | **1.000*** | - |

**Note:** (1) Fargo, ND to Riverside Dam, (2) Riverside Dam to Drayton Dam, (3) Drayton Dam to Border, (4) Border to Morris, MB, (5) Morris, MB to Winnipeg, MB, (6) Assiniboine, (7) City of Winnipeg, (8) Upstream of Lockport, MB, (9) Downstream of Lockport, MB, (10) Netley-Libau Marsh, (11) Lake Winnipeg, (12) Winnipeg River.

Generally, Channel Catfish had a higher retention times in each of the sections compared to Bigmouth Buffalo (Table 6). Section 1 (Fargo, North Dakota to Riverside Dam) had the highest retention time estimates for both species. However, fish were able to freely move upstream from Fargo, ND into the Bois de Sioux and Otter Tail rivers and subsequently out of the range of our receivers, so any potential movements further upstream were not captured by our receiver network. Retention times for Bigmouth Buffalo were likely lower compared to Channel Catfish due to their increased mobility (Table 3) resulting in Bigmouth Buffalo likely spending less time in distinct sections. Fish residency in the Assiniboine River was very similar between the two species with 27.3 and 28.1 d for Bigmouth Buffalo and Channel Catfish, respectively. However, even though the retention times were similar, Bigmouth Buffalo appear to have swum further distance upstream with three individuals detected near the Portage Diversion Dam (near RKM 151) whereas Channel Catfish were not detected at the

Portage Diversion Dam. Movements from fish tagged in the Red River into the Winnipeg River (via Lake Winnipeg) were generally limited, as Channel Catfish were estimated to spend only 1.0 day in the Winnipeg River and no Bigmouth Buffalo moved into the Winnipeg River.

**Table 6.** Predicted mean retention times (in days) in a given river or lake section (see Figure 1) by the CTMM.

| Section | River/Lake Section | Reach Length/Area | Bigmouth Buffalo | Channel Catfish |
|---|---|---|---|---|
| 1 | Fargo, ND to Riverside Dam | 255 km | 137.3 | 134.9 |
| 2 | Riverside Dam to Drayton Dam | 120 km | 11.1 | 104.6 |
| 3 | Drayton Dam to Border | 89 km | 12.8 | 7.7 |
| 4 | Border to Morris, MB | 84 km | 7.1 | 13.4 |
| 5 | Morris, MB to Winnipeg, MB | 75 km | 12.8 | 10.7 |
| 6 | Assiniboine River | 150 km | 27.3 | 28.1 |
| 7 | City of Winnipeg, MB | 40 km | 12.1 | 15.7 |
| 8 | Upstream of Lockport, MB | 21.5 km | 15.1 | 24.6 |
| 9 | Downstream of Lockport, MB | 45 km | 7.6 | 21.1 |
| 10 | Netley-Libau Marsh | 115 km$^2$ | 4.4 | 22.4 |
| 11 | Lake Winnipeg | 2862 km$^2$ | 5.7 | 2.0 |
| 12 | Winnipeg River | 15.5 km | - | 1.0 |

### 3.3. Limitations to Habitat Connectivity in the RED, Winnipeg, and Assiniboine Rivers

In 2016, nine of the tagged Bigmouth Buffalo moved upstream over the Drayton Dam, while no Channel Catfish were detected undergoing an upstream movement over the weir (Table 7). In the following year, a total of twelve Bigmouth Buffalo and 17 Channel Catfish were observed to pass upstream over the Drayton Dam. In 2018, Bigmouth Buffalo completed 13 upstream movements and no Channel Catfish were observed. Downstream passage over the Drayton Dam was observed in 2017 ($n = 4$ Bigmouth Buffalo and $n = 2$ Channel Catfish) and 2018 ($n = 13$ Bigmouth Buffalo). Interestingly, both up and downstream movements over the weir at the Drayton Dam for both species appeared to occur during peak flows or descending hydrographs (Figure 5a,b).

**Table 7.** Number per fish species and year of up and downstream passage over the St. Andrews Lock and Dam, Drayton Dam, and Riverside Dam. In parenthesis the number of fish present in river section downstream for upstream passage and in the river section upstream for downstream passage. Number of fish detected on the receiver downstream (Downstream Presence) of the Portage Diversion Dam and of the Pine Falls Hydroelectric Station that are both impassable for upstream migrating fish.

| Barrier | Passage/Presence | Bigmouth Buffalo | | | Channel Catfish | | |
|---|---|---|---|---|---|---|---|
| | | 2016 | 2017 | 2018 | 2016 | 2017 | 2018 |
| St. Andrews Locks and Dam | Upstream passage | 0 (1) | 3 (22) | 1 (12) | 0 (20) | 0 (10) | 0 (4) |
| | Downstream passage | 1 (13) | 4 (21) | 1 (9) | 5 (15) | 3 (7) | 0 (1) |
| Drayton Dam | Upstream passage | 9 (27) | 12 (25) | 13 (26) | 0 (0) | 17 (23) | 0 (0) |
| | Downstream passage | 0 (5) | 4 (20) | 13 (19) | 0 (0) | 2 (11) | 0 (0) |
| Riverside Dam | Upstream passage | 1 (5) | 0 (20) | 3 (0) | 0 (0) | 0 (11) | 0 (0) |
| | Downstream passage | 0 (1) | 0 (12) | 4 (11) | 0 (0) | 1 (9) | 0 (0) |
| Portage Diversion | Downstream presence | 2 (10) * | 1 (16) * | 0 (4) * | 0 (1) * | 0 (0) * | 0 (0) * |
| Pine Falls Station | Downstream presence | - | - | - | 27 (30) ** | 6 (26) ** | 1 (18) ** |

* In parenthesis the number of fish observed in the Assiniboine River section. ** In parenthesis the number of fish observed in the Winnipeg River section.

Fewer fish passages were observed at the St. Andrews Lock and Dam; three Bigmouth Buffalo passed upstream over the dam in 2017 and one in 2018. Upstream passage of Channel Catfish was not observed. One single downstream passage of Bigmouth Buffalo was detected in 2016 over the dam, four in 2017 and one in 2018 whereas five Channel Catfish passed downstream in 2016 and three in 2017. Similar to the movement patterns observed in Drayton, up and downstream passage at the St. Andrew Lock and Dam appears also to be associated with peak and descending hydrographs in the open water season (Figure 5c,d).

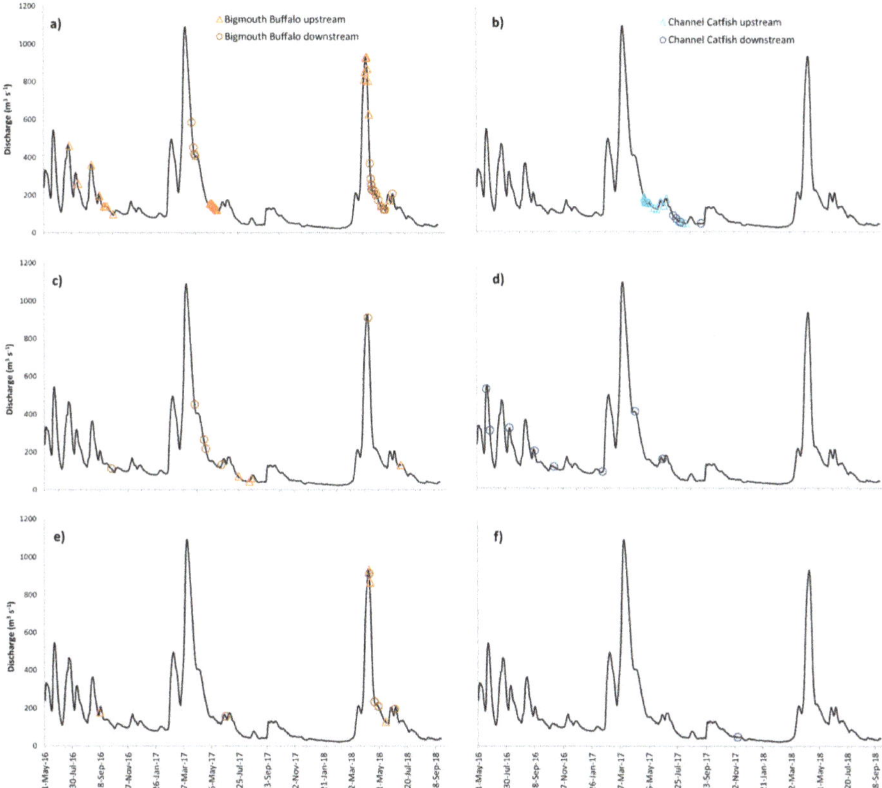

**Figure 5.** Timing of upstream and downstream movements of (**a**) Bigmouth Buffalo and (**b**) Channel Catfish over the St. Andrews Lock and Dam; (**c**) Bigmouth Buffalo and (**d**) Channel Catfish over Drayton Dam, and (**e**) Bigmouth Buffalo and (**f**) Channel Catfish over the Riverside Dam; in relationship to the discharge (daily mean, m$^3$ s$^{-1}$) in the Red River at Emerson, Manitoba (Water Survey of Canada, Hydrometric Station 05OC001).

The Portage Diversion Dam is impassable for upstream fish migration. Tag detections from the receiver closest to the Portage Diversion Dam suggest that the dam blocked the upstream movement of three tagged Bigmouth Buffalo in 2016 ($n = 2$) and 2017 ($n = 1$). The individuals remained below the diversion structure for up to four months in the summer, before returning into the Red River.

The Pine Falls Hydroelectric Station does not provide upstream passage. Tag detections revealed that Channel Catfish tagged in the Lower Winnipeg River moved upstream towards the Pine Falls Hydroelectric Station in each study year. In 2016, 2017, and 2018, 27, six, and one, respectively, of the 30 tagged Channel Catfish that were tagged in the Lower Winnipeg River moved up to the Pine Falls Hydroelectric Station where further upstream movement was impeded by the dam.

## 4. Discussion

The large-scale telemetry study allowed us to gain valuable insights into movement patterns and retention times of fish in the Lake Winnipeg basin and determine bottlenecks for habitat connectivity. Habitat connectivity describes how the environment allows or limits movement between different functional habitats such as feeding, spawning and rearing habitats [33]. Knowledge of species-specific functional connectivity for particular rivers is key given its importance for the persistence of populations. It provides useful perspectives on specific management strategies and is especially valuable in the context of fish passage and barrier removal projects because it can guide decision making to assign restoration priorities [34]. Functional habitat connectivity can be established through fish dispersal and migration patterns using telemetry [35]. By studying fish repeatedly over all seasons and a large geographical area, we observed large scale movement patterns for both, Bigmouth Buffalo and Channel Catfish, in the Lake Winnipeg basin. In particular, Bigmouth Buffalo demonstrated mean annual home ranges of 132.6 to 177.5 km. Our study confirmed the limited information available from other river systems on regular, large-scale movements in Bigmouth Buffalo [36]. In comparison to Bigmouth Buffalo, Channel Catfish had smaller home ranges with mean annual movements ranging from 32.7 to 60.0 km but still completed frequent movements over the geopolitical border. Our findings confirm observations by Siddons et al. [22] that Channel Catfish displayed frequent basin-wide, transboundary movements in the Red River, which is important information for fishery managers from different jurisdictions for the regulation and management of Channel Catfish fisheries.

The inter-specific differences in home range estimates may be influenced by the fact that the majority of Channel Catfish (107 out of 161) were tagged below the St. Andrews Lock and Dam, which may act as a partial barrier to movements [22]. Whereas only 20 of 80 tagged Bigmouth Buffalo were released below the St. Andrews Lock and Dam; thus, their potential for movement may have been less restricted than Channel Catfish. Additionally, Channel Catfish detections decreased each year, either because fish did not move, were not detected on a receiver, lost their tags, suffered natural mortality, were caught in commercial/recreational fisheries and removed from the study system or migrated out of the system. However, evidence in our dataset demonstrated that Bigmouth Buffalo displayed longer and more frequent movements through the available riverine habitats including tributaries (e.g., Seine, La Salle, and Assiniboine rivers) in comparison to Channel Catfish that were mainly observed in the main stems of the Red and Winnipeg rivers.

Understanding the spatial ecology of fishes is of crucial importance to fishery managers as it offers information on how fishes are distributed in both space and time [9]. For example, although Channel Catfish displayed transboundary movements in the Red River, different recreational fishery harvest regulations currently exist for the Manitoban portion of the of the Red River in comparison to the southern reach managed by Minnesota and North Dakota. Subsequently, our results underline the importance of maintaining habitat connectivity throughout the Red River basin and suggest considering a conjoint transboundary fisheries management plan. Similarly, the decline of Bigmouth Buffalo is attributed to the degradation and/or loss of spawning habitat related to water management practices, principally due to the regulation of water levels and channelization [37]. Furthermore, periods of drought, agricultural water demands, the introduction of Common Carp (*Cyprinus carpio*), and commercial fisheries may have also reduced the population size. In addition, the Portage Diversion Dam constructed in 1970, represents a barrier to upstream movement for fish in the Lower Assiniboine River, and coincides with the decline of Bigmouth Buffalo in the Upper Assiniboine and Qu'Appelle rivers that resulted in a commercial fishery closure for Bigmouth Buffalo in Qu'Appelle River in 1983.

In Canada, the Channel Catfish is the only known host species of the endangered Mapleleaf mussel, and the presence of the fish host is one of the key features determining if a given river system supports a healthy mussel population [38]. Among the threats for Mapleleaf populations are aquatic invasive species (e.g., Zebra Mussel (*Dreissena polymorpha*)), habitat loss and degradation, water quality, and siltation, which can negatively impact filter-feeding mussels. Mapleleaf populations can potentially be recovered by their host, the Channel Catfish, as one of the adaptive functions of mussel parasitism of migrant hosts is they can transport glochidia up and downstream. If passage of Channel Catfish is restricted by barriers, it likely poses a constraint on Mapleleaf populations. Being able to observe Channel Catfish movement over multiple years and large distances, the telemetry study allowed monitoring individual movement patterns of Channel Catfish. We observed Channel Catfish move large distances in the system but also pinpointed impediments in the free movement of Chanel Catfish in the Lake Winnipeg basin due to existing barriers that may inhibit the recolonization and recovery potential of Mapleleaf. Consequently, the telemetry study allowed us to gain valuable information for the Recovery Strategy of Mapleleaf and future risk management strategies [39].

The Portage Diversion Dam on the Assiniboine River and the Pine Falls Hydroelectric Station on the Winnipeg River are obvious barriers to upstream fish passage as no fishways are installed, while the St. Andrews Lock and Dam allows for some upstream fish passage through the locks, and a fishway and weir at Drayton is passable at higher water levels. However, the continuous-time Markov model (CTMM) highlighted a low transition probability at the St. Andrews Lock and Dam, suggesting even with the presence of the locks and the fishway, the structures only provide limited passage opportunities for upstream fish movement. It seems that the downstream movement of both, Bigmouth Buffalo and Channel Catfish, considerably supplement the Lower Red River population downstream of the St. Andrews Lock and Dam, given considerably fewer individuals are returning to the Upper Red River. Upstream movement over the St. Andrew's Lock and Dam occurred exclusively in the months of July and August. Our results highlight that the efficiencies to attract and/or pass fish in the St. Andrews Lock and Dam fishway may be limited. Consequently, a more detailed study at St. Andrew's Lock and Dam will be required to analyze attraction and passage efficiencies, as these are the key components to the success of fishways using adequate high resolution telemetry and time to event data analysis.

Due to the small number of fish tagged, the short observation period, and some unforeseen mortality, only a few fish were observed to navigate upstream passage over the rock ramp on the Riverside Dam site. Furthermore, due to the limitations of the receiver network extent, we were not able to accurately quantify downstream movement over the Riverside Dam. Subsequently, we could not fully assess the success and effectiveness of this river restoration project to provide fish passage by converting an existing low head dam to a rapids using a nature-like fish passage design at this time [24]. However, rock ramps are intended to provide a passable slope for fish by building up material on the existing riverbed directly downstream of the dam crest. The approach is particularly applicable for low head dams but has limitations for high head dams. More specifically, a rock arch rapids design was chosen at the Riverside Dam site [40,41]. The configuration has several advantages: It facilitates energy to be dissipated in the center of the rapids whereas the near bank velocities are reduced; boulders within the arch support each other adding stability; and it allows fish passage by providing low velocity eddies and passage is resilient to changing discharges. Further research should be conducted on the ramp to establish its efficiency [14].

The importance of the natural flow regime with its flow variability (i.e., timing, duration, frequency, and rate of change of flows) is well recognized as a driver of ecosystem processes [3,42]. Our telemetry study enabled us to reveal an interesting timing of fish movement in relation to the hydrograph. Movements of Bigmouth Buffalo and Channel Catfish in the Red River seem to be triggered by peak flows and movements were detected close to the peak or during the descending hydrograph limb. This information is useful for approaches such as by Yarnell et al. [43] that focus on retaining specific process-based components of the hydrograph also referred to as functional flows instead of trying to mimic the full natural flow regime. To optimize the functionality of flows, knowledge about which flows trigger fish movement and other life processes are key elements [44].

Anthropogenic instream barriers, such as weirs and dams serve human needs such as hydroelectric generation or flood control, but they may restrict fish movements. Consequently, when barriers are constructed there is concern in regards to changes in fish community assemblages and for potadromous species that are using diverse habitat types at different times of the year and life stages [45]. Truncated distributions, degraded fish assemblages, and changes in age class composition are frequently observed below dams and weirs in Midwestern and Prairie rivers [46–49]. Continued research will be required to study how the two study species are impacted by barriers and how the barriers impact their reproductive success, and what adjustments are needed to increase the fishway attraction and passage efficiencies.

**Author Contributions:** E.E., D.W., and M.P. conceived and designed the study and acquired project funding, C.C., D.W., C.K., D.L., and H.H. conducted the fieldwork, C.C. conducted the data analysis and D.W. contributed to data visualization, E.E. wrote the manuscript, C.C., D.W., C.K., D.L., H.H., and M.P. reviewed and commented on the manuscript.

**Funding:** Financial support for this project was provided by the Fisheries and Oceans Canada's (DFO) Partnership Fund, Species at Risk and Fisheries Protection programs, Manitoba Fish Futures, Inc., the Fish and Wildlife Enhancement Fund, the International Joint Commission, Manitoba Sustainable Development, and the University of Nebraska.

**Acknowledgments:** We wish to thank everybody on the field crews who participated in the field work, in particular Geoff Klein, Jamison Wendel, and Todd Caspers. We would like to thank two anonymous reviewers for the constructive reviews of earlier versions of the manuscript.

**Conflicts of Interest:** The authors declare no conflict of interest. The funders had no role in the design of the study, the data collection and analysis or the interpretation of the data.

## Appendix A

**Table A1.** Number of tagged fish per species and tagging site that were detected in each year of the study.

| Species | Section ID | Section of Tagging Site | Year | n |
|---|---|---|---|---|
| Bigmouth Buffalo | 1 | Fargo, ND to Riverside Dam | 2017 | 12 |
| | | | 2018 | 12 |
| | 3 | Drayton Dam to Border | 2017 | 8 |
| | | | 2018 | 7 |
| | 5 | Morris, MB to Winnipeg, MB | 2016 | 20 |
| | | | 2017 | 20 |
| | | | 2018 | 17 |
| | 7 | City of Winnipeg, MB | 2016 | 20 |
| | | | 2017 | 18 |
| | | | 2018 | 18 |
| | 9 | Downstream of Lockport, MB | 2017 | 19 |
| | | | 2018 | 19 |
| Channel Catfish | 1 | Fargo, ND to Riverside Dam | 2017 | 9 |
| | | | 2018 | 1 |
| | 2 | Riverside Dam to Drayton Dam | 2017 | 12 |
| | | | 2018 | 3 |
| | 3 | Drayton Dam to Border | 2017 | 10 |
| | | | 2018 | 3 |
| | 7 | City of Winnipeg, MB | 2016 | 22 |
| | | | 2017 | 8 |
| | | | 2018 | 2 |
| | 9 | Downstream of Lockport, MB | 2016 | 64 |
| | | | 2017 | 31 |
| | | | 2018 | 14 |
| | 12 | Winnipeg River | 2016 | 30 |
| | | | 2017 | 26 |
| | | | 2018 | 18 |

## References

1. Fuller, M.R.; Doyle, M.W.; Strayer, D.L. Causes and consequences of habitat fragmentation in river networks. *Ann. N. Y. Acad. Sci.* **2015**, *1355*, 31–51. [CrossRef]
2. Nilsson, C.; Reidy, C.A.; Dynesius, M.; Revenga, C. Fragmentation and flow regulation of the world's large river systems. *Science* **2005**, *308*, 405–408. [CrossRef]
3. Poff, N.L.; Allan, J.D.; Bain, M.B.; Karr, J.R. The natural flow regime. *Bioscience* **1997**, *47*, 769–784. [CrossRef]
4. Liermann, C.R.; Nilsson, C.; Robertson, J.; Ng, R.Y. Implications of dam obstruction for global freshwater fish diversity. *Bioscience* **2012**, *62*, 539–548. [CrossRef]
5. Pracheil, B.M.; McIntyre, P.B.; Lyons, J.D. Enhancing conservation of large-river biodiversity by accounting for tributaries. *Front. Ecol. Environ.* **2013**, *11*, 124–128. [CrossRef]

6.  Cairns, D.; Castonguay, M.; Dumont, P.; Caron, F.; Verreault, G.; Mailhot, Y.; de Lafontaine, Y.; Casselman, J. Why has the American Eel, Anguilla anguilla declined dramatically in the St. Lawrence River but not the Gulf? ICES CM 2006/J:33. 2006. Available online: https://www.researchgate.net/publication/230710494_Why_has_the_American_Eel_Anguilla_rostrata_declined_dramatically_in_the_St_Lawrence_River_but_not_the_Gulf. (accessed on 28 May 2019).

7.  Hilborn, R. Ocean and dam influences on salmon survival. *Proc. Natl. Acad. Sci. USA* **2013**, *110*, 6618–6619. [CrossRef] [PubMed]

8.  Cooke, S.J. Biotelemetry and biologging in endangered species research and animal conservation: Relevance to regional, national, and IUCN Red List threat assessments. *Endanger. Species Res.* **2008**, *4*, 165–185. [CrossRef]

9.  Lucas, M.C.; Baras, E. Methods for studying the spatial behaviour of freshwater fishes in the natural environment. *Fish Fish.* **2000**, *1*, 238–316. [CrossRef]

10. Kraus, R.T.; Holbrook, C.M.; Vandergoot, C.S.; Stewart, T.R.; Faust, M.D.; Watkinson, D.A.; Charles, C.; Pegg, M.A.; Enders, E.C.; Krueger, C.C. Evaluation of acoustic telemetry grids for determining aquatic animal movement and survival. *Meth. Ecol. Evol.* **2018**, *9*, 1489–1502. [CrossRef]

11. Fausch, K.D.; Torgersen, C.E.; Baxter, C.V.; Li, H.W. Landscapes to riverscapes: Bridging the gap between research and conservation of stream fishes. *Bioscience* **2002**, *52*, 483–498. [CrossRef]

12. Roni, P.; Beechie, T.; Schmutz, S.; Muhar, S. Prioritization of watersheds and restoration orojects. In *Stream and Watershed Restoration: A Guide to Restoring Riverine Processes and Habitats*; Roni, P., Beechie, T., Eds.; John Wiley & Sons, Ltd.: Hoboken, NJ, USA, 2012; pp. 189–214. [CrossRef]

13. Catalano, M.J.; Bozek, M.A.; Pellett, T.D. Effects of dam removal on fish assemblage structure and spatial distributions in the Baraboo River, Wisconsin. *N. Am. J. Fish. Manag.* **2007**, *27*, 519–530. [CrossRef]

14. Bunt, C.M.; Castro-Santos, T.; Haro, A. Performance of fish passage structures at upstream barriers to migration. *River Res. Appl.* **2012**, *28*, 457–478. [CrossRef]

15. Bunt, C.M.; Castro-Santos, T.; Haro, A. Reinforcement and validation of the analyses and conclusions related to fishway evaluation data from Bunt et al.: 'Performance of Fish Passage Structures at Upstream Barriers to Migration'. *River Res. Appl.* **2016**, *32*, 2125–2137. [CrossRef]

16. Walters, D.M.; Zuellig, R.E.; Crockett, H.J.; Bruce, J.F.; Lukacs, P.M.; Fitzpatrick, R.M. Barriers impede upstream spawning migration of Flathead Chub. *Trans. Am. Fish. Soc.* **2014**, *143*, 17–25. [CrossRef]

17. Williams, J.G.; Gessel, M.H. A history of research to develop guidance systems to divert juvenile salmonids, *Oncorhynchus* spp., from turbines at federal hydroelectric dams on the mainstem Columbia and Snake rivers, U.S.A. *Mar. Fish. Rev.* **2018**, *80*, mfr8023. [CrossRef]

18. Fuller, I.C.; Death, R.G. The science of connected ecosystems: What is the role of catchment-scale connectivity for healthy river ecology? *Land Degrad. Dev.* **2018**, *29*, 1413–1426. [CrossRef]

19. Cooke, S.J.; Paukert, C.; Hogan, Z. Endangered river fish: Factors hindering conservation and restoration. *Endanger. Species Res.* **2012**, *17*, 179–191. [CrossRef]

20. Arthington, A.H.; Dulvy, N.K.; Gladstone, W.; Winfield, I.J. Fish conservation in freshwater and marine realms: Status, threats and management. *Aquat. Conserv. Mar. Freshw. Ecosyst.* **2016**, *26*, 838–857. [CrossRef]

21. Stewart, K.W.; Watkinson, D.A. *Freshwater Fishes of Manitoba*; University of Manitoba Press: Winnipeg, MB, USA, 2004; 278p.

22. Siddons, S.F.; Pegg, M.A.; Klein, G.M. Borders and barriers: Challenges of fisheries management and conservation in open systems. *Riv. Res. Appl.* **2017**, *33*, 578–585. [CrossRef]

23. Manitoba Sustainable Development. Annual Report 2016–2017. 2017; 165p. Available online: https://www.gov.mb.ca/sd/annual-reports/con_reports/sd_annual_report_2016_17.pdf (accessed on 28 May 2019).

24. Aadland, L.P. *Reconnecting Rivers: Natural Channel Design in Dam Removals and Fish Passage*; Minnesota Department of Natural Resources: St Paul, MN, USA, 2010; 196p.

25. Vandergoot, C.S.; Murchie, K.J.; Cooke, S.J.; Dettmers, J.M.; Bergstedt, R.A.; Fielder, D.G. Evaluation of two forms of electroanesthesia and carbon dioxide for short-term anesthesia in Walleye. *N. Am. J. Fish. Manag.* **2011**, *31*, 914–922. [CrossRef]

26. Marty, G.D.; Summerfelt, R.C. Pathways and mechanisms for expulsion of surgically implanted dummy transmitters from Channel Catfish. *Trans. Am. Fish. Soc.* **1986**, *115*, 577–589. [CrossRef]

27. Siegwarth, G.L.; Pitlo, P.M. A modified procedure for surgically implanting radio transmitters in Channel Catfish. *Am. Fish. Soc. Symp.* **1999**, *24*, 287–292.

28.  R Core Development Team. *R: A Language and Environment for Statistical Computing*; R Foundation for Statistical Computing: Vienna, Austria, 2018; Available online: https://www.R-project.org/ (accessed on 28 May 2019).

29.  Wickham, H.; François, R.; Henry, L.; Müller, K. Dplyr: A Grammar of Data Manipulation. R package version 0.7.6. 2018. Available online: https://CRAN.R-project.org/package=dplyr (accessed on 28 May 2019).

30.  Tyers, M. Riverdist: River Network Distance Computation and Applications. R package Version 0.15.0. 2017. Available online: https://CRAN.R-project.org/package=riverdist (accessed on 28 May 2019).

31.  Schnute, J.T.; Boers, N.; Haigh, R. PBSmapping: Mapping Fisheries Data and Spatial Analysis Tools. R package version 2.70.5. 2018. Available online: https://CRAN.R-project.org/package=PBSmapping (accessed on 28 May 2019).

32.  Jackson, C.H. Multi-state models for panel data: The msm package for R. J. Statist. *Software* **2011**, *38*, 1–29. Available online: http://www.jstatsoft.org/v38/i08/ (accessed on 28 May 2019).

33.  Taylor, P.D.; Fahrig, L.; Henein, K.; Merriam, G. Connectivity is a vital element of landscape structure. *Oikos* **1993**, *68*, 571–573. [CrossRef]

34.  Branco, P.; Segurado, P.; Santos, J.M.; Ferreira, M.T. Prioritizing barrier removal to improve functional connectivity of rivers. *J. Appl. Ecol.* **2014**, *5*, 1197–1206. [CrossRef]

35.  Kanno, Y.; Letcher, B.H.; Coombs, J.A.; Nislow, K.H.; Whiteley, A.R. Linking movement and reproductive history of brook trout to assess habitat connectivity in a heterogeneous stream network. *Freshw. Biol.* **2014**, *59*, 142–154. [CrossRef]

36.  Moen, T.E. *Population Trends, Growth, and Movement of Bigmouth Buffalo, Ictiobus Cyprinellus, in Lake Oahe, 1963–70*; Technical Paper 78; U.S. Fish and Wildlife Service: Washington, DC, USA, 1974; 20p. Available online: https://pubs.er.usgs.gov/publication/tp78 (accessed on 28 May 2019).

37.  COSEWIC. *Assessment and update status report on the Bigmouth Buffalo Ictiobus cyprinellus Lakes—Great Lakes—Upper St. Lawrence populations Saskatchewan—Nelson River populations—in Canada*; Committee on the Status of Endangered Wildlife in Canada: Ottawa, ON, Canada, 2009; Volume vii, 40p, Available online: https://www.registrelep-sararegistry.gc.ca/virtual_sara/files/cosewic/sr_bigmouth_buffalo_0809_e.pdf (accessed on 28 May 2019).

38.  COSEWIC. *COSEWIC assessment and status report on the Mapleleaf Quadrula quadrula, Great Lakes—Upper St. Lawrence population and Saskatchewan—Nelson Rivers population, in Canada*; Committee on the Status of Endangered Wildlife in Canada: Ottawa, ON, Canada, 2016; Volume xi. Available online: http://www.registrelep-sararegistry.gc.ca/virtual_sara/files/cosewic/sr_Mapleleaf_2016_e.pdf (accessed on 28 May 2019).

39.  Fisheries and Oceans Canada. *Recovery strategy and action plan for the Mapleleaf (Quadrula quadrula) in Canada (Great Lakes-Upper St. Lawrence population)*; Species at Risk Act Recovery Strategy Series; Fisheries and Oceans Canada: Ottawa, ON, Canada, 2018; Volume vi, 59p. Available online: http://www.registrelep-sararegistry.gc.ca/virtual_sara/files/plans/RsAp-Mapleleaf-v00-2018July-Eng.pdf (accessed on 28 May 2019).

40.  Newbury, R. Rivers and the art of stream restoration. In *Natural and Anthropogenic Influences in Fluvial Geomorphology*; Costa, J.E., Miller, A.J., Potter, K.W., Wilcock, P.R., Eds.; Monograph Series; American Geophysical Union: Washington, DC, USA, 1995; Volume 89, pp. 137–149. [CrossRef]

41.  Rosgen, D.L. *Applied River Morphology*; Wildland Hydrology: Pagosa Springs, CO, USA, 1996; 390p.

42.  Naiman, R.J.; Latterell, J.J.; Pettit, N.E.; Olden, J.D. Flow variability and the biophysical vitality of river systems. *Compt. Rendus Geosci.* **2008**, *340*, 629–643. [CrossRef]

43.  Yarnell, S.M.; Petts, G.E.; Schmidt, J.C.; Whipple, A.A.; Beller, E.E.; Dahm, C.N.; Goodwin, P.; Viers, J.H. Functional flows in modified riverscapes: Hydrographs, habitats and opportunities. *BioScience* **2015**, *65*, 963–972. [CrossRef]

44.  Kiernan, J.D.; Moyle, P.B.; Crain, P.K. Restoring native fish assemblages to a regulated California stream using the natural flow regime concept. *Ecol. Appl.* **2012**, *22*, 1472–1482. [CrossRef] [PubMed]

45.  Landsman, S.J.; Nguyen, V.M.; Gutowsky, L.F.G.; Gobin, J.; Cook, K.V.; Binder, T.R.; Lower, N.; McLaughlin, R.L.; Cooke, S.J. Fish movement and migration studies in the Laurentian Great Lakes: Research trends and knowledge gaps. *J. Great Lakes Res.* **2011**, *37*, 365–379. [CrossRef]

46.  Cooke, S.J.; Bunt, C.M.; Hamilton, S.J.; Jennings, C.A.; Pearson, M.P.; Cooperman, M.S.; Markle, D.F. Threats, conservation strategies, and prognosis for suckers (Catastomidae) in North America: Insights from regional case studies of a diverse family of non-game fishes. *Biol. Conserv.* **2005**, *121*, 317–331. [CrossRef]

47. Wang, L.; Infante, D.; Lyons, J.; Stewart, J.; Cooper, A. Effects of dams in river networks on fish assemblages in non-impoundment sections of rivers in Michigan and Wisconsin, USA. *Riv. Res. Appl.* **2011**, *27*, 473–487. [CrossRef]

48. Pierce, C.L.; Ahrens, N.L.; Loan Wilsey, A.K.; Simmons, G.A.; Gelwicks, G.T. Fish assemblage relationships with physical characteristics and presence of dams in three eastern Iowa rivers. *River Res. Appl.* **2014**, *30*, 427–441. [CrossRef]

49. Enders, E.C.; Watkinson, D.A.; Ghamry, H.; Mills, K.H.; Franzin, W.G. Fish age and size distributions and species composition in a large, hydropeaking prairie river. *Riv. Res. Appl.* **2017**, *33*, 1246–1256. [CrossRef]

*Article*

# Hydropower Development and Fishways: A Need for Connectivity in Rivers of the Upper Paraná Basin

Sergio Makrakis [1,2,*], Ana P. S. Bertão [1,2], Jhony F. M. Silva [1,2], Maristela C. Makrakis [1,2], Fco. Javier Sanz-Ronda [3] and Leandro F. Celestino [1,2,*]

[1]  Grupo de Pesquisa em Tecnologia em Ecohidráulica e Conservação de Recursos Pesqueiros e Hídricos—GETECH, Universidade Estadual do Oeste do Paraná/ Rua da Faculdade, 645, Jd. Santa Maria, Toledo, PR 85903-000, Brazil

[2]  Programa de Pós-graduação em Recursos Pesqueiros e Engenharia de Pesca/ Rua da Faculdade, 645 Jd. Santa Maria, Toledo, PR 85903-000, Brazil

[3]  GEA-ecohidraulica.org, Universidad de Valladolid, E.T.S.II.AA. de Palencia, Avda. Madrid, 44, 34004 Palencia, Spain

*  Correspondence: sergio.makrakis@unioeste.br (S.M.); le_celestino@hotmail.com (L.F.C.)

Received: 31 May 2019; Accepted: 3 July 2019; Published: 9 July 2019

**Abstract:** South American rivers have become intensely affected by the construction of hydroelectric dams that block the river's connectivity for migratory fish species. In order to mitigate the problems caused by dams and to reestablish connections between habitats, fishways are implemented. Fishways are structures that aid fish in overcoming obstacles and help preserve migratory, reproductive, and feeding routes. This study performed an inventory of all hydropower plants—present and future—in the Upper Paraná River, with the objective of identifying fishways unknown to scientific literature, as well as the task of mapping them. By doing so, the current situation of structural connectivity via fishways in the Upper Paraná River Basin was described. Overall, 389 dams along 209 rivers were identified; of these, only 9% (35 dams) have fishways. In addition, an alarming explosion of future medium-sized hydropower plants was observed, with an expectation of an almost 500% increase in relation to those existing. This data reveals a trend of reduction of free-flowing river stretches, which are crucial habitats for Neotropical potamodromous species, and point to a deficiency in the structural connectivity of existing hydropower dams. Furthermore, if the implementations of these expected constructions are associated with limited connectivity as a result of the absence of fishways, the management of fisheries and their resources in the Upper Paraná River may become unsustainable.

**Keywords:** potamodromous fish; fish ladder; fisheries management; free-flowing rivers; Neotropical

---

## 1. Introduction

The great rivers of South America have been highly fragmented by hydroelectric dams in order to meet an ever-growing energy demand [1]. The construction of such dams is one of many anthropic actions which result in the greatest impact on hydrographic basins [2]. Among such impacts is the regulation of the river flow, which alters nutrient dynamics [3], fragments the ecosystem, changes the morphology of water bodies [4], and promotes the longitudinal imbalance of the rivers [5], which greatly decreases biodiversity and leads to the loss of biological and genetic resources [2,6,7]. Furthermore, this interferes in the migration of diadromous and potamodromous fish by blocking the connectivity between feeding, reproduction, and development habitats [8].

Potamodromous fish, which migrate solely in inland waters, are severely impacted by the fragmentation of rivers as a result of dams [9]. The Upper Paraná River Basin is home to approximately 310 fish species [10], of which 15–20 are long-distance migratory species [11–13], which are popularly known as 'piracema' fish [14], and can usually migrate hundreds of kilometers to reproduce [13,15].

Fish generally migrate longitudinally in an upstream spawning movement into the main river channel, followed by a downstream dispersion of eggs and larvae [11]. The intact longitudinal pathways are critical for survival, as well as the lateral migration into tributaries that are often very important for reproduction and can serve as nursery areas for larvae and young fishes [13,16,17]. Most of these species are very important for commercial and recreational fishing and represent a source of income and nourishment for local communities [18]. Several different management strategies, such as stocking programs and construction of fish passes [19], have been attempted in order to improve the sustainability of artisanal fishery and the social conditions of families who are dependent on this resource.

Fishways are constructed in order to reestablish connectivity between upstream and downstream areas [20–22]. By its basic definition, fishways are structures for conducting water by dissipating energy through baffles in order to provide a safe passage for migratory fish, without causing stress, delays, or injuries, and at a low expense of energy [21,23–25]. Fishways can be divided into (i) technical fishways, like fish ladders; (ii) nature-like structures, like bypass channels and rock ramps; and (iii) special-purpose structures, like elevators, fish locks, and eel passes [21,26,27]. Nature-like channels, aside from providing connectivity for fish species, also bear the highest resemblance to the natural habitat, due to the presence of invertebrates and macrophytes [24,28–31]. In contrast, fish ladders can be installed in higher dams and possess a shorter length as a result of the dissipation of water energy through baffles, consequently reducing the cost of installation [21,24,32]. Fish ladders have several designs which vary depending on the different types of baffles (also called weirs). As a result, they are known by different names, such as pool and weir, weir and orifice, vertical slot and Denil, and other mixed systems which present more than one design [21,23,26].

Although several types of fish ladders have been implemented over the last three centuries [33], fishway science is still in its infancy, especially in South America [34]. The first fish passage constructed in Brazil was a fish ladder in the Atibaia River in 1911, at the Salto Grande Hydroelectric Dam in the São Paulo State [35]. This was followed by federal law n° 794 of 1938, enforcing the use of fishways as follows: "Dams in rivers, streams, and creeks must have, as a mandatory complement, constructions that allow the conservation of fluvial fauna by either facilitating the passage of fish, or by installing fish farming". However, this federal law was subsequently revoked by law n° 221 of 1967 as a result of its non-specificity and its enforcement of fishway construction even when not necessary. In the years that followed, some Brazilian states began to require constructions of fishways in waters within their boundaries, provided that the environmental agencies issue favorable technical advice. This was the case for both the states of São Paulo, with law n° 9.798 of 1997, and Minas Gerais, with law n° 12.488 of 1997.

In view of the mandatory nature of fishway implementation, combined with an absence of a specific technological understanding of tropical species, existing fishways in South America have followed a standard of size, hydraulic characteristics, and declivity based on fishways designed for salmonids and other potamodromous species of temperate climates [35,36]. The morphology, physiology, behavior, swimming capacity, and life history of Neotropical species, however, differ from those of temperate regions [14,37–40], for which these fish ladders were originally designed. Thus, the non-specificity of fishway projects undertaken in South America have furthered the installation of low-efficiency systems, which have prompted an overall distrust of their functionality [41–43]. To make matters worse, few hydroelectric enterprises perform adequate evaluations of attraction and efficiency, as well as the necessary monitoring [44]. Attraction is a metric that evaluates the amount of fish from a population that are able to find the fishway entrance, while efficiency is another metric that evaluates how many fish have passed through the fishway, regarding the amount which entered into the fishway—both metrics are generally expressed in percentages [45,46]. In addition, only a small percentage of the resulting fishway evaluation and monitoring reports are divulged to the scientific community [35]. This situation exacerbates the uncertainty regarding the fishways' performance, and in turn becomes an argument for the closure of already constructed fishways [47,48], as well as not implementing them at all in future projects. Nevertheless, ensuring structural connectivity, that is, the presence of physical structures that allow the movement of organisms [49], such as fish ladders, may be the first step in guaranteeing sustainability in fishery resources of regulated rivers [2].

Considering that the Upper Paraná River Basin is the most regulated in South America [1,11], along with the expectation of future hydropower plants, very little is actually known about the quantity, types, and dimensions of existing fishways. This lack of information is further inhibited by the scattering of existing data in different journals and environmental agencies—both state and federal—as well as the difficult access to projects and installations of hydropower plants and the spatial difficulty of visiting all of them in such a vast region. Therefore, the need for a public inventory is crucial for understanding the current and future situation of the Upper Paraná River Basin and to be able to guide the environmental management of the river and local fishery resources. This study quantified the hydropower dams—planned, under construction and in operation—in the Upper Paraná River Basin, identified and characterized the existing fishways, as well as summarized biologic and ecological aspects of the migratory species in the Upper Paraná River Basin.

## 2. Materials and Methods

The Paraná River basin has an extension of $2.8 \times 10^6$ km$^2$ and covers roughly 80% of the Paraná-Paraguay River basin [11]. Also known as La Plata River basin, it includes areas of Brazil, Argentina, Bolivia, and Paraguay (Figure 1). The Paraná River is formed by the confluence of the Paranaíba and Grande rivers and possesses important tributaries such as Tietê, Sucuriú, Verde, Pardo, Aguapeí, Ivinhema, Amambaí, Iguatemi, Paranapanema, Ivaí, Piquiri, and Iguaçu rivers. This fluvial system is distributed through areas of Brazil, Paraguay, and Argentina. The main trajectory of the Paraná River is roughly 4600 km in length, and the annual sediment load is estimated to be over $1.5 \times 10^8$ tons [50]. The mainstem of the Upper Paraná River extends 750 km upstream of the Itaipu Dam (Brazil-Paraguay border) and northeasterly up to the confluence of the Grande and Paranaíba rivers [22], comprising Brazilian and Paraguayan territory. However, due to the lack of information regarding hydroelectric power plants and their respective fishways in Paraguay, this study limits itself to the dams located in the Upper Paraná River Basin within Brazilian territory, including the Iguaçu River and its tributaries. The Upper Paraná River in Brazil partly drains the states of Goiás, Minas Gerais, Mato Grosso do Sul, São Paulo, Paraná, and Santa Catarina.

### 2.1. Data Collection

#### 2.1.1. Spatial Distribution and Classification of Hydroelectric Dams

The coordinates of hydroelectric dams in operation, planned and under construction, were obtained through the Georeferenced Information System of the Electric Sector (SIGEL—https://sigel. aneel.gov.br/portal/home/) of the Brazilian National Hydroelectric Agency (ANEEL), accessed in May of 2018. The data obtained was processed in the Shapefile format by the Geographical Information System (SIG) QGIS 2.18 [51], with the World Geographical System (WGS 84).

The hydropower dams were classified into three groups, according to the guidelines set by the ANEEL, based on the installed power (Megawatt—MW) and the area of the reservoir: (i) small-sized (*Central Geradora Hidrelétrica*—CGH), installed power rated at ≤3 MW, divided into microgenerators, with installed power rated at >0.075 and ≤3 MW [52]; (ii) medium-sized (*Pequena Central Hidrelétrica* —PCH)—installed power rated at >3 MW and ≤30 MW, including a reservoir with an area up to 13 km$^2$ [53]; and (iii) large (*Usina Hidrelétrica de Energia*—UHE)—installed power rated at >30 MW [54].

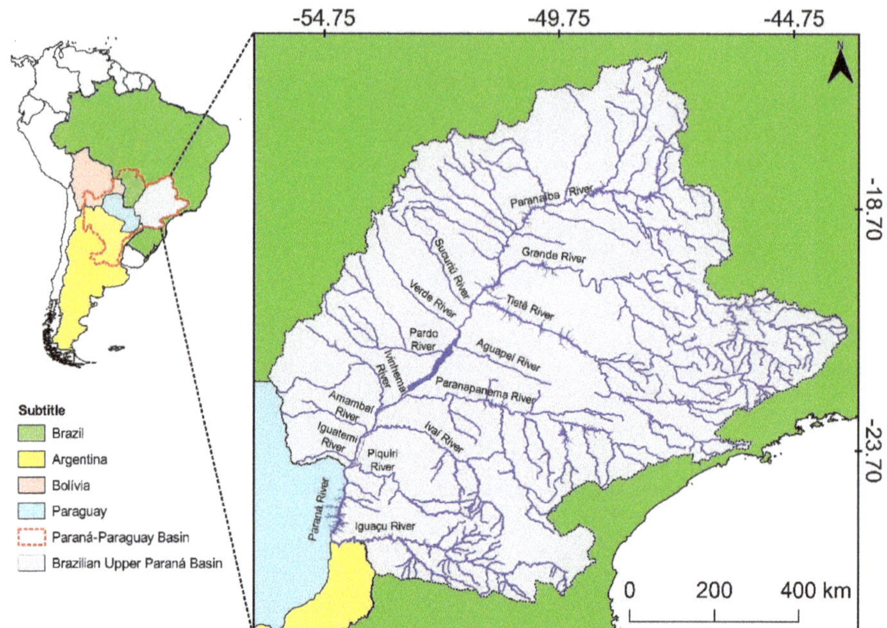


**Figure 1.** The Upper Paraná River Basin in Brazilian territory and its main tributaries.

### 2.1.2. Fishway Survey

The geographical coordinates of the hydropower plants in operation obtained through the ANEEL website were inserted into the Google EarthTM software and were utilized to visually inspect each hydroelectric enterprise to verify the existence of fishways [1]. The information regarding the types of fishways, design, construction date, and materials, were obtained through broad review of available literature, including scientific articles, books, and reports. The origin of the information regarding the fishways were classified into scientific sources (books and articles), reports (from environmental agencies, hydroelectric plants, universities, as well as doctorate thesis), and visual inspection (fishways which did not have any previous mention in the literature and were identified for the first time through Google EarthTM). With a basis in the available data, the fishways (type and design) were categorized according to Clay (1995), Larinier (2001), and Silva et al. (2018) [21,26,27].

### 2.1.3. River Connectivity

A risk map of river connectivity of the Upper Paraná River Basin was constructed. For this, the river stretches were quantified and classified in three categories: (i) blocked—river stretches fragmented by dams without fishways; (ii) connected—river stretches dammed but connected by fishways at dams; (iii) free-flowing river stretches—river stretches with natural flow.

It was considered hydropower plants in operation, however, only in this analysis the Baixo Iguaçu Dam (large-sized), the last one of a cascade of dams along the Iguaçu River, was included, which started the operation in March 2019.

### 2.1.4. Bioecological Aspects and Conservation Status of Migratory Species

A review of the literature was conducted to summarize some biological and ecological aspects of migratory species occurring in the Upper Paraná River Basin: status of conservation [55–57]; maximum standard length [12,58,59], feeding [11,60,61], spawning season [12,58,61–63], habitat [12,61], and migratory movements [13].

## 2.2. Data Analysis

The proportions of the types of hydropower plants in operation (CGH, PCH and UHE), as well as the proportions of fishway types were tested by the chi-square test for equal proportion, while the proportions of fishways by type of hydropower plant were tested by the likelihood test and Fisher's exact test. All the analyses were performed by the PROC FREQ procedure of the SAS University Edition 9.4 (Cary, NC, USA), with a significance level of 5%.

## 3. Results

### 3.1. Hydropower Development

Overall, ANEEL's database provided information regarding 974 hydropower plants, of which 59.2% (n = 577) are planned for construction, 0.8% (n = 8) are under construction, and 39.9% (389 plants) are in operation (Table 1).

The proportion of dams in operation was different between the types of hydroelectric power plants (CGH, PCH, and UHE) (test for equal proportions: $x^2$ = 59.850, df = 2, $p < 0.001$). The proportion of CGH in operation was 51.7% (n = 201) and was greater to the proportion of PCH and UHE, which represented 26.2% (n = 102) and 22.1% (n = 86), respectively (Table 1). The proportions of PCH and UHE did not differ significantly (test for equal proportions: $x^2$ = 1.361, df = 1, $p = 0.243$). The spatial distribution of CGH, PCH and UHE in operation are illustrated in Figure 2.

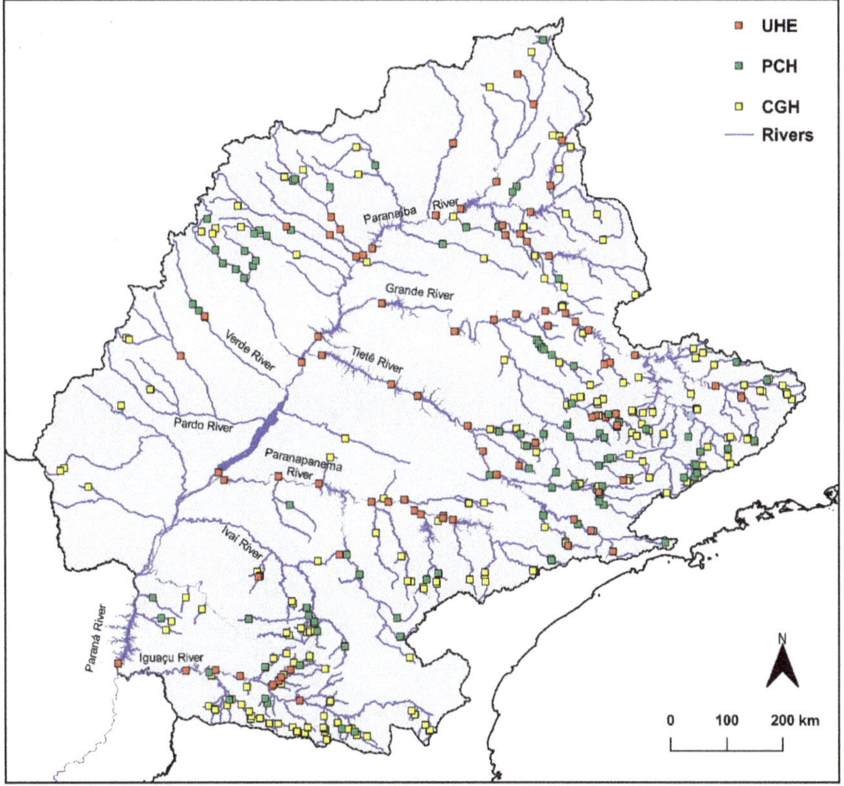


**Figure 2.** Distribution map of hydropower dams in the main rivers from Upper Paraná River Basin, by type of enterprise: UHE = large, PCH = medium, and CGH = small-sized hydropower plants.

**Table 1.** Number of hydropower plants planned, under construction, and in operation according to ANEEL.

| Hydropower Plant Classification | Planned | Under Construction | In Operation | Total |
| --- | --- | --- | --- | --- |
| Large-sized: UHE | 61 | 2 | 86 | **149** |
| Medium-sized: PCH | 516 | 6 | 102 | **623** |
| Small-sized: CGH | - | - | 201 | **201** |
| **Total** | **577** | **8** | **389** | **974** |

In relation to future hydropower plants, the proportion of planned PCHs was of 89.4% (n = 516) and was greater to the proportion of planned UHEs, which represented only 10.6% (n = 61). Therefore, the number of future PCHs (n = 516) represented a five-fold (505%) increase in relation to PCHs in operation, whereas the number of future UHEs represented an increase of 71% (Table 1).

### 3.2. Fishway by Hydropower

A total number of 37 fishways were identified and distributed between 35 dams. The Engenheiro Sérgio Motta Hydropower Plant (UHE) has two fishways—a fish ladder and an elevator, and the São Joaquim (PCH) has two fish ladders (Table 2). The percentage of fishways differed between the CGH, PCH, and UHE (likelihood ratio: $x^2 = 32.074$, df = 2, $p < 0.001$). The proportion of CGHs with fishways (1.5%; n = 3) was significantly lower in relation to PCHs (Fisher's exact test: $p < 0.001$), as well as UHEs (Fisher's exact test: $p < 0.001$). Nevertheless, the proportion of hydropower plants with fishways did not differ significantly between PCHs (20.6%, n = 21) and UHEs (12.8%; n = 11) (Fisher's exact test: $p = 0.176$).

### 3.3. Fishway Information Profile

There was no significant difference in proportion of the type of information regarding fishways, that is, fishways without information (fishway found by visual inspection), with scientific information, and with reports (test for equal proportions: $x^2 = 1.523$, df = 2 $p = 0.469$). Overall, of the 37 fishways encountered, 35.1% (n = 13) did not possess any type of available information and were only identified for the first time through satellite imaging (Figure 3A; Table 2). Alternatively, 40.5% of the fishways (n = 15) displayed scientific information (articles or books) in regard to evaluation of attraction, efficiency, or continuous monitoring. Finally, 24.3% of the fishways (n = 9) had information available through reports, which can best be described as grey literature (Figure 3A). Of the fish passages with available scientific information, the fish ladder at Engenheiro Sérgio Motta hydropower dam (known as Porto Primavera, São Paulo State) stood out as the most studied, having been the subject of seven published scientific articles.

### 3.4. Fishway Types

The proportion of fishway types varied in relation to fish ladders, elevators, and lateral system. Fish ladders were predominant, representing 91.9% (n = 34) of all constructed fishways, while elevators corresponded to 5.4% (n = 2), and lateral system only accounted for 2.7% (n = 1) (Figure 3B).

Of the 34 fish ladders, no information about design was found for 44.1% (n = 15), whereas for 55.9% (n = 19), some information was available. Of these 55.9%, with respect to design, 26.5% (n = 9) were pool and weir, 23.5% (n = 8) were weir and orifice, 2.9% (n = 1) were vertical slot, and 2.9% (n = 1) were of a mixed design (Figure 3C). Photos of the different types and designs of these fishways can be seen in Figure A1.

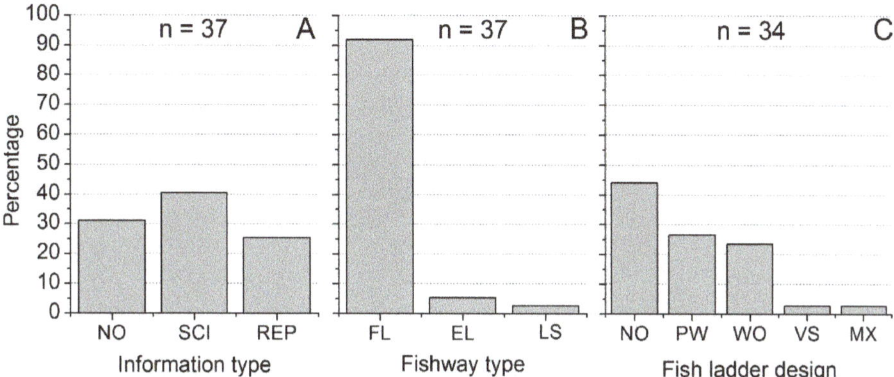

**Figure 3.** Origin of the information regarding encountered fishways: NO = no previous information (first records of the transposition system), SCI = fishway with available scientific information (articles or books), and REP = grey literature information (reports, theses, webpages) (**A**). Percentage of the types of fishways found in the Upper Paraná River basin: FL = fish ladder, EL = elevator, and LS = lateral system (**B**). Designs of encountered fish ladders: NO = no information regarding baffle types, PW = pool and weir, WO = weir and orifice, VS = vertical slot, and MX = mixed designs (more than one design) (**C**).

*3.5. History of Fishways in the Upper Paraná River Basin*

The fishways in the Upper Paraná River Basin that possess information regarding their construction date (n = 19) were built between 1911 and 2015 (a 104-year interval), resulting in an age mean of 47.1 years. In general, 41.1% of these (n = 8) were constructed within the last 18 years (Figure 4A), preceded by 21% (n = 4) from 1940 to 1959, 15.8% (n = 3) from 1900 to 1919, 10.5% (n = 2) from 1980 to 1999, 5.3% (n = 1) from 1920 to 1939, and 5.3% (n = 1) from 1960 to 1979. In regard to fish ladder designs, pool and weir was the most widely utilized until the 1950s. From the 1950s onward, what are thought to be more efficient fish ladder designs gained popularity, predominantly weir and orifice designs (Figure 4B). In addition, the unique fish ladder with vertical slot design was built in the 1980–99 period. Alternatively, mixed designs such as the lateral system, Canal da Piracema, as well as elevators, were constructed exclusively from 2000–2018 (Figure 4B).

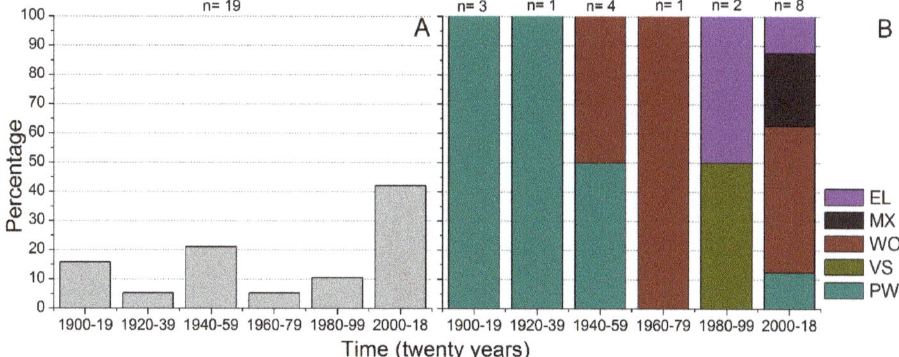

**Figure 4.** Percentage of fishways constructed per 20-year period in the Upper Paraná River Basin (**A**). Fishway type and fish ladder design constructed per 20-year period (**B**). PW = pool and weir, WO = weir and orifice, VS = vertical slot, EL = elevator and MX = mixed system (includes fishways with more than one design).

**Table 2.** List of hydropower plants with fishways and their characteristics. FL = fish ladder, EL = elevator, LS = lateral system, NO = no information available, - = not applicable, * = disabled fishway, ● = fishway with efficiency information. Dam ID represents the spatial positioning of the dam in Figure 5.

| N | Dam ID | Dam Name | Coordinates X, Y | River | Fishway Type | Year | Fishway Design | Baffle Type/Material | References |
|---|---|---|---|---|---|---|---|---|---|
| | | | | Large-Sized—UHE | | | | | |
| 1 | 05 | Igarapava | −19.989255° −47.755567° | Grande | FL● | 1999 | Vertical Slot | -/Concrete | [64,65] |
| 2 | 24 | Canoas I | −22.939658° −50.517983° | Paranapanema | FL*● | 2000 | Weir and Orifice | Bottom orifice and notch/Concrete | [48,62,66,67] |
| 3 | 25 | Canoas II | −22.939505° −50.251990° | Paranapanema | FL*● | 2000 | Weir and Orifice | Bottom orifice and notch/Concrete | [48,62,66,67] |
| 4 | 26 | Ourinhos | −23.070170° −49.838412° | Paranapanema | FL ● | 2005 | Pool and Weir | -/Concrete | [68] |
| 5 | 28 | Piraju I | −23.154237° −49.379971° | Paranapanema | FL | 1971 | Weir and Orifice | Bottom orifice/Concrete | [62,69–71] |
| 6 | 29 | Piraju II | −23.187694° −49.384466° | Paranapanema | FL | NO | Weir and Orifice | Bottom orifice/Concrete | [69] |
| 7 | 06 | São Domingos | −20.083344° −53.174786° | Verde | FL | 2015 | Weir and Orifice | Bottom orifice and notch/Metal | [72] |
| 8 | 22 | Engenheiro Sérgio Motta (Porto Primavera) | −22.483899° −52.957212° | Paraná | FL ● | 2001 | Weir and Orifice | Bottom orifice and notch/Concrete | [22,34,38,39, 73–75] |
| 9 | 22 | Engenheiro Sérgio Motta (Porto Primavera) | −22.483899° −52.957212° | Paraná | EL ● | 1999 | Pipe and Gravity | -/Metal | [76–79] |
| 10 | 35 | Itaipu Binacional | −25.430486° −54.581765° | Paraná | LS ● | 2002 | Semi-natural Vertical Slot Pool and weir | -/Concrete, rocks and excavated | [41,80–82] |
| 11 | 01 | Rochedo | −17.388624° −49.216373° | Meia Ponte | FL | NO | NO | NO/NO | [83] |
| 12 | 12 | Funil | −21.144011° −45.036538° | Grande | EL | 2004 | Flume channel | -/Metal | [44,84,85] |

**Table 2.** *Cont.*

| N | Dam ID | Dam Name | Coordinates X, Y | River | Fishway Type | Year | Fishway Design | Baffle Type/Material | References |
|---|---|---|---|---|---|---|---|---|---|
| | | | | Medium-sized—PCH | | | | | |
| 13 | 34 | Salto Mauá | −24.058585° −50.712232° | Tibagi | FL | 1943 | Pool and Weir | -/Concrete | [62,69,71,86] |
| 14 | 04 | Salto Moraes | −18.950109° −49.382568° | Tijuco | FL ● | 1951 | Weir and Orifice | Bottom orifice/Concrete | [62,71,87] |
| 15 | 14 | João Baptista de Lima Figueiredo | −21.584954° −46.746961° | Pardo | FL | NO | NO | NO/NO | [83] |
| 16 | 21 | Corumbataí | −22.480472° −47.592333° | Corumbataí | FL | NO | Pool and Weir | NO/Concrete | [88] |
| 17 | 07 | Retiro | −20.436132° −47.889945° | Sapucaí | FL | NO | NO | NO/NO | [83] |
| 18 | 08 | Anhanguera | −20.493856° −47.858010° | Sapucaí | FL | NO | NO | NO/NO | [83] |
| 19 | 09 | Palmeiras | −20.550638° −47.813264° | Sapucaí | FL | NO | NO | NO/NO | [83] |
| 20 | 10 | São Joaquim | −20.581831° −47.780019° | Sapucaí | FL | NO | NO | NO/NO | [89] |
| 21 | 10 | São Joaquim | −20.581831° −47.780019° | Sapucaí | FL | 1911 | Pool and Weir | -/Concrete | [69,89] |
| 22 | 11 | Dourados | −20.666837° −47.654310° | Sapucaí | FL | 1926 | Pool and Weir | -/Concrete | [69] |
| 23 | 13 | Itaipava | −21.413491° −47.334915° | Pardo | FL | 1911 | Pool and Weir | -/Concrete | [69,71] |
| 24 | 27 | San Juan | −23.149462° −47.793913° | Sorocaba | FL | NO | NO | NO/NO | [83] |
| 25 | 23 | Salto Grande | −22.933744° −46.896052° | Atibaia | FL | 1911 | Pool and Weir | -/Concrete | [35,83,90] |

**Table 2.** *Cont.*

| N | Dam ID | Dam Name | Coordinates X, Y | River | Fishway Type | Year | Fishway Design | Baffle Type/Material | References |
|---|---|---|---|---|---|---|---|---|---|
| 26 | 18 | Cachoeira de Emas | −21.926500° −47.366357° | Mogi-Guaçu | FL • | 1922 1943 | Pool and Weir | -/Concrete | [71,91–93] |
| 27 | 20 | Mogi-Guaçu | −22.379299° −46.900892° | Mogi-Guaçu | FL | NO | Pool and Weir | -/Concrete | [83,94] |
| 28 | 19 | São José | −21.938628° −46.816218° | Jaguari Mirim | FL | NO | NO | NO | [83] |
| 29 | 33 | Cachoeira Poço Preto II | −24.047850° −49.457784° | Itararé | FL | NO | NO | NO | [83] |
| 30 | 32 | Cachoeira Poço Preto I | −24.036877° −49.462776° | Itararé | FL | NO | NO | NO | [83] |
| 31 | 03 | Jataí | −17.943613° −51.726355° | Claro | FL | NO | NO | NO | [83] |
| 32 | 02 | Ypê | −17.725705° −50.451810° | Verdão | FL | NO | NO | NO | [83] |
| 33 | 15 | Gavião-Peixoto | −21.847628° −48.489485° | Jacaré-Guaçu | FL | 1913 1987 1995 2007 | Mixed system: Pool and weir and excavated rock | -/Concrete and rocks | [43,95] |
| 34 | 16 | Capão Preto | −21.895113° −47.814506° | Quilombo | FL | NO | NO | NO/NO | [83,96] |
| Small-sized—CGH | | | | | | | | | |
| 35 | 17 | Quatiara | −21.951352° −50.929426° | Do Peixe | FL | 1949 | Weir and Orifice | Notch/concrete | [69] |
| 36 | 30 | Santa Adélia | −23.327529° −47.768980° | Sorocaba | FL | NO | NO | NO/NO | [83] |
| 37 | 31 | Do Túnel | −23.414886° −50.452437° | Laranjinha | FL | NO | NO | NO/NO | [83] |

From all 37 fishways found, 24.3% (n = 9) have some kind of information regarding efficiency, while only 5.4% (n = 2) were rebuilt or structurally modified to improve their efficiency. The first to undergo modifications/reconstruction was the Cachoeira das Emas fish ladder in the Mogi-Guaçu River, São Paulo State, built in 1922 and rebuilt in 1943 (Table 2). Additionally, the Gavião-Peixoto fish ladder in the Jacaré-Guaçu River was built in 1913 and was rebuilt in 1987, 1995, and 2007 (Table 2).

### 3.6. Structural Connectivity by Fishways

Regarding all the hydropower plants currently in operation in the Upper Paraná River Basin (n = 389), only 9% (n = 35) have some type of fishway to provide structural connectivity (Figure 5). Furthermore, at least two fish ladders are inactive, both are fish ladders in the Paranapanema River at the Canoas I and II dams (Table 2).

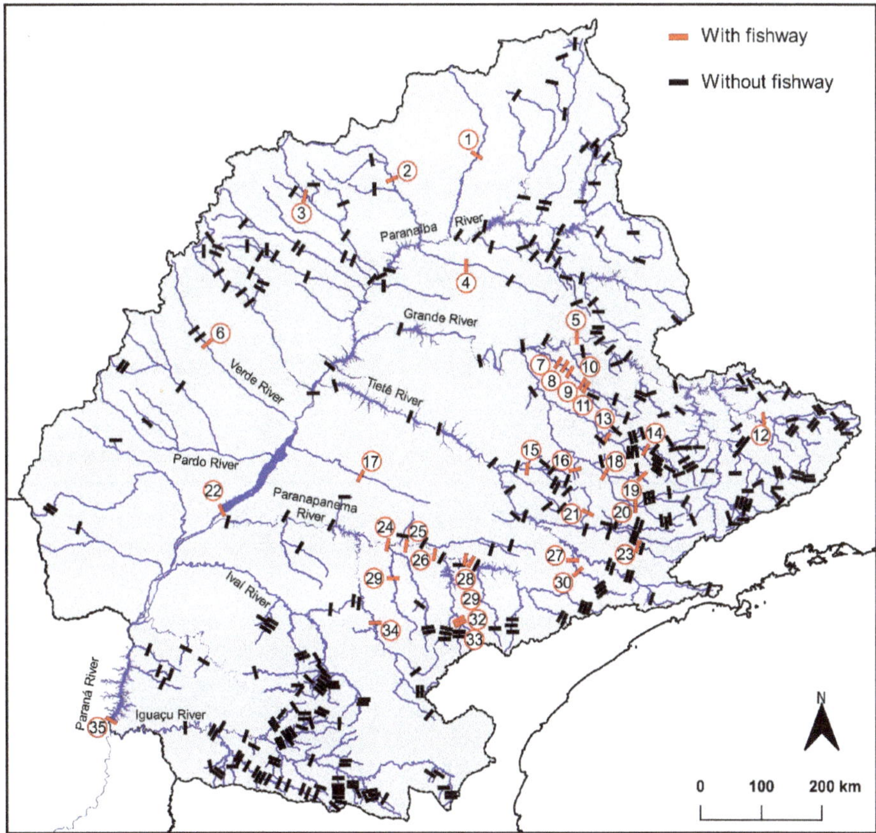


**Figure 5.** Distribution of hydropower dams in the main rivers of the Paraná River Basin. Black dashes indicate dams without fishways, and red dashes indicate dams with fishways. Numbers within the red circles represent the dam ID with fishways and can be found in Table 2.

The Upper Paraná River Basin is composed of a fluvial network that extends through 154,608 km, where 209 rivers (32,490.4 km) of this network have at least one hydropower dam, which corresponds to 0.0119 dam km$^{-1}$. However, only 10% (n = 21) of these rivers have some type of fishway to provide structural connectivity (Figure 6). The main water courses of the Upper Paraná River Basin are composed of a 31,454 km river network, where only 9% (2788 km) are connected by fishways, 48% (15,187 km) are blocked, and 43% (13,598 km) are free-flowing (Figure 6).

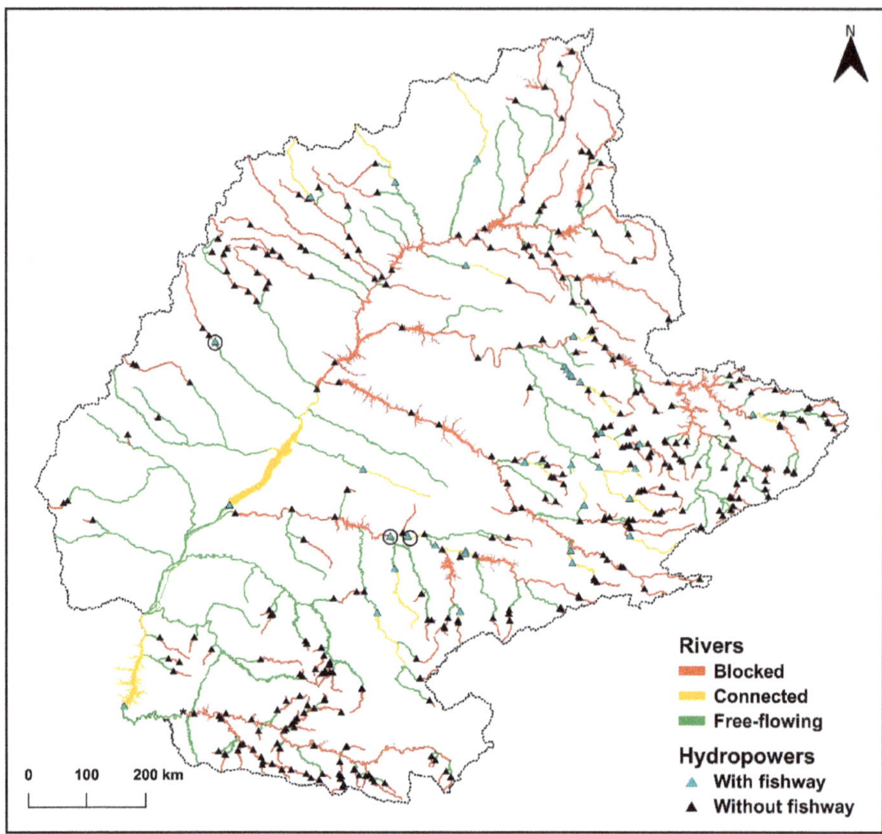



**Figure 6.** Risk map of river connectivity in the Upper Paraná Basin illustrating hydropower dams and the stretches of rivers: blocked, connected by fishways, and free-flowing. The circles indicate disabled fishways, and the asterisk indicates Baixo Iguaçu Dam.

The Paranapanema and Sapucaí rivers present the highest number of dams with fishways (n = 5), followed by the Grande, Itararé, Paraná, Pardo, Sorocaba, and Mogi-Guaçu rivers (n = 2 each river), while the other rivers have only one dam with a transposition system (Table 3).

Within a geopolitical distribution, 71.4% of dams with fishways (n = 25) reside within the state of São Paulo, followed by the states of Minas Gerais, Goiás, and Paraná, with 8.6% (n = 3) each, and the state of Mato Grosso do Sul with 2.9% (n = 1) (Table 3).

**Table 3.** Rivers of the Upper Paraná Basin with fishways and their respective length, slope, quantities of dams per river, dams with fishways, and the percentage of dams that provide connectivity via fishway. Dam location corresponds to the state in which the dam is located. * Indicates that the river flows on the border of a neighboring state, but the dam with the fishway follows the legislation of the state marked by the asterisk. # Indicates a fishway in each state.

| N | River | River Length (km) | River Slope (%) | Number of Dams per River | Number of Dams per River with Fishway | % of Dams with Fishway per River | Dam Location (State) |
|---|---|---|---|---|---|---|---|
| 1 | Atibaia | 181.5 | 0.23 | 2 | 1 | 50.0 | São Paulo |
| 2 | Claro | 353.8 | 0.14 | 4 | 1 | 25.0 | Goiás |
| 3 | Corumbataí | 84.7 | 0.29 | 1 | 1 | 100.0 | São Paulo |
| 4 | Peixe | 92.1 | 1.01 | 1 | 1 | 100.0 | São Paulo |
| 5 | Grande | 1156.9 | 0.08 | 12 | 2 | 16.7 | Minas Gerais * |
| 6 | Itararé | 197.8 | 0.21 | 2 | 2 | 100.0 | São Paulo |
| 7 | Jacaré-Guaçu | 171.7 | 0.21 | 2 | 1 | 50.0 | São Paulo |
| 8 | Jaguari-mirim | 98.3 | 0.21 | 2 | 1 | 50.0 | São Paulo |
| 9 | Laranjinha | 257.8 | 0.26 | 1 | 1 | 100.0 | Paraná |
| 10 | Meia Ponte | 434.9 | 0.16 | 1 | 1 | 100.0 | Goiás |
| 11 | Mogi-Guaçu | 406.1 | 0.24 | 6 | 2 | 33.3 | São Paulo |
| 12 | Paraná | 802.4 | 0.02 | 4 | 2 | 50.0 | São Paulo and Paraná# |
| 13 | Paranapanema | 718.3 | 0.08 | 11 | 5 | 45.5 | São Paulo * |
| 14 | Pardo | 481.0 | 0.18 | 7 | 2 | 28.6 | São Paulo |
| 15 | Quilombo | 35.7 | 0.56 | 1 | 1 | 100.0 | São Paulo |
| 16 | Sapucaí | 305.8 | 0.17 | 5 | 5 | 100.0 | São Paulo |
| 17 | Sorocaba | 204.1 | 0.19 | 5 | 2 | 40.0 | São Paulo |
| 18 | Tibagi | 336.8 | 0.11 | 2 | 1 | 50.0 | Paraná |
| 19 | Tijuco | 274.8 | 0.15 | 2 | 1 | 50.0 | Minas Gerais |
| 20 | Verdão | 384.3 | 0.15 | 1 | 1 | 100.0 | Goiás |
| 21 | Verde | 386.3 | 0.07 | 3 | 1 | 33.3 | Mato Grosso do Sul |
| | Total | 7365.1 | - | 75 | 35 | 46.7 | - |

## 3.7. Bioecological Aspects and Conservation Status of Migratory Species

Eighteen migratory fish species are listed in Table 4, which occur in the Upper Paraná River Basin. Some species are rare in this region, such as *Salminus hilarii*, *Steindachneridion scriptum* and *Zungaro jahu*. Regarding the status of conservation, six species are threatened, among them one near threatened, NT, (*Pseudoplatystoma corruscans*); two vulnerable, VU, (*Salminus brasiliensis*, *Rhinelepis aspera*); and three endangered species, EN, (*Brycon orbignyanus*, *S. scriptum* and *Z. jahu*). However, *B. orbignyanus* is also classified as critically endangered, CR, in the State of São Paulo, Brazil. The migratory fish are medium and large-sized species (characiformes, catfish and armored catfish) and exhibit varied feeding, with short reproductive periods (ranging from two to four months) between October and March. Most species are restricted to stretches of free-flowing rivers (FFRS) of dams, in the mainstem and/or tributaries, while some species also inhabit reservoir-dammed areas (*Prochilodus lineatus*, *Pimelodus maculatus*, *Pinirampus pirinampu*, *Pterodoras granulosus*, and *Rhaphiodon vulpinus*). Migratory species move longitudinally in the mainstem river, upstream and/or downstream, as well as most species perform lateral movements in dam-free tributaries.

Particularly in the Lower Iguaçu River, upstream of the Iguaçu Falls, there is an endemic and endangered species, and possibly migratory, *Steindachneridion melanodermatum*, similarly to *S. scriptum*. This species is restricted to a river stretch of 190 km, dam-free, and its knowledge is still limited.

**Table 4.** Migratory fish species in the Upper Paraná River Basin. CR = critically endangered, EN = endangered, VU = vulnerable, NT = near threatened, maximum standard length (SL), FFRS = free-flowing river stretches of dam, RES = reservoir, LON = longitudinal, UP = upstream, DO = downstream, LAT = lateral, * = rare species, [1] = MMA (2014) [56]; [2] = São Paulo (2018) [55], [3] = Abilhoa & Duboc (2004) [57].

| Family and Species | Common Name | Status of Conservation | SL (cm) | Feeding | Spawning Season | Habitat | Migratory Movements |
|---|---|---|---|---|---|---|---|
| **CHARACIFORMES** | | | | | | | |
| **Anostomidae** | | | | | | | |
| *Megaleporinus obtusidens* | Piapara | | 51.7 | Omnivorous | Dec-Jan | FFRS | LON-UP |
| *Megaleporinus piavussu* | Piapara | | 40.0 | Omnivorous | Nov-Jan | FFRS | |
| *Megaleporinus macrocephalus* | Piavuçu | | 50.0 | Omnivorous | - | FFRS | |
| **Bryconidae** | | | | | | | |
| *Brycon orbignyanus* | Piracanjuba | EN[1], CR[2] | 62.5 | Insectivorous | Oct-Jan | FFRS | LONG-UP |
| *Salminus brasiliensis* | Dourado | VU[3] | 85.9 | Piscivorous | Oct-Jan | FFRS | LON-UPDO, LAT |
| *Salminus hilarii** | Tabarana | | 34.0 | Piscivorous | Nov-Jan | FFRS | |
| **Cynodontidae** | | | | | | | |
| *Rhaphiodon vulpinus* | Dourado-cachorro | | 78.0 | Piscivorous | Out-Jan | FFRS, RES | - |
| **Prochilodontidae** | | | | | | | |
| *Prochilodus lineatus* | Curimba | | 54.2 | Iliophagous | Oct-Jan | FFRS, RES | LON-UPDO, LAT |
| **Serrasalmidae** | | | | | | | |
| *Piaractus mesopotamicus* | Pacu | | 52.6 | Omnivorous | Oct-Jan | FFRS | LON-UP, LAT |
| **SILURIFORMES** | | | | | | | |
| **Doradidae** | | | | | | | |
| *Pterodoras granulosus* | Armado | | 63.5 | Omnivorous | Jan-Mar | FFRS, RES | LON-UPDO, LAT |
| *Rhinelepis aspera* | Cascudo-preto | VU[3] | 49.0 | Iliophagous | Oct-Jan | FFRS | |
| **Pimelodidae** | | | | | | | |
| *Hemisorubim platyrhynchos* | Jurupoca | | 51.4 | Piscivorous | Dec-Jan | FFRS | LON-UP |
| *Pimelodus maculatus* | Mandi amarelo | | 36.0 | Omnivorous | Nov-Jan | FFRS, RES | LON-UPDO, LAT |
| *Pinirampus pirinampu* | Barbado | | 68.0 | Piscivorous | Dec-Jan | FFRS, RES | LON-UPDO, LAT |
| *Pseudoplatystoma corruscans* | Pintado | NT[3] | 140.0 | Piscivorous | Nov-Feb | FFRS | LON-UP, LAT |
| *Sorubim lima* | Jurupensem | | 60.5 | Piscivorous | Nov-Dec | FFRS | |
| *Steindachneridion scriptum** | Surubim | EN[1] | 64.0 | Piscivorous | Dec-Jan | FFRS | |
| *Zungaro jahu** | Jaú | EN[2] | 83.0 | Piscivorous | Dec-Feb | FFRS | |

## 4. Discussion

The fragmentation of rivers caused by dams is one of the leading factors of population decline in fish species [4]. Despite the negative effects on migratory and resident fish species, there are still ongoing plans to develop new hydropower dams, mainly in regions of high hydropower potential that have yet been unexploited, much like the basins of the Congo, Mekong, and Amazon rivers [97,98]. Therefore, the future of these free-flowing stretches of the Upper Paraná River Basin may be compromised, due to an expected 505% increase in medium-sized dams (PCHs). It is possible that this elevated number of medium-sized dams is the result of an exhaustion of the great hydraulic potentials caused by the construction of large-sized hydropower plants (UHEs) within the last 50 years [1]. This scenario jeopardizes migratory fish to an even greater risk than they face with existing dams.

Overall, 389 hydropower dams were identified, and only 35 (9%) of them have fishways in the Brazilian Upper Paraná Basin (some dams have more than one fishway, totaling 37 fishways overall). Although 37 fishways is a proportionally small number (9.5%) in comparison to 389 dams, the number of fishways encountered in this study alone was greater than the 25 fishways listed in all South America according to the most recent fishway inventory performed by Lira et al. [44]. These authors, however, only utilized information derived from scientific articles. By contrast, the results encountered in this study were derived from the integration of different research methods, such as scientific literature, grey literature, and satellite imaging, which provided the description of 13 new fishways in the Upper Paraná River Basin.

The number of fishways in this study was much higher in the state of São Paulo than in the five other Brazilian states drained by the Upper Paraná River. This fact may be related to current state legislation which requires the construction of fishways, when the position of the state environmental agency is favorable [99]. The state with the second largest number of fishways was Minas Gerais, which also has legislation regarding fishway implementation [100], while the other states that are part of Upper Paraná Basin do not have specific legislation regarding fishways. Despite some Brazilian states possessing legislation that describes the implementation of fishways whenever pertinent [99,100], there is no legislation that enforces the evaluation or monitoring of these structures. Once built, the evaluation and monitoring of the fishway depends entirely on requests from the environmental agencies and how it affects their granting of operating licenses for each hydropower plant. Otherwise, the necessary evaluation and monitoring will depend on the good faith of the developers. It is important to remember that the mere existence of a fishway will not guarantee that it will function as desired, especially when considering that the observed efficiency of existing structures hardly ever surpasses 30%.

This reality certainly explains the absence of scientific literature regarding fishways in the region this study took place. For the most part, fishways with scientific information are a result of partnerships with the hydropower companies and universities which utilize these fishways as a subject of study [44]. This is the case of the fish ladder at Engenheiro Sérgio Motta Hydropower Plant (Porto Primavera, SP), which is the subject of the highest number of published scientific articles [34,38,39,73–75], followed by Itaipu Binacional—Canal da Piracema [41,80–82], Canoas I and II [48,66], but it is also the reality of other fishways in the Tocantins River, Amazon Basin [101–103]. Many fishways in South America that have been evaluated are deemed selective or inefficient [41,80,104,105], and despite efficiency problems being commonplace, only two fishways have been modified or reconstructed in an aim to improve their efficiency, as were the cases of the Cachoeira das Emas [71,91] and the Gavião-Peixoto fish ladders [43,95].

During the data collection for this study, there was a predominance of fish ladders in relation to other types of fishways (elevators and lateral system). The small number of recorded elevators in this study could be a result of negative experiences with this fishway type in South American rivers. Evaluations conducted in some elevators showed low biomass transposition, high operational cost, and electromechanical problems [42,78,79]. Furthermore, the transposition of fish in this system is unidirectional (from downstream to upstream), not allowing the return of fish from upstream to downstream [106]. These factors may have contributed to the non-installation of this type of fishway in the other dams.

The lateral system found in this study, the Canal da Piracema, located at the Itaipu Dam, is more than 10 km long, and is considered the longest fishway in the world, comprising different designs such as semi-natural, vertical slot, and pool and weir [80,82]. Nature-like systems possess characteristics similar to those found in nature, facilitating the passage of fish, as they can offer habitats for aquatic organisms [31] and provide restored environments when compared to technical fishways [28]. Additionally, in some cases, nature-like systems and technical fishways can be favored over other structures such as elevators and fish locks, as they may allow for bidirectional connectivity [22,29]. A large part of fish ladders constructed in the Upper Paraná River until 1960 were of pool and weir design, and although they are encountered in a much higher frequency, they are considered to be fairly selective, as they favor jumper fish and limit the passage of fish which swim at the bottom [24,107]. Later, however, more adequate designs were implemented, such as weir and orifice and vertical slot, which permit the passages of benthic as well as pelagic fish species.

Fishways should obligatorily be implemented when: (i) there are native migratory species that require passage through the area where the axis of the dam is located; (ii) there is no impassable natural barrier that separates habitats prior to the construction of the dam (e.g., waterfalls); (iii) migratory species present a spatial distribution in both stretches that are to be fragmented by the dam, that is, where gene flow is present; and (iv) there are habitats favorable for spawning upstream from the axis of the future dam [108]. It is very likely that many of the 354 dams without fishways in the Upper Paraná River Basin would fit these four criteria proposed by Makrakis et al. [108], which would justify the implementation of fishways. Considering that the majority of hydropower dams do not permit a connection to the stretch of river blocked by the dam, it can be assumed that vast stretches have been highly fragmented, and consequently, gene flow in the Upper Paraná River has been compromised. Rivers which are prominently inhabited by migratory species are considerably fragmented, such as the Grande River, which hosts 12 dams, and only 2 of them (16.7%) provide fishways; the Pardo River, with 2 dams (28.6%); the Mogi-Guaçu River with 2 dams (33.2%); and the Paranapanema, a large tributary which spans roughly 1,000 km, with only 5 dams (45.5%) (Table 3). The Paranapanema River is especially notable, as the Rosana Hydropower Plant, which does not offer a fishway, is 25 km from where it meets the Paraná River, thereby severely hindering connectivity with the mainstream. Studies have shown that populations of long-distance migratory species in this tributary have been reduced [109], and this fact may be associated with the lack of connectivity at Rosana Dam.

New dams are expected and they could drastically reduce the number of free-flowing stretches, which are important areas for the maintenance of potamodromous species [110]. In the Paraná River mainstem, the preservation of the free-flowing river stretches, the floodplain of Upper Paraná River, which extends about 230 km from the Guaíra municipality up to downstream of the Porto Primavera Dam, is essential for the maintenance of fish diversity (Figure 6). Regarding the main tributaries in the Upper Paraná River Basin, notably the Aguapeí, Ivinhema, Iguatemi, Ivaí and Piquiri and Capanema (a tributary of Iguaçu River) rivers are completely free-flowing rivers and must be preserved (Figure 6). Additionally, other rivers that are partially dammed and have long free-flowing stretches require special attention because they are important areas of spawning and nursery for migratory fish species, e.g., the Pardo, Verde and Lower Iguaçu rivers [17,110–112].

The role of fishways in connectivity for some long-distance migratory species in Neotropical rivers has been widely studied, and its efficacy has been proven, as was the case of the Porto Primavera fish ladder [17,22,34,39,73,74,113]. A long period of monitoring the Porto Primavera fish ladder has shown that the fish ladder can provide bidirectional connectivity for some fish species, such as *Prochilodus lineatus* [22] and *Megaleporinus obtusidens* [34], and the results of the connectivity have been corroborated with the maintenance of *P. lineatus* gene flow in the whole Upper Paraná Basin [113]. In this sense, it is fundamental that hydropower plants meet the necessary requirements [108], and should seek the installation of fishways, especially fish ladders, which allow for the voluntary and bidirectional passage of fish species [22].

This study conducted a data survey, a revision, and an update to technical scientific information of existing fishways in the Upper Paraná River Basin within Brazilian borders. Thus, hydropower plants, both current and future, in attending to the requirements of implementation, must enable the connectivity of critical habitats to migratory species via fish ladders. The designs of these fish ladders must meet the biological demands related to the behaviors and swimming capabilities of Neotropical migratory fish species. Although fish ladders are not a panacea to solve the environmental distresses caused by impoundments, they can mitigate impacts [22]. A feasible implementation with adequate designs of fish ladders in hydropower dams of the Upper Paraná River Basin is one of the fundamental requisites for the viability of maintaining connectivity, and therefore, the conservation of Neotropical long-distance migratory species.

**Author Contributions:** Conceptualization, S.M., M.C.M., L.F.C., A.P.S.B., J.F.M.S. and F.J.S.-R.; methodology, S.M., M.C.M., L.F.C., A.P.S.B., and J.F.M.S.; software, L.F.C. and J.F.M.; validation, S.M., M.C.M., L.F.C., A.P.S.B., J.F.M.S. and F.J.S.-R.; formal analysis, L.F.C. and J.F.M.; investigation, S.M., M.C.M., L.F.C., A.P.S.B., J.F.M.S.; data curation, L.F.C., J.F.M. and A.P.S.B.; writing—original draft preparation, S.M., M.C.M., L.F.C. and A.P.S.B.; writing—review and editing, S.M., M.C.M., L.F.C., J.F.M.S. and F.J.S.-R.; supervision, S.M., M.C.M.

**Funding:** This research received no external funding.

**Acknowledgments:** We are grateful to CNPq for the Productivity Grant in Technological Development and Innovative Extension—DT (SM), and Coordination for the Improvement of Higher Level Personnel—CAPES for the Doctoral scholarship (LFC) and Master scholarship (APSB). We are also grateful to the Araucária Foundation for technical fellowship (JFMS). Any use of trade, firm, or product names is for descriptive purposes only and does not imply endorsement.

**Conflicts of Interest:** The authors declare no conflict of interest.

## Appendix A

**Figure A1.** Weir and orifice fish ladder constructed in metal at the UHE São Domingos—Photo: [61] (**A**). Pool and weir fish ladder constructed in the Cachoeira das Emas PCH—Photo: [82] (**B**). Pool and weir fish ladder constructed at the São Joaquim PCH—Photo: [102] (**C**). Weir and orifice fish ladder of the Canoas I UHE, without water—Photo: [58] (**D**). Weir and orifice ladder of the Canoas II UHE with water—Photo: [103] (**E**). lateral system (Canal da Piracema) of ITAIPU without water—Photo: Leandro F. Celestino (**F**). Example of a mixed system (Canal da Piracema) of ITAIPU with water—Photo: [104] (**G**). Weir and orifice fish ladder of the Engenheiro Sérgio Motta UHE without water—Photo: Leandro F. Celestino (**H**). Weir and orifice fish ladder of the Engenheiro Sérgio Motta with water—Photo: Leandro F. Celestino (**I**). Elevator of Engenheiro Sérgio Motta UHE—Photo: Leandro F. Celestino (**J**). Example of a vertical slot at Igarapava UHE—Photo: [105] (**K**). Pool and weir as part of a mixed ladder at Galvão-Peixoto PCH—Photo: [106] (**L**), Pool and weir fish ladder constructed at Salto Grande PCH—Photo: [107] (**M**).

## References


1. Zarfl, C.; Lumsdon, A.E.; Berlekamp, J.; Tydecks, L.; Tockner, K. A global boom in hydropower dam construction. *Aquat. Sci.* **2015**, *77*, 161–170. [CrossRef]
2. Albins, M.; Evans, A.; Ismail, G.; Neilsen, B.; Pusack, T.; Schemmel, E.; Smith, W.; Stoike, S.; Li, H.W.; Noakes, D.L.G. Can humans coexist with fishes? *Environ. Biol. Fishes* **2013**, *96*, 1301–1313. [CrossRef]
3. Timpe, K.; Kaplan, D. The changing hydrology of a dammed Amazon. *Sci. Adv.* **2017**, *3*, 1–14. [CrossRef] [PubMed]
4. Agostinho, A.; Pelicice, F.; Gomes, L. Dams and the fish fauna of the Neotropical region: Impacts and management related to diversity and fisheries. *Brazi. J. Biol.* **2008**, *68*, 1119–1132. [CrossRef]
5. Ward, J.V.; Stanford, J.A. Ecological connectivity in alluvial river ecosystems and its disruption by flow regulation. *Regul. Rivers Res. Manag.* **1995**, *11*, 105–119. [CrossRef]
6. Barletta, M.; Jaureguizar, A.J.; Baigun, C.; Fontoura, N.F.; Agostinho, A.A.; Almeida-Val, V.M.F.; Val, A.L.; Torres, R.A.; Jimenes-Segura, L.F.; Giarrizzo, T.; et al. Fish and aquatic habitat conservation in South America: A continental overview with emphasis on neotropical systems. *J. Fish Biol.* **2010**, *76*, 2118–2176. [CrossRef]
7. Gouskov, A.; Reyes, M.; Wirthner-Bitterlin, L.; Vorburger, C. Fish population genetic structure shaped by hydroelectric power plants in the upper Rhine catchment. *Evol. Appl.* **2016**, *9*, 394–408. [CrossRef]
8. Lucas, M.C.; Baras, E. *Migration of Freshwater Fishes*; Wiley Online Library: Oxford, UK, 2001.
9. Moraes, P.; Deverat, F. *An Introduction to Fish Migration*; Moraes, P., Deverat, F., Eds.; Taylor & Francis Group: London, UK, 2016; ISBN 978-1-4987-1874-5.
10. Langeani, F.; Oyakawa, O.T.; Shibatta, A.O.; Pavanelli, C.S.; Casatti, L. Diversidade da ictiofauna do Alto Rio Paraná: Composição atual e perspectivas futuras. *Biota Neotrop.* **2007**, *7*, 1–18. [CrossRef]
11. Agostinho, A.A.; Gomes, L.C.; Suzuki, H.I.; Júlio-Júnior, H.F. Migratory fishes of the Upper Paraná River Basin, Brazil. In *Migratory Fishes of South America*; Carolsfeld, J., Harvey, B., Ross, C., Baer, A., Eds.; World Bank: Washington, DC, USA, 2003; ISBN 0-9683958-2-12.
12. Suzuki, H.I.; de Vazzoler, A.E.A.M.; Marques, E.E.; Lizama, M.L.A.P.; Inada, P. Reproductive Ecology of the Fish Assemblages. In *Upper Paraná River and its Floodplain: Physical Aspects, Ecology and Conservation*; Thomaz, S.M., Agostinho, A.A., Hahn, N.S., Eds.; Backhuys Publishers: Leiden, The Netherlands, 2004; pp. 271–292.
13. Makrakis, M.C.; Miranda, L.E.; Makrakis, S.; Fontes, H.M.; Morlis, W.G.; Dias, J.H.P.; Garcia, J.O. Diversity in migratory patterns among Neotropical fishes. *J. Fish Biol.* **2012**, *81*, 866–881. [CrossRef]
14. Godinho, A.L.; Kynard, B. Migratory fishes of Brazil: Life history and fish passage needs. *River Res. Appl.* **2008**, *25*, 702–712. [CrossRef]
15. Godoy, M.P. *Peixes do Brasil, Subordem Characoidei: Bacia do Rio Mogí Guassú*; Franciscana: Piracicaba, Brazil, 1975.
16. Cowx, I.G.; Welcomme, R.L. *Rehabilitation of Rivers for Fish*; Food & Agriculture Organization: Rome, Italy, 1998.
17. Da Silva, P.S.; Makrakis, M.C.; Miranda, L.E.; Makrakis, S.; Assumpção, L.; Paula, S.; Dias, J.H.P.; Marques, H. Importance of reservoir tributaries to spawning of migratory fish in the upper Paraná River. *River Res. Appl.* **2015**, *31*, 313–322. [CrossRef]
18. Quirós, R. The Paraná river basin development and the changes in the lower basin fisheries. *Interciencia* **1990**, *15*, 442–451.
19. Hoeinghaus, D.J.; Agostinho, A.A.; Gomes, L.C.; Pelicice, F.M.; Okada, E.K.; Latini, J.D.; Kashiwaqui, E.A.L.; Winemiller, K.O. Effects of river impoundment on ecosystem services of large tropical rivers: Embodied energy and market value of artisanal fisheries. *Conserv. Biol.* **2009**, *23*, 1222–1231. [CrossRef] [PubMed]
20. Porcher, J.P.; Travade, F. Fishways: Biological basis, limits and legal considerations. *Bull. Français la Pêche la Piscic.* **2002**, *364*, 9–20. [CrossRef]
21. Clay, C.H. *Design of Fishways and Other Fish Facilities*, 2nd ed.; CRC Press: Boca Raton, FL, USA, 1995.
22. Celestino, L.F.; Sanz-Ronda, F.J.; Miranda, L.E.; Makrakis, M.C.; Dias, J.H.P.; Makrakis, S. Bidirectional connectivity via fish ladders in a large Neotropical river. *River Res. Appl.* **2019**, *35*, 236–246. [CrossRef]
23. Larinier, M. Fishways-General consideration. *Bull. Français La Pêche La Piscic.* **2002**, *364*, 21–27. [CrossRef]
24. DVWK. *Fish Passages—Design, Dimension and Monitoring*; FAO—Food and Agriculture Organization of the United Nations: Rome, Italy, 2002; ISBN 92-5-104894-0.

25. Castro-Santos, T.; Cotel, A.; Webb, P. Fishway evaluations for better bioengineering: An Integrative approach a framework for fishway. *Am. Fish. Soc.* **2009**, *69*, 557–575.

26. Silva, A.T.; Lucas, M.C.; Castro-Santos, T.; Katopodis, C.; Baumgartner, L.J.; Thiem, J.D.; Aarestrup, K.; Pompeu, P.S.; O'Brien, G.C.; Braun, D.C.; et al. The future of fish passage science, engineering, and practice. *Fish Fish.* **2018**, *19*, 340–362. [CrossRef]

27. Larinier, M. Environmental Issues, Dams and Fish Migration. In *Dams, Fish and Fisheries: Opportunities, Challenges and Conflict Resolution*; Marmulla, G., Ed.; Fisheries Techinical Paper; FAO Food and Agriculture Organization: Rome, Italy, 2001; pp. 45–90, ISBN 92-5-104694-8.

28. Gustafsson, S.; Österling, M.; Skurdal, J.; Schneider, L.D.; Calles, O. Macroinvertebrate colonization of a nature-like fishway: The effects of adding habitat heterogeneity. *Ecol. Eng.* **2013**, *61*, 345–353. [CrossRef]

29. Dodd, J.R.; Cowx, I.G.; Bolland, J.D. Efficiency of a nature-like bypass channel for restoring longitudinal connectivity for a river-resident population of brown trout. *J. Environ. Manag.* **2017**, *204*, 318–326. [CrossRef]

30. Calles, O.; Greenberg, L. Connectivity is a two-way street-the need for a holistic approach to fish passage problems in regulated rivers. *River Res. Appl.* **2009**, *25*, 1268–1286. [CrossRef]

31. Pander, J.; Mueller, M.; Geist, J. Ecological functions of fish bypass channels in streams: Migration corridor and habitat for rheophilic spceies. *River Res. Appl.* **2013**, *29*, 441–450. [CrossRef]

32. Rajaratnam, N.; Van der Vinne, G.; Katopodis, C. Hydraulics of Vertical Slot Fishways. *J. Hydraul. Eng.* **1986**, *112*, 909–927. [CrossRef]

33. Katopodis, C.; Williams, J.G. The development of fish passage research in a historical context. *Ecol. Eng.* **2012**, *48*, 8–18. [CrossRef]

34. Gutfreund, C.; Makrakis, S.; Castro-Santos, T.; Celestino, L.F.; Dias, J.H.P.; Makrakis, M.C. Effectiveness of a fish ladder for two Neotropical migratory species in the Paraná River. *Mar. Freshw. Res.* **2018**, 1–9. [CrossRef]

35. Quirós, R. *Structures Assisting the Migrations of Non-Salmonid Fish: Latin America*; Quirós, R., Ed.; Techinical Paper. N°5; COPESCAL: Rome, Italy, 1989; ISBN 92-5-102683-1.

36. Fernandez, D.R.; Agostinho, A.A.; Bini, L.M.; Pelicice, F.M. Diel variation in the ascent of fishes up an experimental fish ladder at Itaipu Reservoir: Fish size, reproductive stage and taxonomic group influences. *Neotrop. Ichthyol.* **2007**, *5*, 215–222. [CrossRef]

37. De Santos, H.A.; Viana, E.M.F.; Pompeu, P.S.; Martinez, C.B. Optimal swim speeds by respirometer: An analysis of three neotropical species. *Neotrop. Ichthyol.* **2012**, *10*, 805–811. [CrossRef]

38. Assumpção, L.; Makrakis, M.C.; Makrakis, S.; Wagner, R.L.; Da Silva, P.S.; De Lima, A.F.; Kashiwaqui, E.A.L. The use of morphometric analysis to predict the swimming efficiency of two Neotropical long-distance migratory species in fish passage. *Neotrop. Ichthyol.* **2012**, *10*, 797–804. [CrossRef]

39. Bido, A.F.; Urbinati, E.C.; Makrakis, M.C.; Celestino, L.F.; Serra, M.; Makrakis, S. Stress indicators for *Prochilodus lineatus* (Characiformes: Prochilodontidae) breeders during passage through a fish ladder. *Mar. Freshw. Res.* **2018**, 1–8. [CrossRef]

40. Castro-Santos, T.; Haro, A. Fish Guidance and Passage at Barriers. In *Fish Locomotion: An Eco-Ethological Perspective*; Domenici, P., Kapoor, B.G., Eds.; Science Publisher: Enfield, NH, USA, 2010; pp. 62–89.

41. Fontes, H.M., Jr.; Castro-santos, T.; Makrakis, S.; Gomes, L.C.; Latini, J.D. A barrier to upstream migration in the fish passage of Itaipu Dam (Canal da Piracema), Paraná River basin. *Neotrop. Ichthyol.* **2012**, *10*, 697–704. [CrossRef]

42. Oldani, N.O.; Baigún, C.R.M. Performance of a fishway system in a major South American dam on the Parana River (Argentina-Paraguay). *River Res. Appl.* **2002**, *18*, 171–183. [CrossRef]

43. Almeida, L.E.G.; Mattos, M.E.; Tanaka, H.R. Modernização da PCH Gavião Peixoto: Gestão, controle ambiental e mecanismo de desenvolvimento limpo. *Semin. Nac. PRODUÇÃO E Transm. Energ. ELÉTRICA* **2010**, 7.

44. Lira, N.A.; Pompeu, P.S.; Agostinho, C.S.; Agostinho, A.A.; Arcifa, M.S.; Pelicice, F.M. Fish passages in South America: An overview of studied facilities and research effort. *Neotrop. Ichthyol.* **2017**, *15*, 1–14. [CrossRef]

45. Cooke, S.J.; Hinch, S.G. Improving the reliability of fishway attraction and passage efficiency estimates to inform fishway engineering, science, and practice. *Ecol. Eng.* **2013**, *58*, 123–132. [CrossRef]

46. Castro-Santos, T.; Haro, A.; Walk, S. A passive integrated transponder (PIT) tag system for monitoring fishways. *Fish. Res.* **1996**, *28*, 253–261. [CrossRef]

47. Energy, D. ABC da Energia. 2013. Available online: https://abc-da-energia.webnode.com/ (accessed on 5 July 2019).

48. Britto, S.G.D.C.; Carvalho, E.D. Reproductive migration of fish and movement in a series of reservoirs in the Upper Parana River basin, Brazil. *Fish. Manag. Ecol.* **2013**, *20*, 426–433. [CrossRef]

49. Auffret, A.G.; Plue, J.; Cousins, S.A.O. The spatial and temporal components of functional connectivity in fragmented landscapes. *Ambio* **2015**, *44*, 51–59. [CrossRef] [PubMed]

50. Orfeo, O.; Stevaux, J. Hydraulic and morphological characteristics of middle and upper reaches of the Paraná River (Argentina and Brazil). *Geomorphology* **2002**, *44*, 309–322. [CrossRef]

51. QGIS Development Team. *Development Geographic Information System User Guide. Open Source Geospatial Foundation Project 2019*; QGIS Development Team: Grut, Switzerland, 2019.

52. ANEEL—Agência Nacional de Energia Elétrica. *Resolução Normativa N° 687, de 24 de Novembro de 2015*; Gência Nacional de Energia Elétrica: Brasilia, Brazil, 2015.

53. *ANEEL—Agência Nacional de Energia Elétrica Resolução Normativa N° 673, de 4 de Agosto de 2015*; Gência Nacional de Energia Elétrica: Brasilia, Brazil, 2015.

54. ANEEL. *Atlas de Energia Elétrica do Brasil*, 3rd ed.; de Elétrica, A.N.E., Ed.; Agência Nacional de Energia Elétrica: Brasilia, Brazil, 2008; ISBN 978-85-87491-10-7.

55. São Paulo. Decreto N° 63.853, de 27 de Novembro de 2018. São Paulo. 2018. Available online: https://www.al.sp.gov.br/repositorio/legislacao/decreto/2018/decreto-63853-27.11.2018.html (accessed on 5 July 2019).

56. Brasil Portaria do Ministério do Meio Ambiente (MMA). *445, de 17 de Dezembro de 2014. Reconhece Como Espécies de Peixes e Invertebrados Aquáticos da Fauna Brasileira Ameaçadas de Extinção Aquelas Constantes da Lista Nacional Oficial de Espécies da Fauna Ameaçada*; Brasil Portaria do Ministério do Meio Ambiente: Brasilia, Brazil, 2014; pp. 126–130.

57. Da Ingenito, L.F.S.; Duboc, L.F.; Abilhoa, V. Contribuição ao conhecimento da ictiofauna da bacia do alto rio Iguaçu, Paraná, Brasil. *Arq. Ciências Veterinárias e Zool. da Unipar* **2004**, *7*, 23–36.

58. De Vazzoler, A.E.A.M. *Biologia da Reprodução de Peixes Teleósteos: Teoria e Prática*; Eduem: Maringá, Brazil, 1996; ISBN 85-85545-16-X.

59. Ota, R.R.; de Deprá, G.C.; da Graça, W.J.; Pavanelli, C.S. Peixes da planície de inundação do alto rio Paraná e áreas adjacentes: Revised, annotated and updated. *Neotrop. Ichthyol.* **2018**, *16*, 1–111. [CrossRef]

60. Hahn, N.S.; Fugi, R.; Andrian, I.F. Trophic Ecology of Fish Assemblages. In *The Upper Paraná River Floodplain: Physical Aspects, Ecology and Conservation*; Thomaz, S.M., Agostinho, A.A., Hahn, N.S., Eds.; Backhuys Publishers: Leiden, The Netherlands, 2004; pp. 247–269.

61. De Resende, E.K. Migratory Fishes of the Paraguay-Paraná Basin. In *Migratoty Fishes of South America: Biology, Fisheries, and Conservation Status*; Carolsfeld, J., Harvey, B., Ross, C., Baer, A., Eds.; World Fisheries Trust: Washington, DC, USA, 2003; pp. 98–155.

62. Agostinho, A.A.; Gomes, L.C.; Pelicice, F.M. *Ecologia e Manejo de Recursos Pesqueiros em Rereservatórios do Brasil*; EDUEM: Maringá, Brazil, 2007.

63. Godinho, A.L.; Lamas, I.R.; Godinho, H.P. Reproductive ecology of Brazilian freshwater fishes. *Environ. Biol. Fishes* **2010**, *87*, 143–162. [CrossRef]

64. Bowen, M.D.; Marques, S.; Silva, L.G.M.; Vono, V.; Godinho, H.P. Comparing on site human and video counts at Igarapava fish ladder, Southeastern Brazil. *Neotrop. Ichthyol.* **2006**, *4*, 291–294. [CrossRef]

65. Bizzotto, P.M.; Godinho, A.L.; Vono, V.; Kynard, B.; Godinho, H.P. Influence of seasonal, diel, lunar, and other environmental factors on upstream fish passage in the Igarapava Fish Ladder, Brazil. *Ecol. Freshw. Fish* **2009**, *18*, 461–472. [CrossRef]

66. Lopes, C.M.; De Almeida, F.S.; Orsi, M.L.; Britto, S.G.D.C.; Sirol, R.N.; Sodré, L.M.K. Fish passage ladders from Canoas Complex—Paranapanema River: Evaluation of genetic structure maintenance of *Salminus brasiliensis* (Teleostei: Characiformes). *Neotrop. Ichthyol.* **2007**, *5*, 131–138. [CrossRef]

67. Britto, S.G.D.C. A estratégia reprodutiva dos peixes migradores frente às escacda do complexo canoas (rio Paranapanema, bacia do alto Paraná). Ph.D. Thesis, Universidade Estadual Paulista, Sao Paolo, Brazil, 2009.

68. Arcifa, M.S.; Luiz, A.; Esguícero, H. The fish fauna in the fish passage at the Ourinhos Dam, Paranapanema River. *Neotrop. Ichthyol.* **2012**, *10*, 715–722. [CrossRef]

69. Martins, S.L. Sistemas Para a Transposição de Peixes Neotropicais Potamódromos. Ph.D. Thesis, Universidade Estadual Paulista, Sao Paolo, Brazil, 2005.

70. De Viana, E.M.F. Mapeamento do Campo de Velocidades em Mecanismos de Transposição de Peixes do Tipo Slot Vertical em Diferentes Escale. Ph.D. Thesis, Universidade Federal de Minas Gerais, Belo Horizonte, Brazil, 2005.

71. Godoy, M.P. *Aqüicultira—Atividade Multidisciplinar e Outras Facilidades Para Passagens de Peixes, Estações de Piscicultura*; Eletrosul: Florianópolis, Brazil, 1985.

72. Eletrosul. *5° Relatório Semestral de Andamento dos Programas e Planos Ambietais*; Fase de Operação UHE São Domingos; Eletrosul: Florianópolis, Brazil, 2015; Volume 1.

73. Makrakis, S.; Makrakis, M.C.; Wagner, R.L.; Dias, J.H.P.; Gomes, L.C. Utilization of the fish ladder at the Engenheiro Sergio Motta Dam, Brazil, by long distance migrating potamodromous species. *Neotrop. Ichthyol.* **2007**, *5*, 197–204. [CrossRef]

74. Wagner, R.L.; Makrakis, S.; Castro-Santos, T.; Makrakis, M.C.; Dias, J.H.P.; Belmont, R.F. Passage performance of long-distance upstream migrants at a large dam on the Paraná River and the compounding effects of entry and ascent. *Neotrop. Ichthyol.* **2012**, *10*, 785–795. [CrossRef]

75. Volpato, G.L.; Barreto, R.E.; Marcondes, A.L.; Andrade Moreira, P.S.; de Barros Ferreira, M.F. Fish ladders select fish traits on migration–still a growing problem for natural fish populations. *Mar. Freshw. Behav. Physiol.* **2009**, *42*, 307–313. [CrossRef]

76. CESP. *Programa de Manejo Pesqueiro: Plano de Trabalho 2000–2001*; CESP: São Paulo, Brazil, 2000.

77. CESP. *Programa de Manejo Pesqueiro: Plano de Trabalho 2002–2003*; CESP: São Paulo, Brazil, 2002.

78. CESP. *Programa de Manejo Pesqueiro: Plano de Trabalho 2003–2004*; CESP: São Paulo, Brazil, 2003.

79. Makrakis, S.; Makrakis, M.C.; Da Silva, P.S.; Celestino, L.F. *Monitoramento do Ictioplâncton em Tributários do Reservatório de Porto Primavera e Monitoramento da Transposição*; Unioeste: Toledo, Brazil, 2015.

80. Makrakis, S.; Gomes, L.C.; Makrakis, M.C.; Fernandez, D.R.; Pavanelli, C.S. The Canal da Piracema at Itaipu Dam as a fish pass system. *Neotrop. Ichthyol.* **2007**, *5*, 185–195. [CrossRef]

81. Hahn, L.; English, K.; Carosfeld, J.; Gustavo, L.; Latini, J.D.; Agostinho, A.A.; Fernandez, D.R. Preliminary study on the application of radio-telemetry techniques to evaluate movements of fish in the Lateral canal at Itaipu Dam, Brazil. *Neotrop. Ichthyol.* **2007**, *5*, 103–108. [CrossRef]

82. Makrakis, S.; Miranda, L.E.; Gomes, L.C.; Makrakis, M.C.; Junior, H.M.F.J. Ascent of neotropical fish in the Itaipu reservoir fish pass. *River Res. Appl.* **2011**, *27*, 511–519. [CrossRef]

83. Google Maps. Available online: https://www.google.com.br/maps (accessed on 15 February 2018).

84. Suzuki, F.M.; Pires, L.V.; Pompeu, P.S. Passage of fish larvae and eggs through the Funil, Itutinga and Camargos Reservoirs on the upper Rio Grande (Minas Gerais, Brazil). *Neotrop. Ichthyol.* **2011**, *9*, 617–622. [CrossRef]

85. Aliança Usina de Funil. Available online: https://aliancaenergia.com.br/br/nossas-usinas/usina-de-funil (accessed on 24 April 2019).

86. Shibatta, O.A.; Gealh, A.M.; Bennemann, S.T. Ictiofauna dos trechos alto e médio da bacia do rio Tibagi, Paraná, Brasil. *Biota Neotrop.* **2007**, *7*, 125–134. [CrossRef]

87. Godinho, H.P.; Godinho, A.L.; Formagio, P.S.; Torquato, V.C. Fish ladder efficiency in a southeastern Brazilian river. *Cienc. Cult.* **1991**, *43*, 63–67.

88. EMAE—Fundação de Energia e Saneamento. *PCH Corubataí, Rio claro—SP.*; Fundação de Energia e Saneamento: Sao Paulo, Brazil, 2014.

89. Ricardi, A.; Meneguello, C.; Bueno, E.; Santos, G.; Magalhães, G.; Limnos, G.; Bizello, M.L.; Xavier, M.; Diniz, R.; Furlan, S.; et al. *Relaório Técnico da 8° Expedição: Usinas Lobo, São Joaquim, Buritis*; Esmeril: São Paulo, Brazil, 2015; Volume 8.

90. Mortati, D.; Bueno, E.; Dizzio, F.; Limnios, G.; Magalhães, G.; Midori, M.; Lima, N.; Gazoni, P.; Diniz, R.; Carvalho, T. *Relatório Técnico da 2° Expedição: Usinas de Porto Goes, Lavras, Salto Grande, Cariobinha, Carioba, America, Jaguari*; Esmeril: São Paulo, Brazil, 2014; Volume 2.

91. Capeleti, A.R.; Petrere Jr, M. Migration of the curimbatá *Prochilodus lineatus* (Valenciennes, 1836) (Pisces, Prochilodontidae) at the waterfall "Cachoeira de Emas" of the Mogi-Guaçu river—São Paulo, Brazil. *Brazilian J. Biol.* **2006**, *66*, 651–659. [CrossRef]

92. Peressin, A.; Gonçalves, S.; Manoel, F.; Braga, D.S. Reproductive strategies of two Curimatidae species in a Mogi Guaçu impoundment, upper Paraná River basin, São Paulo, Brazil. *Neotrop. Ichthyol.* **2012**, *10*, 847–854. [CrossRef]

93.  Migliorini, P.C.P. *Repotenciação da Pequena Central Hidrelétrica de Emas "Nova"—Pirassununga—SP: Aspectos Técnicos, Socioambientais e Economicos*; Universidade de São Paulo: Sao Paulo, Brazil, 2011.

94.  Gonçalves, C.D.S.; Braga, F.M.D.S. Diversidade e ocorrência de peixes na área de influência da UHE Mogi Guaçu e lagoas marginais, bacia do alto rio Paraná, São Paulo, Brasil. *Biota Neotrop.* **2008**, *8*, 103–114. [CrossRef]

95.  Esguícero, A.L.H.; Arcifa, M.S. Fragmentation of a Neotropical migratory fish population by a century-old dam. *Hydrobiologia* **2010**, *638*, 41–53. [CrossRef]

96.  Ricardi, A.; Argollo, A.; Freitas, C.; Barbanti, C.; Mortati, D.; Geribello, D.; Bueno, E.; Drizzo, F.; Santos, G.; Magalhães, G.; et al. Relatório técnico 5° Expedição: Usinas de Corumbataí, Capão Pretro, Monjolinho, Lia-Marmelos I, II, III e Isabl. 2014. Available online: http://eletromemoria.fflch.usp.br/sites/eletromemoria.fflch.usp.br/files/relatorio_da_5a_expedicao.pdf (accessed on 5 July 2019).

97.  Winemiller, K.O.; McIntyre, P.B.; Castello, L.; Fluet-Chouinard, E.; Giarrizzo, T.; Nam, S.; Baird, I.G.; Darwall, W.; Lujan, N.K.; Harrison, I.; et al. Balancing hydropower and biodiversity in the Amazon, Congo, and Mekong. *Science* **2016**, *351*, 128–129. [CrossRef] [PubMed]

98.  Latrubesse, E.M.; Arima, E.Y.; Dunne, T.; Park, E.; Baker, V.R.; D'Horta, F.M.; Wight, C.; Wittmann, F.; Zuanon, J.; Baker, P.A.; et al. Damming the rivers of the Amazon basin. *Nature* **2017**, *546*, 363–369. [CrossRef]

99.  ALESP. *LEI N° 9.798, de 07 de Outubro de 1997*; Assembleia Legislativa do Estado de São Paulo: Sao Paulo, Brazil, 1997.

100. Minas Gerais. Lei N° 12.488, de 9 de Abril de 1997. 1997. Available online: http://www.ctpeixes.ufmg.br/html/conteudo/lei_12488.htm (accessed on 5 July 2019).

101. Agostinho, A.A.; Marques, E.E.; Agostinho, C.S.; Almeida, D.A.; Oliveira, R.J.; Rodrigues, J.; Melo, B. Fish ladder of Lajeado Dam: Migrations on one-way routes? *Neotrop. Ichthyol.* **2007**, *5*, 121–130. [CrossRef]

102. Agostinho, C.S.; Pelicice, F.M.; Marques, E.E.; Soares, A.B.; de Almeida, D.A.A. All that goes up must come down? Absence of downstream passage through a fish ladder in a large Amazonian river. *Hydrobiologia* **2011**, *675*, 1–12. [CrossRef]

103. Pelicice, F.M.; Agostinho, C.S. Deficient downstream passage through fish ladders: The case of Peixe Angical Dam, Tocantins River, Brazil. *Neotrop. Ichthyol.* **2012**, *10*, 705–713. [CrossRef]

104. Baigún, C.R.M.; Nestler, J.M.; Minotti, P.; Oldani, N.O. Fish passage system in an irrigation dam (Pilcomayo River basin): When engineering designs do not match ecohydraulic criteria. *Neotrop. Ichthyol.* **2012**, *10*, 741–750.

105. Oldani, N.O.; Baigún, C.R.M.; Nestler, J.M.; Goodwin, R.A. Is fish passage technology saving fish resources in the lower La Plata River basin? *Neotrop. Ichthyol.* **2007**, *5*, 89–102. [CrossRef]

106. Croze, O.; Bau, F.; Delmouly, L. Efficiency of a fish lift for returning Atlantic salmon at a large-scale hydroelectric complex in France. *Fish. Manag. Ecol.* **2008**, *15*, 467–476. [CrossRef]

107. Larinier, M.; Marmulla, G. Fish passes: Types, principles and geographical distribution—An overview. In Proceedings of the Second International Symposium on the Management of Large Rivers for Fisheries; Welcomme, R.L., Petr, T., Eds.; FAO Fisheries: Bangkok, Thailand, 2004; pp. 183–206, ISBN 974-7946-65-3.

108. Makrakis, S.; Dias, J.H.P.; Lopes, J.D.M.; Fontes-Junior, H.M.; Godinho, A.L.; Martinez, C.B.; Makrakis, M.C. Premissas e Critérios Mínimos para Implantação, Avaliação e Monitoramento de Sistemas de Transposição para Peixes. *Bol. Soc. Bras. Ictiol.* **2015**, *114*.

109. Pelicice, F.M.; Azevedo-Santos, V.M.; Esguícero, A.L.H.; Agostinho, A.A.; Arcifa, M.S. Fish diversity in the cascade of reservoirs along the Paranapanema River, southeast Brazil. *Neotrop. Ichthyol.* **2018**, *16*, 1–18. [CrossRef]

110. Da Silva, P.S.; Miranda, L.E.; Makrakis, S.; Assumpção, L.; Dias, J.H.P.; Makrakis, M.C. Tributaries as biodiversity preserves: An ichthyoplankton perspective from the severely impounded Upper Paraná River. *Aquat. Conserv. Mar. Freshw. Ecosyst.* **2019**, 258–269. [CrossRef]

111. Marques, H.; Dias, J.H.P.; Perbiche-Neves, G.; Kashiwaqui, E.A.L.; Ramos, I.P. Importance of dam-free tributaries for conserving fish biodiversity in Neotropical reservoirs. *Biol. Conserv.* **2018**, *224*, 347–354. [CrossRef]

112. Assumpção, L.; Makrakis, S.; Sarai, P.; Makrakis, M.C. Espécies de peixes ameaçadas de extinção no Parque Nacional do Iguaçu. *Biodiversidade Bras.* **2017**, *7*, 4–17.
113. Ferreira, D.G.; Souza-Shibatta, L.; Shibatta, O.A.; Sofia, S.H.; Carlsson, J.; Dias, J.H.P.; Makrakis, S.; Makrakis, M.C. Genetic structure and diversity of migratory freshwater fish in a fragmented Neotropical river system. *Rev. Fish Biol. Fish.* **2017**, *27*, 209–231. [CrossRef]

*Communication*

# Habitat Enhancement Solutions for Iberian Cyprinids Affected by Hydropeaking: Insights from Flume Research

**Maria João Costa *, António N. Pinheiro and Isabel Boavida**

Civil Engineering Research and Innovation for Sustainability, Instituto Superior Técnico, Universidade de Lisboa, 1049-001 Lisboa, Portugal; antonio.pinheiro@tecnico.ulisboa.pt (A.N.P.); isabelboavida@tecnico.ulisboa.pt (I.B.)
* Correspondence: mariajcosta@tecnico.ulisboa.pt

Received: 17 October 2019; Accepted: 3 December 2019; Published: 7 December 2019

**Abstract:** Due to peak electricity demand, hydropeaking introduces rapid and artificial flow fluctuations in the receiving river, which alters the river hydromorphology, while affecting the downstream ecological integrity. The impacts of hydropeaking have been addressed in flumes and in rivers. However, few studies propose mitigation solutions based on fish responses. The objective of this communication was to assemble the methods and outputs of flume research focused on Iberian cyprinids and to present recommendations to be used by freshwater scientists and hydropower producers. Emphasis was given to the critical role of integrating ecology and hydraulics to find the causal pathway between a flow change and a measurable fish response. The use of diverse behaviour quantification methods, flow sensing technologies, and statistical tools were decisive to strengthen the validity of the findings and to identify fish-fluid relationships, according to flow events. This communication encourages further research to identify flow thresholds for key life-cycle stages and complementary river studies to design and assess mitigation solutions for hydropeaking. Although the research focused on an Iberian cyprinid, the methods suggested have the potential to be extended to other fish species affected by hydropeaking.

**Keywords:** hydropeaking; Iberian cyprinids; flow refuges; hydropower; flow variability; fish physiology; fish behaviour; ecohydraulics

---

## 1. Introduction

In response to daily peaks of electricity demand, the hydropower plants (HPP) control the flow through the turbines very rapidly, which results in downstream hydropeaking [1]. These rapid flow fluctuations result from the distinct stages of hydropeaking: base-flow discharge (no electricity production), increasing discharge or up-ramping (powering-on of the turbines), continuous high discharge (peak electricity demand), and decreasing discharge or down-ramping (powering-off of the turbines) [2,3]. This continuous flow variability alters the downstream river morphological and hydrological processes [4–9], with consequences to the ecological integrity of the river system [10]. Notwithstanding, new hydropower plants are planned or under construction [11,12]. In addition, the estimates of global hydropower potential are convenient for producers [13], therefore, hydropower production continues to expand [14]. Given these and the predicted impacts of climate change, it is expected that the artificial flow fluctuations will be particularly severe for Iberian rivers affected by a Mediterranean climate [15], especially in the summer when water is scarce and the flow downstream is low, even if an ecological flow is provided by the HPP.

Fish responses to hydropeaking range from organism-level adjustments (neuroendocrine and metabolic) [16,17], to changes in diel activities (e.g., foraging, finding a suitable habitat, or avoiding predators), and alterations in key life-cycle events (e.g., reproductive migration, recruitment, and survival) [18]. This range of responses is mostly related to the spatiotemporal scale of the effects. The study of population or community dynamics is to larger spatiotemporal scale effects, whereas organism-level responses are to finer spatial scales and short-term effects. Different experimental approaches have been adopted to address those biological responses, namely studies conducted in riverine conditions [19], in artificial flumes [20–22], and numerical modelling studies [23,24]. In rivers, changes in fish assemblages along the longitudinal gradient [25,26], in diel activity [27], and in life-history traits (growth and longevity) [28] have been associated with hydropeaking. In addition, smaller [19,29–32] to wider spatial-scale movements [33–37] have been reported. This high variability has been mainly attributed to the difficulty in isolating external factors, to the presence and use of natural velocity refuges, or to inter-individual and intra-individual variability [19,30]. It is clearly challenging to find a causal pathway between flow variability and a measurable fish response [38], explained by the overlapping effects of other physical and biological factors that are difficult to isolate in rivers [19,31].

Inter-individual and intra-individual changes associated with flow fluctuations are mostly addressed in flume conditions because it is possible to control the flow events and to isolate potential confounding factors. Physiological responses, such as elevations in corticosteroid levels and changes in glucose, have been associated with simulated rapid flow fluctuations in experimental flumes [16]. A high swimming effort [20], no changes in social interactions or growth rates associated with down-ramping [39], and diverse preference patterns according to substrate and refuges [16,22,25,40,41] have been reported in experimental flumes. Although some of these findings clearly illustrate that rapid flow fluctuations affected fish performance, others are inconsistent and even contradictory. It is still challenging to extrapolate the findings regarding specific changes in a flow component (e.g., magnitude or frequency) to population or community-level responses because these flow changes occur at timescales much shorter than population responses [31]. This is justified by the difficulty to isolate each flow component in river conditions. Although the impacts of hydropeaking in population and community dynamics are recognized, it is still complex to attribute a higher effect to magnitude or frequency. This aspect is crucial to implement operational or morphological mitigation strategies. Either in rivers or in flumes, it is possible to characterize the hydraulic conditions by using diverse flow sensing technologies. The measurements are then incorporated in hydraulic numerical models to either characterize hydropeaking and propose habitat enhancement measures [23,42,43], or to propose adaptations in the operational schemes [44,45] to predict potential scenarios, and to develop conceptual frameworks to serve as grounds for hydropeaking studies [2,8,18].

Even recognizing the research effort that has been directed to address the hydropeaking problematic, few studies have proposed mitigation strategies. Specifically referring to habitat enhancement solutions, lateral refuges (deflectors) were effectively used for flow-refuging by Iberian barbels in simulated hydropeaking conditions [21]. T-shaped structures [46] and lateral refuges were suggested as potential flow-refuges for brown trout [22] and young grayling [47]. However, the proposed artificial refuges were not tested in natural conditions. Most of the projected structures were designed to offer flow refuge for salmonids based on changes in the swimming activity [22,46], or the interaction with conspecifics in flume conditions [48,49]. However, considering flow as the abiotic factor that determines fish movement, it is expected that it has considerably different effects on the movement behaviour of fish species, especially considering the morphological adaptations and the diverse swimming modes resulting from millions of years of evolution [50]. For this research, we considered the existing knowledge about habitat enhancement strategies to mitigate the effects of rapid flow fluctuations for salmonids and to bridge the knowledge gap for cyprinids.

Prior to the work that resulted in this communication, the effects of hydropeaking on cyprinids had been scarcely addressed [26,28]. Moreover, the application of habitat mitigation measures for cyprinids was inexistent. Remarkably, this group is the most abundant and dispersed in rivers worldwide [51]. Grounded on these knowledge gaps and on the previous findings for salmonid species, the effects of hydropeaking events were assessed for an Iberian cyprinid and the potential of artificial flow-refuges were evaluated at an indoor flume. The main objective of this communication is to assemble the major findings of those previous experiments by presenting the adopted multidisciplinary approach and a set of practical guidelines for the design of flow-refuges to freshwater scientists and hydropower producers. Particular emphasis is given to the critical role of integrating ecology and hydraulics to find the causal pathway between a specific flow change and a measurable fish response. Lastly, the limitations of this research are enumerated, and future research opportunities are discussed.

## 2. Research Approach

### 2.1. Problem Identification

The first step to implement flow-refuges as a mitigation measure for hydropeaking is the clear identification of the changes in magnitude, duration, and frequency of a flow event likely to occur in rivers affected by hydropower production. Thus, the simulated flow events were based on the operational scheme of Portuguese hydropower plants operating in hydropeaking regime [45,52]. Recognizing the impracticability of dam removal, the potential constraints inherent to the hydropower facility, and the difficulty of convincing hydropower producers to adapt the operational scheme, the research focused on the design of potential habitat enhancement solutions, given as flow-refuges. The methods adopted incorporated a multidisciplinary approach conducted at an indoor flume where physiological and behavioural responses were measured. The resultant hydraulic conditions were characterized (Figure 1).

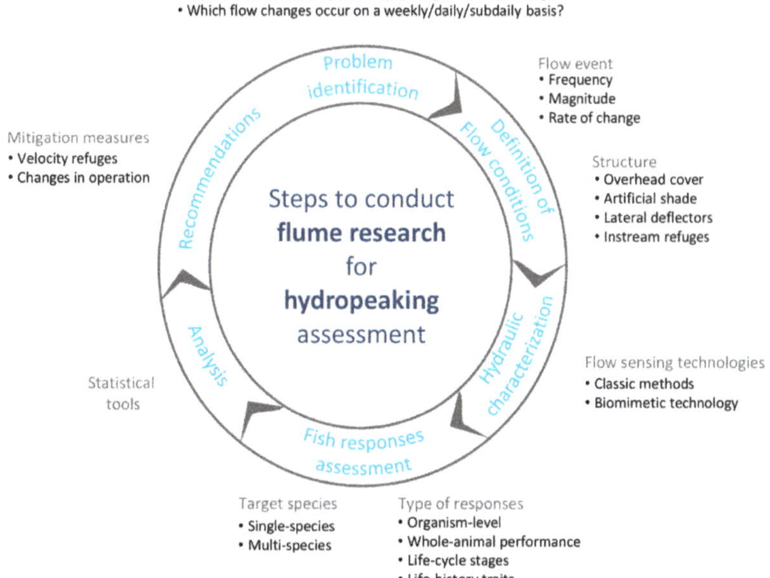


**Figure 1.** Circular flow chart displaying a potential sequence of steps to study the impacts of hydropeaking in flume conditions and to propose mitigation measures accordingly. HPP-Hydropower Plant.

With this approach, it was possible to relate physiological adjustments, changes in motion, and structure use, with the hydraulic conditions created by flow fluctuations and the presence of structures. Although it was adopted to assess the hydropeaking problematic for an Iberian cyprinid, this research approach has the potential to be adapted to other flume-based research and fish species.

### 2.2. Experimental Design

The experiments were conducted at an indoor flume with a rectangular cross-section, constructed on a steel frame with glass panels on both sides (Figure 2). The flume length was shortened to 6.5 m using two perforated metallic panels. The discharge and the water level were controlled by a sluice gate upstream and by a flap gate downstream and the maximum discharge was set to 60 L·s$^{-1}$. This flume is equipped with a false bottom, which means it is highly versatile when testing a wide range of structures.

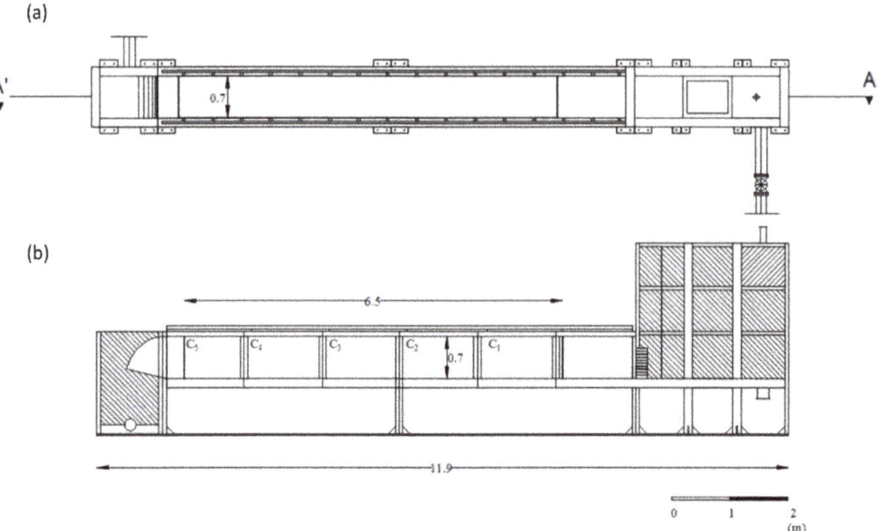

**Figure 2.** Top (**a**) and lateral (**b**) view of the indoor experimental flume, respective dimensions (m). C1-C5—Behaviour observation areas.

A diverse set of flow events was tested, where magnitude, peak duration, peak frequency, and event duration were changed (Figure 3). For all experiments, the base-flow conditions consisted of a continuous 7 L·s$^{-1}$ flow event (Figure 3, blue line). The peak discharges tested were 20, 40, and 60 L·s$^{-1}$ and were adapted during the progress of the experiments based on the findings of each setup (Figure 3).

The peak events were initially studied in the presence of lateral deflectors (Figure 3, Setups 1 and 2) [21,53]. For these structures, two configurations were tested-meandered, and single-sided-in two experimental setups termed setup 1 and 2, respectively (Figure 3). The reasoning behind the selection of lateral structures was: (i) their broad use in restoration actions [54], (ii) velocity preferences of the selected cyprinid species [55,56], (iii) velocity refuges proposed for other species [22], (iv) habitat modelling studies based on suitability curves [23], and (v) in studies aiming at designing effective fishways for this species [57].

Instream structures, namely v-shaped structures and solid triangular pyramids, were tested in the third experimental setup (Figure 3, Setup 3) [58]. The instream structures were designed according to: (i) the results from the previous experiments with lateral deflectors, (ii) fishway experiments with this species [59], and (iii) the potential effect of these structures on the hydrodynamic conditions (e.g., formation of vortices and shaded areas) and, consequently, on specific fish motion patterns.

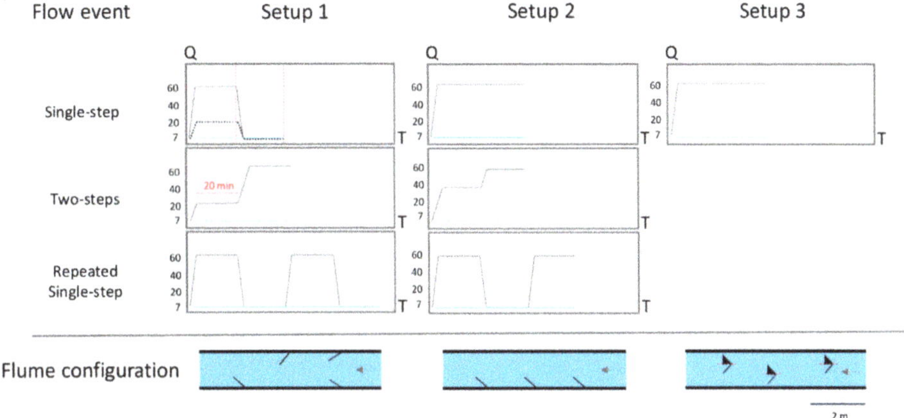

**Figure 3.** Flow events and flume configurations for each experimental setup. The tested structures were: for Setup 1, lateral meandered deflectors, for Setup 2, lateral single-sided deflectors, and, for Setup 3, v-shaped structures and solid triangular pyramids, respectively. Q—Discharge (L·s$^{-1}$); T—Time (minutes).

## 2.3. Hydraulic Characterization

To characterize the hydraulic conditions and the hydrodynamic environment, two measuring technologies were adopted: (i) velocity measurements using acoustic Doppler velocimetry (ADV), followed by a calibration to a theoretical 3D numerical model, and (ii) a novel technology using an artificial lateral line probe (LLP), which measures the pressure gradients over the probe body, which enables a characterization of fluid-body interactions [58]. For both, we selected a measurement grid that covered the areas that would be most affected by the structures and by the peak event. Acoustic Doppler Velocimetry (ADV) technology (Nortek-AS Vectrino 10 MHz, 100 Hz sampling rate) was used to characterize the velocity field in the flume when equipped with meandered deflectors for 7, 20, and 60 L·s$^{-1}$. The ADV output was used to calibrate a numerical model, extending its measurements to the single-sided deflector configuration, and to predict the flow field, according to the imposed changes [21]. One disadvantage of ADV technology is that it only performs point flow measurements, which neglects the fluid-body interactions that fish use to process flow. This and other conventional techniques often measure at a sampling rate that is below (1–50 Hz) the sensory range of lateral lines (10s–100s Hz) [60]. Artificial lateral lines have emerged to provide a closer representation of the hydrodynamic changes that are actually perceived by fish [61,62]. The artificial lateral line used for this research consisted of a 0.22 m length standard NACA 0025 airfoil shape, which is most similar to a cross section of a fish. This sensor measures pressure gradients over its body, mimicking the natural analogue, by using six differential pressure sensors, two in the front (snout) and two at each side of the probe body [58]. The pressure measurements were performed for two depths, using a 200 Hz sampling rate. The selected flow metrics derived from the LLP (i.e., mean pressure, mean front fluctuations, and mean pressure asymmetry) result from previous measurements conducted in an experimental vertical slot fishway where those and other pressure variables were compared with fish behaviour [63]. Those pressure metrics are quadratically related with flow velocity, turbulence, and cyclic flow patterns [63].

## 2.4. Assessment of Fish Responses

The selected cyprinid species was the Iberian barbel, *Luciobarbus bocagei* Steindachner, 1864 (hereafter, *L. bocagei*). This potamodromous cyprinid is endemic to the Iberian Peninsula, and it has specific habitat requirements and velocity preferences during its ontogeny [51]. Thus, it is possible to study the effects of hydropeaking for different life-cycle stages of this species. Furthermore, the swimming mode of *L. bocagei* differs from the widely studied salmonids. Although often identified as slower swimmers, by using the posterior half of the body for propulsion, these fish can be more efficient in turning and accelerating abilities [50].

Two types of fish responses were quantified: organism-level and whole-animal performance. With this approach, it was possible to capture immediate homeostatic adjustments and behavioural patterns, according to the imposed flow changes. To find whether they presented a stressor for *L. bocagei*, the selected organism-level responses were blood glucose and lactate. The reasons to select glucose were: (i) it generally increases after the cortisol response through the cortisol-mediated gluconeogenesis [64,65], which means there is a time latency between cortisol and the glucose response, and (ii) it has been broadly used in flow variability studies with reported increments comparatively to pre-stress levels [19,66]. Following the scientific recommendation that more than one response should be quantified [16], the other physiological response selected was lactate. Lactate is a metabolite of muscle activity, and is expected to increase during continual swimming [67]. Glucose and lactate level adjustments have been widely used as secondary physiological indicators of stress to flow variability [38]. Hence, both physiological responses may represent reliable surrogates of a stress response to flow variability.

Whole animal performance was analysed by quantifying changes in motion related to the structures tested and the occurrence of the peak event. The behavioural metrics were divided into two categories: (1) structure use and (2) swimming activity in the flume. Successful structural use was considered when a single fish or a group of two to five fish were observed in the immediate downstream area of the structure. The swimming activity metrics were separated into two types: fish sprints, defined as a swimming activity lasting a few seconds and characterized by several tail beats, and fish drifts, defined as downstream fish displacements driven by passive advection of the body in the flow direction. The frequency of behaviour was defined as the number of occurrences, in absolute frequency, over the duration of the flow event.

The biological results were analysed and interpreted according to the characterization of the hydraulic and hydrodynamic conditions. Thus, it was possible to relate inter-individual and intra-individual responses with the hydraulic conditions created in the flume and to find the most advantageous hydraulic conditions given by the flow event and the flow-refuge for *L. bocagei*.

## 3. Discussion of the Major Findings

### 3.1. Is There an Effect of the Simulated Peak Events?

After initially testing two peak discharges, i.e., 20 and 60 L·s$^{-1}$, under the highest discharge in the presence of meandered deflectors, (i) the onset of a glucose response occurred sooner, (ii) the use of deflectors as flow refuges was inhibited, and (iii) the sprinting and drifting activity was higher [21]. Additionally, event duration and peak frequency dictated a sustained glucose response in the 60 L·s$^{-1}$ flow events, and the unpredictability of the two-steps event (i.e., 20 L·s$^{-1}$ followed by 60 L·s$^{-1}$) resulted in elevated glucose levels [21]. From these findings, it would be tempting to suggest that there was a clear adverse effect of the hydropeaking events. However, signs of exhaustion were undetected and corroborated by the absence of a lactate response. Fish were able to cope with water velocities > 60 cm·s$^{-1}$ by hiding in the downstream area of the deflectors [21]. Still, the ability of *L. bocagei* to cope with the hydraulic conditions created under 60 L·s$^{-1}$ was lower and the swimming activity decreased especially after peak repetition.

The unpredictability of the events when more than one of the flow components (e.g., magnitude and frequency) was altered resulted in the most visible responses. Longer peak flows, with higher magnitudes and repeated peaks, resulted in glucose increments, increased swimming effort, and difficulty to find the available refuges [21]. Similar findings occurred in the presence of the single-sided deflectors [53]. This cumulative effect is recognized in the literature [18,68]. On the other hand, when a hydropeaking event was tested in the absence of these structures, the lowest physiological responses were obtained [53]. This was explained by the highly predictable and homogenous flow conditions.

Regarding the instream structures tested in the third experimental setup, the effects of the same flow event differed significantly between the tested configurations [58]. After testing the instream structures, it was clear that the combined effect of the flow event and structure created local hydrodynamic changes that dictated the range of responses shown by *L. bocagei*, and not solely by the peak event itself [58].

### 3.2. Which Flow Event Structure Is the Most Ecologically Conclusive?

The most ecologically relevant findings regarding the hydropeaking problematic for this species are not solely those that identify the 60 L·s$^{-1}$ flow events as the most adverse, but mostly the exceptions to this trend. In one hand, the presence of deflectors created a more heterogeneous flow environment, yet *L. bocagei* were still able to use them for flow-refuging. Nonetheless, the scarcer search for deflectors under 60 L·s$^{-1}$ in comparison with the 20 L·s$^{-1}$ flow events and the increased overall swimming activity suggest that the critical hydraulic conditions created in the vicinity of the deflectors reduced *L. bocagei* ability to find the low flow areas downstream of the deflectors [21]. On the other hand, the combination of a flow event with the presence of instream structures altered the distribution of velocity, turbulent fields, and pressure fluctuations, which generates distinct behavioural responses [58].

The movement behaviour of *L. bocagei* was not solely related to the unpredictable flow conditions resulting from the peak discharge (i.e., 60 L·s$^{-1}$). Specifically referring to the instream structures, under the single-step flow event, the individual and group use were higher for the solid triangular pyramids [58]. The lower velocities expected inside the v-shaped structures, which would favour *L. bocagei* for flow-refuging, were masked by the complex flow conditions created by these structures. Even if the swimming activity was more pronounced under 60 L·s$^{-1}$, the physiological responses were not always indicative of stress or exhaustion.

From all experiments, it was possible to demonstrate that the most conclusive responses occurred in the events where more than one flow component (given as magnitude, frequency, or duration) changed in the presence of the structures that created the most complex flow conditions. These were: (i) the single-step and two-steps events for 60 L·s$^{-1}$ for meandered deflectors [21], (ii) the two-steps and the repeated single-step events for one-sided deflectors [53], and (iii) the single-step event for v-shaped structures [58] (see Figure 3).

### 3.3. Which Hydraulic Conditions Are the Most Beneficial?

From the analysis of the three experimental setups, the most favourable hydraulic conditions were those where flow magnitude, peak frequency, and peak duration were lower. These events were the base-flow events in the presence of deflectors and v-shaped structures [53,58], and the single-step events where the lowest peak discharge, i.e., 20 L·s$^{-1}$, was tested [21]. For these events, there were no physiological adjustments, the swimming effort was low demonstrated by the lower frequency of individual sprints and the balanced frequency of individual and group behaviour), and fish could easily access the refuges.

The movement behaviour results indicated that *L. bocagei* were using the deflectors more frequently than the instream structures, and that, under moderate peaks, their use was enhanced [21,53,58]. However, the physiological responses were not so clear. Glucose increments occurred in the presence of both deflector configurations [21,53], whereas lactate increments occurred only in the presence of v-shaped structures [58]. With these findings, a question emerged: does the higher use of deflectors imply that these structures are the most adequate for *L. bocagei* under hydropeaking conditions? By combining ecological and hydraulic perspectives, it was possible to assert that the hydraulic conditions created by the presence of deflectors motivated their use. It was clear that *L. bocagei* were able to use the meandered and the single-sided deflectors as flow refuges more effectively than the two triangular-shaped structures. Under these conditions, the swimming activity (frequency of sprints and drifts) was enhanced.

Analysing the numerical models for deflectors [21,53], and the pressure distribution maps of the Lateral Line Probe (LLP) for the instream structures [58], it was evident that, rather than just velocity, it was the combination of local scale hydrodynamic features, given by mean front pressure fluctuations and mean pressure asymmetry that determined the movement patterns. These explained the difficulty for *L. bocagei* to use the triangular structures as velocity refuges, but also their ability to use those changes in hydrodynamic features to find other areas for flow-refuging [69], move upstream (favouring rheotactic behaviour) [70,71], or to benefit from group behaviour [72].

*3.4. Flow Sensing Technologies: Acoustic Doppler Velocimetry and Lateral Line Probe*

From the Acoustic Doppler Velocimetry (ADV) measurements and the hydraulic models, it was possible to define a flow threshold that represented the resting state for *L. bocagei*. For lateral structures (i.e., deflectors), velocity magnitudes ranging from near 0 to 0.41 m·s$^{-1}$ allowed *L. bocagei* to use the available flume area. This range enhanced *L. bocagei* swimming performance, without any physiological response [21,58]. The hydraulic characterization given by the FLOW-3D$^{®}$ models was relevant to explain deflector use and swimming activity patterns. However, the information that the velocity ranges provided was not sufficient to explain the diversity of *L. bocagei* organism level responses.

The results from the Lateral Line Probe (LLP) provided unique findings on the role of the hydrodynamic conditions as primary triggers of movement patterns by *L. bocagei* [58]. After analysing the fish responses together with the derived pressure variables, it was possible to define pressure thresholds referring to the mean front pressure, mean pressure fluctuations, and mean pressure asymmetry that provided the most and the least favourable hydrodynamic conditions for *L. bocagei* [58]. Asymmetry was the pressure variable that was the most related with *L. bocagei* responses, with a favourable asymmetry window being observed for the base-flow event with v-shaped structures [58]. On the other hand, the highest-pressure asymmetry and the high mean front pressure and pressure fluctuations measured in the single-peak event for the same structures hindered the fish's ability to cope with the hydrodynamic conditions and resulted in lactate adjustments [58]. These pressure metrics were identified for the first time as potential surrogates of local hydrodynamic stimuli, according to the observed fish responses. Rather than just velocity, it is the combination of local hydrodynamic changes (i.e., mean front pressure, mean fluctuations, and asymmetry) that lead fish to select the flow conditions that ensure the lowest energetic costs. These regions are not only represented by the low flow areas inside the refuge, but also represented by regions that favour the formation of vortices and turbulent areas that fish can take advantage of [58]. Considering the range of pressure readings by the lateral line system [72], the LLP represents a potential tool to assess the distributed sensing capacity of fish. Thus, in comparison with the ADV, the LLP provided a better representation of how specific hydrodynamic changes affect the movement behaviour of fish. That study was the first to use this technology to understand smaller-scale movement behavioural patterns of fish associated with flow variability. The results are promising considering the role that local-scale hydrodynamic changes have on the swimming performance of fish, and represent a step forward to understand the ecological significance of the pressure field around the fish and its movement patterns.

## 4. Recommendations for Hydropower Producers

The research effort to find mitigation solutions to hydropeaking consequences is expanding. However, it is urgent to bridge the gap between this scientific knowledge and hydropower producers. With this flume-based research, it was possible to propose potential design guidelines for flow-refuges and recommendations to adapt the operational schemes in a hydropower plant. Specifically referring to the physiological findings, with this research, there was no unequivocal evidence that supported the existence of a flow threshold representing the resting state for *L. bocagei*. However, the lowest physiological responses were observed when *L. bocagei* was subjected to 7 and 20 L·s$^{-1}$ in the presence of lateral deflectors, and in the absence of instream structures. For lateral deflectors, this absence of a physiological response, corresponding to a velocity threshold of 0–0.41 m·s$^{-1}$, which is in accordance with velocity preferences for this species [73–75], represents a further step regarding the definition of the resting state with respect to flow variability. Nonetheless, contrary to what would be expected, the physiological responses were not consistently lower under base flow conditions (continuous 7 L·s$^{-1}$), but fluctuated, according to the flow event and structural configuration. With the exception of the hydropeaking event with v-shaped structures, the levels of blood lactate did not differ significantly between events, and the movement patterns of *L. bocagei* were not indicative that fish were exhausted [21,53]. These results emphasize the importance of habitat heterogeneity as a decisive factor for the definition of the most adequate conditions for this species. The concomitant analysis of physiological and movement behavioural responses was essential to find the most suitable flow event-structure configuration for *L. bocagei*. For any study aiming to assess the potential of mitigation measures to hydropeaking based on quantified fish responses, we recommend selecting more than one level of an ecological organization (e.g., individual and population), and then to quantify different levels of organism-level responses (e.g., metabolic changes and whole-animal performance metrics likely to be associated with rapid flow fluctuations). Considering the absence of a universal threshold for a physiological response that represents the resting state for flow variability, the validity of a physiological result is significant only if compared with a base-flow or control condition. Lastly, in riverine conditions, there is a myriad of environmental variables that are difficult to isolate. Therefore, it is critical to isolate confounding external factors that may mask the effect of the flow changes. This can be partially outweighed by conducting experiments in outdoor and indoor flumes where it is possible to control flow variability and isolate external variables, and to replicate the conditions.

The metrics to describe movement behaviour were selected for their likelihood to be expressed under the simulated peak events. Remarkably, the fish responses were not always indicative that the proposed lateral or instream structures would be effective for flow-refuging. In rivers affected by hydropeaking, it is substantially more challenging to relate specific motion patterns and organism-level responses with the hydrodynamic changes. To implement lateral refuges in natural conditions, the most decisive factors for their design are the hydraulic and hydrodynamic changes that result from hydropeaking, which, in turn, determine the choice of the type, number, and positioning of structures. First, it is necessary to characterize the rapid flow fluctuations and the changes in water depth, velocity, and wetted profile before the design and implementation of a lateral or instream refuge [6]. Afterward, the proposed structures can be tested through numerical modelling to understand whether the added habitat heterogeneity provides velocity refuging areas, or creates unstable hydraulic conditions for fish [43,76]. According to our findings, both classical velocity measurements and novel biomimetic technologies provided sound results regarding the characterization of the flow patterns. Feasibility, cost, and post-processing are aspects to consider regarding the choice of a flow sensing technology. Yet, to assess the potential of habitat enhancement measures, it is more important to consider the fluid changes and their impact on fish organism responses and the motion patterns of fish. Thus, ADV technology provides a solid representation of the flow field, whereas the LLP illustrates fluid-body interactions, which provides a better representation of the pressure field that fish would perceive.

Specifically referring to recommendations to implement the artificial flow-refuges in natural conditions, the mitigation structures to be proposed should assure velocity refuges during up-ramping and lateral connectivity with the main channel during the down-ramping stage. To implement lateral refuges (deflectors), their length, angle, and height are crucial design factors [77,78]. Thus, the opening angle should be at least in the same order of magnitude as the fish body length [79], favour group behaviour, guide the flow, prevent clogging [22], and create or maintain the flow dynamics of the pool-riffle (sediment deposition-transport) systems while avoiding bank erosion [78]. The opening angle, excessive wide angles, in relation to the river bank, are discouraged. To avoid fish stranding, especially during down-ramping, the area behind the deflectors should avoid potential stranding zones, or assure a minimum water depth of 0.5 m [22,76]. Although a heterogeneous habitat is recommended, adding refuge areas is not always the most effective solution [43,79]. There must be a trade-off between existing river habitat conditions and the effectiveness of adding new structures, while remembering that, in natural conditions, each case is unique and generalizations have to be made with caution.

One final recommendation refers to an unexpected finding regarding the fish responses to the two-step flow event (i.e., 20 followed by 60 L·s$^{-1}$). The high glucose levels, the inability to find the deflectors particularly during the transition between discharges, and the increased swimming activity in the second step, demonstrated that a stepped flow event is not favourable for *L. bocagei* [21]. The explanation for these results refers to the difficulty of *L. bocagei* to cope with the unpredictable flow conditions. This finding may be convenient for hydropower producers, since the economic benefit of a two-step event, would likely be lower than, for example, a continuous high peak single-step event. We recommend that the duration of the base-flow events between two high peaks is extended to enable fish to recover from the effort required to cope with the continuous high peak. Nevertheless, prior to the implementation of any operational scheme, it is crucial to quantify the impacts in river conditions.

## 5. Identify Limitations to Find Research Opportunities

In flumes, it is difficult to simulate rapid flow fluctuations of the same order of magnitude and rate of change as in some regulated rivers due to pumping capacity or to limitations inherent with the flow control [68]. Although it was possible to simulate rapid flow fluctuations, the maximum discharge of our flume was limited to 60 L·s$^{-1}$. This construction limitation was partially overcome by reducing the flume cross-section with two parallel deflectors that were installed in the upstream flume area for the meandered deflector's setup. With this solution, it was possible to get higher velocity ranges during the hydropeaking events [21]. The deflectors were clearly used for flow-refuging, but, at the same time, the more complex flow environment reduced the success for *L. bocagei* to use these areas in comparison with the 20 L·s$^{-1}$ flow events [21]. Thus, the effect of the rapid flow fluctuations was still evident. To achieve the most comprehensive results, the construction limitations inherent in the facilities must be recognized and alternatives to surpass potential limitations have to be considered.

Experimental flume research has been suggested as a valuable tool to understand the effects of rapid flow fluctuations in downstream fish communities [18,68]. A wide diversity of studies conducted in laboratory conditions discovered the behavioural diversity that is found in nature [80]. Although experimental flume research is reliable to study behavioural patterns, extrapolations to natural conditions are discouraged and its transferability should be further substantiated. Nonetheless, the behaviour patterns observed (e.g., use of available refuges, longitudinal movements, take advantage of flow changes to minimize energetic costs) have been observed in studies conducted in rivers [17,19], in nature-like channels [20], and in artificial streams [66]. Ideally, complementary river studies should be conducted to assess the ecological consequences of those patterns, and, particularly, in the context of hydropeaking.

One of the biggest problems of ecological studies conducted in controlled conditions is the difficulty to replicate the experiments and their findings. These studies are usually constrained by the number of organisms that is possible to obtain from the natural conditions, which reduces the robustness of the findings. To outweigh this limitation, the recommendations are to use a diverse set of behavioural quantification tools (e.g., physiology and movement behaviour) complemented with a detailed characterization of the hydraulic conditions, and the use of statistical tools that are adequate for small-sized samples. With the analysis of distinct levels of fish responses, it is possible to identify trends and to establish relationships, according to flow variability.

Lastly, although the present recommendations are mostly directed for young adults of *L. bocagei*, they can be extended to other Iberian cyprinid species. Nonetheless, single-species studies are difficult to extrapolate to communities, or the river ecosystem. Although the single-species limitation was recognized, this work represents a step forward to understand and mitigate the impacts of hydropeaking for cyprinids.

## 6. Future Hydropeaking Research: Recommendations for Freshwater Scientists

The findings that motivated this communication present the initial steps to propose effective mitigation measures for hydropeaking consequences for cyprinids inhabiting rivers affected by the Mediterranean climate. Further research opportunities are described afterward.

Long duration and repeated flow peaks produced physiological adjustments as well as a reduction of the swimming activity during the second peak with the same magnitude. Thus, we hypothesized whether the swimming effort was higher in the second peak, which resulted in a reduction of sprinting, drifting, and in an inability to find the velocity refuges, or if *L. bocagei* were getting adapted to the peak. To address these questions, the simulation of more than one peak repetition is recommended.

There was not enough evidence supporting that a specific flow change resulted in an actual stress response. Thus, the identification of a flow threshold that represented the resting state for *L. bocagei* was not straightforward and deserves further investigation. Fish have evolved adaptive mechanisms to cope with natural extremes. However, under hydropeaking conditions, the flow changes are significantly more severe and frequent than in a free-flowing river. The energy-cost that is associated with those rapid and artificial changes is likely proportional. Although fish are able to adapt to novel conditions, the associated energy expenditure has considerable drawbacks. It will certainly reduce the energy available for diel activities or to successfully complete life-cycle stages. If a biomarker for flow variability is established, it will be possible to infer whether the flow fluctuations in natural conditions represent a potential stressor, or, otherwise, initiate behavioural strategies to compensate for the adverse flow fluctuations and maintain homeostasis.

Given the particular role of event predictability for *L. bocagei*, the study of alternative up-ramping and down-ramping rates is strongly advised. It is predicted that the slower onset of the up-ramping stage will enable fish to gradually adapt not only to the severity of a peak flow, but to the water level change. In the same way, slower down-ramping rates have the potential to reduce the stranding probability since the gradual flow decrease may function as a cue for fish to escape from potential isolated pools. A particular focus should be addressed to the critical down-ramping stage since this is usually associated with reduced chances of survival and high mortality rates.

The water velocity preferences of *L. bocagei* vary during its ontogeny. Given that most literature focuses on salmonids, the design of flow refuges according to velocity and habitat preferences for *L. bocagei*, and other cyprinid species, will likely differ. Thus, more studies focusing on other fish species (with distinct swimming modes) and habitat guilds are encouraged. In addition, it is necessary to further the investigation to specific life-cycle stages, namely larvae and juvenile survival, foraging, reproductive migrations, or spawning. Specific aspects of the life-cycle of salmonids have been recently proposed to be included during the development of mitigation solutions for salmonids affected by hydropeaking [81]. However, such guidelines are inexistent for cyprinids. In addition, flow regulations specifically addressed to hydropower plants operating in hydropeaking exist for a few regions in European countries, however, they focus exclusively on salmonid species [82]. Thus, it is critical to study the flow requirements for other fish families, cyprinids in particular, and for critical bottlenecks to ensure the sustainability of fish communities affected by hydropeaking, and to increase the awareness of policymakers to include flow recommendations in national legal instruments.

Local hydrodynamic changes were related to specific movement behaviour and physiological responses [58]. These findings represent the first step of a novel perspective to study behavioural patterns related to rapid flow fluctuations. Thus, it is still premature to generalize. To strengthen the findings regarding fluid-body interactions under rapid flow fluctuations and habitat heterogeneity, it is necessary to establish a quantitative relationship. With it, it would be possible to predict the magnitude of specific movement patterns, relate them with local hydrodynamic changes, and interpret them in terms of swimming effort.

Although it was possible to define guidelines for the design of deflectors and to propose alternative operational measures, it was not possible to upscale them in the river system. Thus, it should be reinforced that, prior to the implementation of any mitigation measure based on flume research, it is necessary to perform experimental tests in riverine conditions. In addition, it is strongly recommended that future research combines flume experiments with research conducted in rivers.

## 7. Conclusions

The motivation to write this communication emerged from the need to assemble the methods and the main findings obtained from flume research, with the purpose that this knowledge could be used by freshwater scientists and hydropower producers. Although the scientific findings were obtained for an Iberian cyprinid inhabiting river affected by the Mediterranean climate, the methods that were adopted have the potential to be extended to other river systems affected by hydropeaking, fish species, and particular bottlenecks. Despite the limitations inherent in indoor flumes, the possibility to control the flow changes and to quantify fish responses provided unique clues regarding fluid-body interactions, given by pressure thresholds, which are difficult to observe in natural conditions. The refined scale of the research not only identified the potential negative aspects of hydropeaking, but also the opportunities that can emerge from them, and the ability of fish to adapt and compensate under adverse flow conditions. The use of diverse behavioural quantification methods (e.g., different levels of fish responses), flow sensing technologies and statistical tools were decisive to strengthen the validity of the findings and to identify fish-fluid relationships, according to flow variability. Emphasis is given to habitat heterogeneity as a decisive factor to propose solutions to mitigate hydropeaking. This communication encourages further research to identify flow thresholds for key life-cycle stages and complementary river studies to design and assess mitigation solutions to hydropeaking.

**Author Contributions:** Conceptualization, M.J.C.; Methodology, M.J.C., A.N.P., I.B.; Validation, M.J.C.; Formal Analysis, M.J.C.; Investigation, M.J.C., I.B.; Resources, M.J.C., A.N.P., I.B.; Data curation, M.J.C.; Writing–Original Draft, M.J.C.; Preparation, M.J.C.; Writing–Review & Editing, M.J.C., A.N.P., I.B.; Visualization, M.J.C.; Supervision, A.N.P., I.B.; Project Administration, A.N.P., I.B.; Funding Acquisition, A.N.P., I.B.

**Funding:** Maria João Costa received funding from the FLUVIO—River Restoration and Management Doctoral Programme from the Fundação para a Ciência e Tecnologia, https://www.fct.pt/, Portugal (grant No. SFRH/BD/52517/2014). This project has received funding from the European Union's Horizon 2020 research and innovation programme under grant agreement No. 727830, http://www.fithydro.eu/.

**Conflicts of Interest:** The authors declare no conflict of interest.

## References

1. Cushman, R.M. Review of ecological effects of rapidly varying flows downstream from hydroelectric facilities. *N. Am. J. Fish. Manag.* **1985**, *5*, 330–339. [CrossRef]
2. Bruder, A.; Tonolla, D.; Schweizer, S.P.; Vollenweider, S.; Langhans, S.D.; Wüest, A. A conceptual framework for hydropeaking mitigation. *Sci. Total Environ.* **2016**, *568*, 1204–1212. [CrossRef] [PubMed]
3. Tonolla, D.; Bruder, A.; Schweizer, S. Evaluation of mitigation measures to reduce hydropeaking impacts on river ecosystems—A case study from the Swiss Alps. *Sci. Total Environ.* **2017**, *574*, 594–604. [CrossRef] [PubMed]
4. Bejarano, M.D.; Jansson, R.; Nilsson, C. The effects of hydropeaking on riverine plants: A review. *Biol. Rev.* **2018**, *93*, 658–673. [CrossRef] [PubMed]
5. Greimel, F.; Schülting, L.; Graf, W.; Bondar-Kunze, E.; Auer, S.; Zeiringer, B.; Hauer, C. Hydropeaking Impacts and Mitigation. In *Riverine Ecosystem Management*; Springer: Cham, Switzerland, 2018; pp. 91–110, ISBN 978-3-319-73250-3.
6. Schmutz, S.; Bakken, T.H.; Friedrich, T.; Greimel, F.; Harby, A.; Jungwirth, M.; Melcher, A.; Unfer, G.; Zeiringer, B. Response of fish communities to hydrological and morphological alterations in hydropeaking rivers of Austria. *River Res. Appl.* **2015**, *31*, 919–930. [CrossRef]
7. Shen, Y.; Diplas, P. Modeling Unsteady Flow Characteristics of Hydropeaking Operations and Their Implications on Fish Habitat. *J. Hydraul. Eng.* **2010**, *136*, 1053–1066. [CrossRef]
8. Zimmerman, J.K.H.; Letcher, B.H.; Nislow, K.H.; Lutz, K.A.; Magilligan, F.J. Determining the effects of dams on subdaily variation in river flows at a whole-basin scale. *River Res. Appl.* **2010**, *26*, 1246–1620. [CrossRef]

9. Zolezzi, G.; Siviglia, A.; Toffolon, M.; Maiolini, B. Thermopeaking in Alpine streams: Event characterization and time scales. *Ecohydrology* **2011**, *4*, 564–576. [CrossRef]

10. Bunn, S.E.; Arthington, A.H. Basic principles and ecological consequences of altered flow regimes for aquatic biodiversity. *Environ. Manag.* **2002**, *30*, 492–507. [CrossRef]

11. Nilsson, C.; Reidy, C.A.; Dynesius, M.; Revenga, C. Fragmentation and flow regulation of the world's large river systems. *Science* **2005**, *308*, 405–408. [CrossRef]

12. Zarfl, C.; Lumsdon, A.E.; Berlekamp, J.; Tydecks, L.; Tockner, K. A global boom in hydropower dam construction. *Aquat. Sci.* **2015**, *77*, 161–170. [CrossRef]

13. Hoes, O.A.C.; Meijer, L.J.J.; Van Der Ent, R.J.; Van De Giesen, N.C. Systematic high-resolution assessment of global hydropower potential. *PLoS ONE* **2017**, *12*, e0171844. [CrossRef] [PubMed]

14. *Hydropower Status Report: Section Trends and Insights*; IHA: London, UK, 2019.

15. Teotónio, C.; Fortes, P.; Roebeling, P.; Rodriguez, M.; Robaina-Alves, M. Assessing the impacts of climate change on hydropower generation and the power sector in Portugal: A partial equilibrium approach. *Renew. Sustain. Energy Rev.* **2017**, *74*, 788–799. [CrossRef]

16. Flodmark, L.E.W.; Urke, H.A.; Halleraker, J.H.; Arnekleiv, J.V.; Vollestad, L.A.; Poléo, A.B.S. Cortisol and glucose responses in juvenile brown trout subjected to a fluctuating flow regime in an artificial stream. *J. Fish Biol.* **2002**, *60*, 238–248. [CrossRef]

17. Taylor, M.K.; Cook, K.V.; Hasler, C.T.; Schmidt, D.C.; Cooke, S.J. Behaviour and physiology of mountain whitefish (Prosopium williamsoni) relative to short-term changes in river flow. *Ecol. Freshw. Fish* **2012**, *21*, 609–616. [CrossRef]

18. Young, P.; Cech, J.; Thompson, L. Hydropower-related pulsed-flow impacts on stream fishes: A brief review, conceptual model, knowledge gaps, and research needs. *Rev. Fish Biol. Fish.* **2011**, *21*, 713–731. [CrossRef]

19. Krimmer, A.N.; Paul, A.J.; Hontela, A.; Rasmussen, J.B. Behavioural and physiological responses of brook trout Salvelinus fontinalis to midwinter flow reduction in a small ice-free mountain stream. *J. Fish Biol.* **2011**, *79*, 707–725. [CrossRef]

20. Auer, S.; Zeiringer, B.; Fuhrer, S.; Tonolla, D.; Schmutz, S. Effects of river bank heterogeneity and time of day on drift and stranding of juvenile European grayling (Thymallus thymallus L.) caused by hydropeaking. *Sci. Total Environ.* **2017**, *575*, 1515–1521. [CrossRef]

21. Costa, M.; Boavida, I.; Almeida, V.; Cooke, S.; Pinheiro, A. Do artificial velocity refuges mitigate the physiological and behavioural consequences of hydropeaking on a freshwater Iberian cyprinid? *Ecohydrology* **2018**, *11*, e1983. [CrossRef]

22. Ribi, J.-M.; Boillat, J.-L.; Peter, A.; Schleiss, A.J. Attractiveness of a lateral shelter in a channel as a refuge for juvenile brown trout during hydropeaking. *Aquat. Sci.* **2014**, *76*, 527–541. [CrossRef]

23. Boavida, I.; Santos, J.M.; Ferreira, M.T.; Pinheiro, A.N. Barbel habitat alterations due to hydropeaking. *J. Hydro-Environ. Res.* **2015**, *9*, 237–247. [CrossRef]

24. Person, E.; Bieri, M.; Peter, A.; Schleiss, A.J. Mitigation measures for fish habitat improvement in Alpine rivers affected by hydropower operations. *Ecohydrology* **2014**, *7*, 580–599. [CrossRef]

25. Vehanen, T.; Jurvelius, J.; Lahti, M. Habitat utilisation by fish community in a short-term regulated river reservoir. *Hydrobiologia* **2005**, *545*, 257–270. [CrossRef]

26. Alexandre, C.M.; Almeida, P.R.; Neves, T.; Mateus, C.S.; Costa, J.L.; Quintella, B.R. Effects of flow regulation on the movement patterns and habitat use of a potamodromous cyprinid species. *Ecohydrology* **2016**, *9*, 326–340. [CrossRef]

27. Vehanen, T.; Louhi, P.; Huusko, A.; Mäki-Petäys, A.; Meer, O.; Orell, P.; Huusko, R.; Jaukkuri, M.; Sutela, T. Behaviour of upstream migrating adult salmon (Salmo salar L.) in the tailrace channels of hydropeaking hydropower plants. *Fish. Manag. Ecol.* **2019**. [CrossRef]

28. Alexandre, C.M.; Ferreira, M.T.; Almeida, P.R. Life history of a cyprinid species in non-regulated and regulated rivers from permanent and temporary Mediterranean basins. *Ecohydrology* **2015**, *8*, 1137–1153. [CrossRef]

29. Scruton, D.A.; Ollerhead, L.M.N.; Clarke, K.D.; Pennell, C.; Alfredsen, K.; Harby, A.; Kelley, D. The behavioural response of juvenile Atlantic salmon (Salmo salar) and brook trout (Salvelinus fontinalis) to experimental hydropeaking on a Newfoundland (Canada) River. *River Res. Appl.* **2003**, *19*, 577–587. [CrossRef]

30. Taylor, M.K.; Hasler, C.T.; Findlay, C.S.; Lewis, B.; Schmidt, D.C.; Hinch, S.G.; Cooke, S.J. Hydrologic correlates of bull trout (Salvelinus confluentus) swimming activity in a hydropeaking river. *River Res. Appl.* **2013**, *30*, 756–765. [CrossRef]

31. Taylor, M.K.; Hasler, C.T.; Hinch, S.G.; Lewis, B.; Schmidt, D.C.; Cooke, S.J. Reach-scale movements of bull trout (Salvelinus confluentus) relative to hydropeaking operations in the Columbia River, Canada. *Ecohydrology* **2014**, *7*, 1079–1086. [CrossRef]

32. Thompson, L.C.; Cocherell, S.A.; Chun, S.N.; Cech, J.J.; Klimley, A.P. Longitudinal movement of fish in response to a single-day flow pulse. *Environ. Biol. Fishes* **2011**, *90*, 253–261. [CrossRef]

33. Boavida, I.; Harby, A.; Clarke, K.D.; Heggenes, J. Move or stay: Habitat use and movements by Atlantic salmon parr (Salmo salar) during induced rapid flow variations. *Hydrobiologia* **2017**, *785*, 261–275. [CrossRef]

34. Burnett, N.J.; Hinch, S.G.; Braun, D.C.; Casselman, M.T.; Middleton, C.T.; Wilson, S.M.; Cooke, S.J. Burst swimming in areas of high flow: Delayed consequences of anaerobiosis in wild adult sockeye salmon. *Physiol. Biochem. Zool.* **2014**, *87*, 587–598. [CrossRef]

35. Capra, H.; Plichard, L.; Bergé, J.; Pella, H.; Ovidio, M.; McNeil, E.; Lamouroux, N. Fish habitat selection in a large hydropeaking river: Strong individual and temporal variations revealed by telemetry. *Sci. Total Environ.* **2017**, *578*, 109–120. [CrossRef] [PubMed]

36. De Vocht, A.; Baras, E. Effect of hydropeaking on migrations and home range of adult Barbel (Barbus barbus) in the river Meuse. In *Aquatic Telemetry: Advances and Applications, Proceedings of the Fifth Conference on Fish Telemetry, Ustica, Italy, 9–13 June 2003*; FAO-COISPA: Rome, Italy, 2005; pp. 35–44.

37. Harvey-Lavoie, S.; Cooke, S.J.; Guénard, G.; Boisclair, D. Differences in movements of northern pike inhabiting rivers with contrasting flow regimes. *Ecohydrology* **2016**, *9*, 1687–1699. [CrossRef]

38. Costa, M.; Lennox, R.; Katopodis, C.; Cooke, S. Is there evidence for flow variability as an organism-level stressor in fluvial fish? *J. Ecohydraulics* **2017**, *2*, 68–83. [CrossRef]

39. Flodmark, L.E.W.; Forseth, T.; L'Abée-Lund, J.H.; Vøllestad, L.A. Behaviour and growth of juvenile brown trout exposed to fluctuating flow. *Ecol. Freshw. Fish* **2006**, *15*, 57–65. [CrossRef]

40. Chun, S.N.; Cocherell, S.A.; Cocherell, D.E.; Miranda, J.B.; Jones, G.J.; Graham, J.; Klimley, A.P.; Thompson, L.C.; Cech, J.J., Jr. Displacement, velocity preference, and substrate use of three native California stream fishes in simulated pulsed flows. *Environ. Biol. Fishes* **2011**, *90*, 43–52. [CrossRef]

41. Vilizzi, L.; Copp, G.H. An analysis of 0+ barbel (Barbus barbus) response to discharge fluctuations in a flume. *River Res. Appl.* **2005**, *21*, 421–438. [CrossRef]

42. Hauer, C.; Holzapfel, P.; Leitner, P.; Graf, W. Longitudinal assessment of hydropeaking impacts on various scales for an improved process understanding and the design of mitigation measures. *Sci. Total Environ.* **2017**, *575*, 1503–1514. [CrossRef]

43. Hauer, C.; Unfer, G.; Holzapfel, P.; Haimann, M.; Habersack, H. Impact of channel bar form and grain size variability on estimated stranding risk of juvenile brown trout during hydropeaking. *Earth Surf. Process. Landf.* **2014**, *39*, 1622–1641. [CrossRef]

44. Casas-Mulet, R.; Alfredsen, K.; Killingtveit, A. Modelling of environmental flow options for optimal Atlantic salmon, Salmo salar, embryo survival during hydropeaking. *Fish. Manag. Ecol.* **2014**, *21*, 480–490. [CrossRef]

45. Pragana, I.; Boavida, I.; Cortes, R.; Pinheiro, A. Hydropower Plant Operation Scenarios to Improve Brown Trout Habitat. *River Res. Appl.* **2017**, *33*, 364–376. [CrossRef]

46. Vehanen, T.; Bjerket, P.L.; Heggenes, J.; Huusko, A.; Mäki-Petäys, A. Effect of fluctuating flow and temperature on cover type selection and behaviour by juvenile brown trout in artificial flumes. *J. Fish Biol.* **2000**, *56*, 923–937. [CrossRef]

47. Valentin, S.; Sempeski, P.; Souchon, Y.; Gaudin, P. Short-term habitat use by young grayling, Thymallus thymallus L., under variable flow conditions in an experimental stream. *Fish. Manag.* **1994**, *1*, 57–65. [CrossRef]

48. Sloman, K.A.; Gilmour, K.M.; Taylor, A.C.; Metcalfe, N.B. Physiological effects of dominance hierarchies within groups of brown trout, Salmo trutta, held under simulated natural conditions. *Fish Physiol. Biochem.* **2000**, *22*, 11–20. [CrossRef]

49. Sloman, K.A.; Taylor, A.C.; Metcalfe, N.B.; Gilmour, K.M. Effects of an environmental perturbation on the social behaviour and physiological function of brown trout. *Anim. Behav.* **2001**, *61*, 325–333. [CrossRef]

50. Sfakiotakis, M.; Lane, D.M.; Davies, J.B.C. Review of fish swimming modes for aquatic locomotion. *IEEE J. Ocean. Eng.* **1999**, *24*, 237–252. [CrossRef]

51. Kottelat, M.; Freyhof, J. *Handbook of European Freshwater Fishes*; Publications Kottelat, 2007; ISBN 9782839902984.

52. Boavida, I.; Caetano, L.; Pinheiro, A.N. E-flows to reduce the hydropeaking impacts on the Iberian barbel (Luciobarbus bocagei) habitat. An effectiveness assessment based on the COSH Tool application. *Sci. Total Environ.* **2020**, *699*, 134209. [CrossRef]

53. Costa, M.; Ferreira, M.; Pinheiro, A.; Boavida, I. The potential of lateral refuges for Iberian barbel under simulated hydropeaking conditions. *Ecol. Eng.* **2019**, *127*, 567–578. [CrossRef]

54. Pretty, J.L.; Harrison, S.S.C.; Shepherd, D.J.; Smith, C.; Hildrew, A.G.; Hey, R.D. River rehabilitation and fish populations: Assessing the benefit of instream structures. *J. Appl. Ecol.* **2003**, *40*, 251–265. [CrossRef]

55. Ferreira, M.T.; Oliveira, J.; Caiola, N.; de Sostoa, A.; Casals, F.; Cortes, R.; Economou, A.; Zogaris, S.; Garcia-Jalon, D.; Ilhéu, M.; et al. Ecological traits of fish assemblages from Mediterranean Europe and their responses to human disturbance. *Fish. Manag. Ecol.* **2007**, *14*, 473–481. [CrossRef]

56. Oliveira, J.M.; Ferreira, A.P.; Ferreira, M.T. Intrabasin variations in age and growth of Barbus bocagei populations. *J. Appl. Ichthyol.* **2002**, *18*, 134–139. [CrossRef]

57. Branco, P.; Santos, J.M.; Katopodis, C.; Pinheiro, A.N.; Ferreira, M.T. Effect of flow regime hydraulics on passage performance of Iberian chub (Squalius pyrenaicus) (Günther, 1868) in an experimental pool-and-weir fishway. *Hydrobiologia* **2013**, *714*, 145–154. [CrossRef]

58. Costa, M.; Fuentes-Pérez, J.; Boavida, I.; Tuhtan, J.; Pinheiro, A. Fish under pressure: Examining behavioural responses of Iberian barbel under simulated hydropeaking with instream structures. *PLoS ONE* **2019**, *14*, e0211115. [CrossRef]

59. Santos, J.M.; Branco, P.; Katopodis, C.; Ferreira, M.T.; Pinheiro, A.N. Retrofitting pool-and-weir fishways to improve passage performance of benthic fishes: Effect of boulder density and fishway discharge. *Ecol. Eng.* **2014**, *73*, 335–344. [CrossRef]

60. van Netten, S.; McHenry, M. The Biophysics of the Fish Lateral Line. In *The Lateral Line System*; Springer: New York, NY, USA, 2013; Volume 48, pp. 99–119. ISBN 978-1-4614-8850-7.

61. Tuhtan, J.; Fuentes-Pérez, J.; Toming, G.; Kruusmaa, M. Flow velocity estimation using a fish-shaped lateral line probe with product-moment correlation features and a neural network. *Flow Meas. Instrum.* **2017**, *54*, 1–8. [CrossRef]

62. Fuentes-Pérez, J.; Kalev, K.; Tuhtan, J.; Kruusmaa, M. Underwater vehicle speedometry using differential pressure sensors: Preliminary results. In Proceedings of the IEEE/OES AUV, Tokyo, Japan, 6–9 November 2016; p. 6.

63. Fuentes-Pérez, J.; Eckert, M.; Tuhtan, J.; Ferreira, M.; Kruusmaa, M.; Branco, P. Spatial preferences of Iberian barbel in a vertical slot fishway under variable hydrodynamic scenarios. *Ecol. Eng.* **2018**, *125*, 131–142. [CrossRef]

64. Bracewell, P.; Cowx, I.G.; Uglow, R.F. Effects of handling and electrofishing on plasma glucose and whole blood lactate of Leuciscus cephalus. *J. Fish Biol.* **2004**, *64*, 65–71. [CrossRef]

65. Pankhurst, N.W. The endocrinology of stress in fish: An environmental perspective. *Gen. Comp. Endocrinol.* **2011**, *170*, 265–275. [CrossRef]

66. Arnekleiv, J.V.; Urke, H.A.; Kristensen, T.; Halleraker, J.H.; Flodmark, L.E.W. Recovery of wild, juvenile brown trout from stress of flow reduction, electrofishing, handling and transfer from river to an indoor simulated stream channel. *J. Fish Biol.* **2004**, *64*, 541–552. [CrossRef]

67. Kieffer, J.D. Limits to exhaustive exercise in fish. *Comp. Biochem. Physiol. Part A Mol. Integr. Physiol.* **2000**, *126*, 161–179. [CrossRef]

68. Harby, A.; Noack, M. Rapid flow fluctuations and impacts on fish and the aquatic ecosystem. In *Ecohydraulics—An Integrated Approach*; Maddock, I., Harby, A., Kemp, P., Wood, P., Eds.; Wiley Blackwell: Hoboken, NJ, USA, 2013; pp. 323–335.

69. Liao, J.C. A review of fish swimming mechanics and behaviour in altered flows. *Philos. Trans. R. Soc. Lond. B Biol. Sci.* **2007**, *362*, 1973–1993. [CrossRef] [PubMed]

70. Bak-Coleman, J.; Court, A.; Paley, D.A.; Coombs, S. The spatiotemporal dynamics of rheotactic behavior depends on flow speed and available sensory information. *J. Exp. Biol.* **2013**, *216*, 4011–4024. [CrossRef]

71. Kanter, M.J. Rheotaxis and prey detection in uniform currents by Lake Michigan mottled sculpin (Cottus bairdi). *J. Exp. Biol.* **2003**, *206*, 59–70. [CrossRef]

72. Kalmijn, A.J. Detection of weak electric fields. In *Sensory Biology of Aquatic Animals*; Atema, J., Fay, R.R., Popper, A.N., Tavolga, W.N., Eds.; Springer: New York, NY, USA, 1988; pp. 151–186, ISBN 978-1-4612-8317-1, 978-1-4612-3714-3.

73. Martínez-Capel, F.; García de Jalón, D. Desarrollo de curvas de preferencia de microhábitat para Leuciscus pyrenaicus y Barbus bocagei por buceo en el río Jarama (Cuenca del Tajo). *Limnetica* **1999**, *17*, 71–83.

74. Mateus, C.S.; Quintella, B.R.; Almeida, P.R. The critical swimming speed of Iberian barbel Barbus bocagei in relation to size and sex. *J. Fish Biol.* **2008**, *73*, 1783–1789. [CrossRef]

75. Santos, J.M.; Rivaes, R.; Boavida, I.; Branco, P. Structural microhabitat use by endemic cyprinids in a Mediterranean-type river: Implications for restoration practices. *Aquat. Conserv. Mar. Freshw. Ecosyst.* **2017**, *28*, 26–36. [CrossRef]

76. Almeida, R.; Boavida, I.; Pinheiro, A.N. Habitat modeling to assess fish shelter design under hydropeaking conditions. *Can. J. Civ. Eng.* **2017**, *44*, 9098. [CrossRef]

77. Pagliara, S.; Hassanabadi, L.; Kurdistani, S.M. Clear water scour downstream of log deflectors in horizontal channels. *J. Irrig. Drain. Eng.* **2015**, *141*, 04015007. [CrossRef]

78. Carré, D.; Biron, P.; Gaskin, S. Flow dynamics and bedload sediment transport around paired deflectors for fish habitat enhancement: A field study in the Nicolet River. *Can. J. Civ. Eng.* **2007**, *34*, 761–769. [CrossRef]

79. Tuhtan, J.A.; Noack, M.; Wieprecht, S. Estimating stranding risk due to hydropeaking for juvenile European grayling considering river morphology. *KSCE J. Civ. Eng.* **2012**, *16*, 197–206. [CrossRef]

80. Mittelbach, G.G.; Ballew, N.G.; Kjelvik, M.K.; Fraser, D. Fish behavioral types and their ecological consequences. *Can. J. Fish. Aquat. Sci.* **2014**, *71*, 927–944. [CrossRef]

81. Hayes, D.S.; Moreira, M.; Boavida, I.; Haslauer, M.; Unfer, G.; Zeiringer, B.; Greimel, F.; Auer, S.; Ferreira, T.; Schmutz, S. Life stage-specific hydropeaking flow rules. *Sustaianbility* **2019**, *11*, 1547. [CrossRef]

82. Moreira, M.; Hayes, D.; Boavida, I.; Schletterer, M.; Schmutz, S.; Pinheiro, A. Ecologically-based criteria for hydropeaking mitigation: A review. *Sci. Total Environ.* **2019**, *657*, 1508–1522. [CrossRef] [PubMed]

*Review*

# Life Stage-Specific Hydropeaking Flow Rules

**Daniel S. Hayes** [1,2,*], **Miguel Moreira** [3], **Isabel Boavida** [3], **Melanie Haslauer** [1], **Günther Unfer** [1], **Bernhard Zeiringer** [1], **Franz Greimel** [1], **Stefan Auer** [1], **Teresa Ferreira** [2] and **Stefan Schmutz** [1]

[1] Department of Water, Atmosphere and Environment, Institute of Hydrobiology and Aquatic Ecosystem Management, University of Natural Resources and Life Sciences, Vienna (BOKU), 1180 Vienna, Austria; melanie.haslauer@boku.ac.at (M.H.); guenther.unfer@boku.ac.at (G.U.); bernhard.zeiringer@boku.ac.at (B.Z.); franz.greimel@boku.ac.at (F.G.); stefan.auer@boku.ac.at (S.A.); stefan.schmutz@boku.ac.at (S.S.)

[2] Forest Research Center (CEF), Instituto Superior de Agronomia, University of Lisbon, 1349-017 Lisbon, Portugal; terferreira@isa.utl.pt

[3] Center for Engineering Research and Innovation for Sustainability (CERIS), Instituto Superior Técnico, University of Lisbon, 1049-001 Lisbon, Portugal; miguelrmoreira@tecnico.ulisboa.pt (M.M.); isabelboavida@tecnico.ulisboa.pt (I.B.)

* Correspondence: daniel.hayes@boku.ac.at; Tel.: +43-147-654-81-223

Received: 15 February 2019; Accepted: 8 March 2019; Published: 14 March 2019

**Abstract:** Peak-operating hydropower plants are usually the energy grid's backbone by providing flexible energy production. At the same time, hydropeaking operations are considered one of the most adverse impacts on rivers, whereby aquatic organisms and their life-history stages can be affected in many ways. Therefore, we propose specific seasonal regulations to protect ecologically sensitive life cycle stages. By reviewing hydropeaking literature, we establish a framework for hydrological mitigation based on life-history stages of salmonid fish and their relationship with key parameters of the hydrograph. During migration and spawning, flows should be kept relatively stable, and a flow cap should be implemented to prevent the dewatering of spawning grounds during intragravel life stages. While eggs may be comparably tolerant to dewatering, post-hatch stages are very vulnerable, which calls for minimizing or eliminating the duration of drawdown situations and providing adequate minimum flows. Especially emerging fry are extremely sensitive to flow fluctuations. As fish then grow in size, they become less vulnerable. Therefore, an 'emergence window', where stringent thresholds on ramping rates are enforced, is proposed. Furthermore, time of day, morphology, and temperature changes must be considered as they may interact with hydropeaking. We conclude that the presented mitigation framework can aid the environmental enhancement of hydropeaking rivers while maintaining flexible energy production.

**Keywords:** sustainable hydropower; sub-daily flow fluctuations; peak-load energy production; pulsed flows; environmental flow; biologically sensitive periods; salmonids; *Salmo salar*; *Salmo trutta*; *Thymallus thymallus*

---

## 1. Introduction

Mountainous rivers are often subjected to sub-daily flow variations caused by peak-operating hydropower plants, which run their turbines according to the demand of the electricity market. These hydropower plants allow high flexibility in energy production, making them an essential part of the current and future electric grid as they can buffer periods of low energy availability of other renewables, such as wind or solar energy [1,2]. At the same time, hydropeaking entails numerous adverse ecological consequences and has therefore been described as "one of the most significant impacts on rivers downstream of dams" [3]. Fish communities, in particular, are severely threatened by hydropeaking [4]. Fish can be affected by changes in various components of the hydrograph, whereby the most common responses—stranding, drift, and dewatering of spawning grounds—are mostly related to up- and downramping rates [5,6], peak flow magnitude [5], and baseflow duration [7].

Considering the large capacity of existing storage hydropower plants [8], as well as new ones that are currently being planned and installed [9], it is imperative to develop appropriate and transferable management measures to mitigate these ecological impacts. Many structural (e.g., constructing retention basins) and operational (e.g., reducing flow fluctuation rates) mitigation measures have been proposed [10,11], but implementation remains difficult, among other issues, because of significant reductions in the energy yield when setting ecological thresholds [2,12]. Therefore, well-targeted mitigation measures have to be developed to avoid energy losses and to guarantee ecological efficiency.

Freeman et al. [13] argue that adverse effects can be minimized by either restoring vital features of the natural flow regime or by implementing a flow management scheme which avoids hydropower-induced habitat bottlenecks. Regarding the latter, multiple studies point out the need to identify critical flows, which include seasonal and diel considerations when determining operational mitigation strategies in rivers affected by hydropeaking [5,13–16]. To maintain self-sustaining fish populations in regulated water bodies, river management must take all life-history stages into account, especially during ecologically sensitive periods, when designing and implementing customized flow regimes [7]. This study aims to establish a framework for hydrological mitigation in rivers affected by hydropeaking, based on life cycle stages of fish and their relationship to key parameters of the hydrograph. We, therefore, describe the impacts of hydropeaking on each life cycle stage and propose critical aspects which must be considered when defining mitigation rules. We focus on salmonid fish as they are the most studied and most affected fish family regarding hydropeaking [17,18]. Nevertheless, many aspects of the presented mitigation concept can potentially be transferred to or adapted to the requirements of other taxa.

## 2. Life Cycle Stage-Specific Mitigation Approach

Within the life cycle of many salmonids, upstream spawning migrations are followed by the deposition of fertilized eggs in the gravel bed. In the following weeks, the eggs develop within the gravel substrate where, after hatching, the larvae (alevins) also stay until their yolk sack is absorbed. Afterwards, the fry emerge from the riverbed to find nursery habitats, e.g., along the shoreline for feeding and growth. As fish increase in size, they use different habitats. Once they reach sexual maturity, their life cycle starts over again [19,20].

The literature indicates that each of these life cycle stages can show a distinct sensitivity to different aspects of the hydropeaking hydrograph, whereby reproduction and early life stages seem to be the most sensitive ones [15]. Thus, the key parameters for flow restoration will vary between the life cycle stages. In the following subchapters, we will therefore discuss the ecological effects of hydropeaking for each sensitive life stage, as well as establish a specific mitigation framework approach for each life cycle stage (Figure 1).

### 2.1. Migration and Spawning

River flow is a crucial factor for spawning-related activities [21]. In hydropeaking rivers, highly variable flows can influence, among other factors, migration, nest digging, and spawning behavior. Studies have shown that migration patterns of lake trout (*Salmo trutta lacustris*), as well as Chinook (*Oncorhynchus tshawytscha*) and pink salmon (*O. gorbuscha*), were correlated with the hydropeaking regime. Fish avoided migrating during changing flow magnitudes and they were able to cover greater distances on the days with peaking operations of lower amplitudes [22,23].

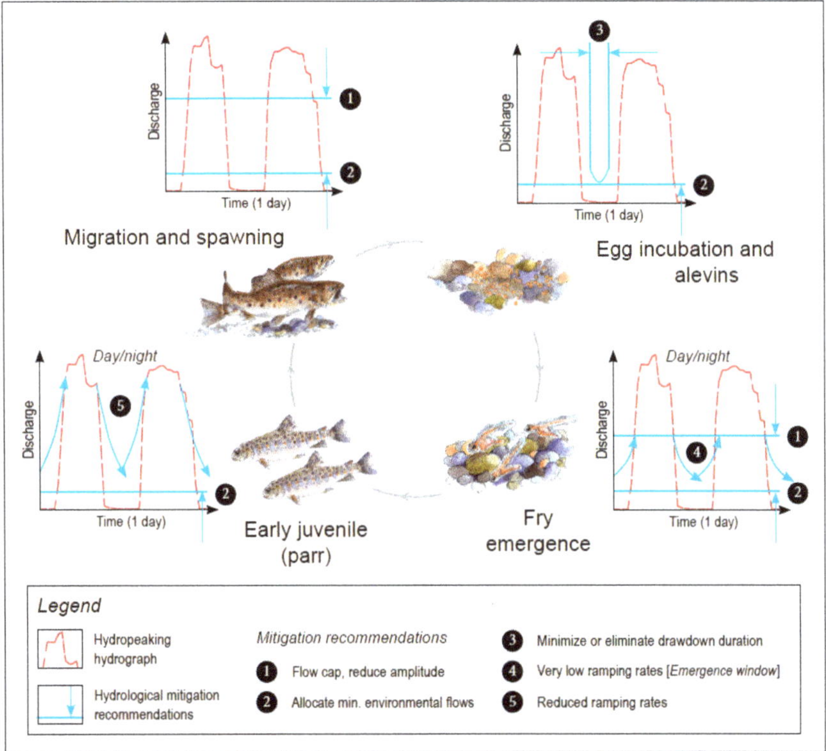



**Figure 1.** Life stage-specific hydropeaking flow rules: Conceptual framework for the sensitive life stage approach to mitigate the adverse impacts of hydropeaking. The dashed red lines represent a schematic daily hydropeaking hydrograph (two peaks and a baseflow phase), whereas the solid blue lines depict recommendations for hydrological restrictions to aid the environmental enhancement of hydropeaking rivers. "Day/night" indicates that restrictions might differ with time of day. For a detailed description, see Sections 2.1–2.4 (salmonid illustrations by DAB graphics, used with permission from The Wild Trout Trust Ltd, Waterlooville, England, www.wildtrout.org).

When fish are able to reach suitable reproduction areas, spawning can be interrupted by rapid flow fluctuations [16,24–26]. During downramping events, brown trout (*S. trutta fario*) stop preparing their redds [24] and brown trout and Atlantic salmon (*S. salar*) leave the spawning area if hydraulic conditions become unsuitable [26]. As soon as flows increase again, Atlantic salmon are highly motivated to continue redd preparation, and resume spawning once flows have returned to more stable conditions [24,25], whereas brown trout take more time to re-engage in spawning-related activities [24].

To mitigate the effects of hydropeaking on migrating fish, it is therefore advantageous to release higher flows during migration periods [21]. Furthermore, during spawning, it is suggested that flows are kept relatively stable [16], at least for a sufficient duration to allow females to complete nest preparation and oviposition [26], which can take multiple hours to days for one spawning bed [27,28], but several weeks for the entire population. The flow magnitude to enable these ecological functions can differ among geomorphic habitat units as it depends on the hydraulic conditions of the reach [21,25] and the species present [26]. By allocating a steady environmental flow release during the peak spawning period (e.g., ca. four weeks for *S. trutta*), the ecological conditions can already be improved [29].

Although some salmonids tend to spawn during discharges greater than the median [30], the release of too high flows is not recommended if these water levels cannot be sustained throughout the incubation period [28]. At that stage, higher discharges could encourage nest-building in areas that will be at a higher risk of being dewatered between the hydropeaks later in the season. Instead, flow caps, i.e., upper peak limits, should be implemented [21,26,31–33], as the survival of intragravel life stages is linked to redd site selection during the spawning period.

*2.2. Intragravel Life Cycle Stages: Egg Incubation and Alevins*

In hydropeaking rivers, intragravel life cycle stages can be predominantly influenced through either siltation of sediments [34,35], scouring [21,35], or dewatering of spawning grounds [21,31,32]. In the winter, mountainous rivers are characterized by an extended low baseflow period. During this season, hydropeaking rivers, however, often exhibit relatively high flow fluctuations. This operational scheme leads to higher flow amplitudes in winter compared to summer [36,37].

As salmonids deposit their eggs in the gravel bed, siltation may reduce hatching success by affecting interstitial water flow [38,39]. It has been suggested that hydropeaking is a governing factor regarding fine sediment dynamics in gravel-bed rivers, as hydropeaking can change the fine sediment composition of both surface and subsurface layers [40]. Indeed, a recent study [34] found that ramping zones exhibited significant surface clogging due to a continuous accumulation of fines. In contrast, permanently inundated areas contained little or no fine sediment infiltration into the riverbed's surface layers as fines are subject to transport. Interestingly, the hydropeak magnitude itself, expressed as the peaking ratio, was not related to fine sediment infiltration rates [34]. It seems, therefore, that the spawning ground position in the riverbed (see Section 2.1) can influence hatching success through the effect of sediment sizes.

It is expected that, due to nest-building of female salmonids, spawning ground stability is increased as sediments are sorted and redds, therefore, feature coarser substrate surfaces than unspawned beds through winnowing of fines. Despite this coarsening effect, however, studies [41] have demonstrated that redds are more unstable than unspawned beds. For example, redds exhibited a 12–37% lower grain resistance to motion, as well as a 13–41% higher boundary shear stress for the same flows in comparison to unspawned beds. Also, bed-average shear stress was significantly reduced [41]. Although studies indicate that salmonid spawning usually takes place in locations with less excess shear stress [21], hydropower peak flows may enhance the risk of embryo mortality as redds are more likely to be eroded than the surrounding gravel bed [21,41]. However, although scour has been cited as a potentially adverse effect, no study has yet quantified its impact magnitude. Furthermore, the scouring potential also depends on the peak flow magnitude. Unfer et al. [42] demonstrated that only flow magnitudes larger than half the size of mean annual high flow events were able to substantially erode sediments in the Alpine Ybbs River, Austria. The risk of egg erosion due to hydropeaking may, therefore, be rather case-specific [35] and may only occur in rivers with a high peaking magnitude.

If spawning occurs during peak flow periods, the drawdown to winter baseflows between peaks could lead to the dewatering of spawning grounds, which is a significant concern as it can result in the mortality of eggs and alevins [31,43]. It has been demonstrated that salmonid eggs are generally rather robust to dewatering and can survive extended time frames without inundation, provided that they are kept moist, are not subjected to extreme temperatures (i.e., freezing or heating) exceeding incubation tolerances, and receive sufficient oxygen through the influx of air into the interstitial spaces [31,43–46]. In contrast, newly hatched alevins (eleutheroembryos) are less tolerant to redd desiccation and may die within 4–12 h of dewatering, whereas pre-emergent alevins are considered the most sensitive intragravel life stage [44–46]. Since pre-emergent alevins depend on gills for respiration, dewatering events <1 h lead to very high (>96%) mortality rates [44], which can have profound impacts on fish populations. However, groundwater upwelling might attenuate apparent adverse effects where available [21,43,47].

To prevent the mortality of intragravel life stages, it is recommended that power production is adapted to discourage fish from spawning in shallow water which will later fall dry (see Section 2.1). Additionally, the duration of hydropower production stops should be minimized. Also, the provision of enhanced minimum flows during this critical development period can help to reduce the difference between incubation and spawning flows [32,43,48,49]. During the egg incubation, limited redd dewatering through the hydropower plant operation might not entail complete losses in some cases [16,32], such as in the presence of local groundwater upwellings and at temperatures above freezing [43]. However, considering that spawning can occur over an extended period and that multiple species can be present in the same river, allowing limited redd dewatering is not recommended since alevins, which require continuous inundation [7], may be present throughout the entire time [16].

### 2.3. Fry Emergence and Early Juvenile Development

After alevins have absorbed the major portion of their yolk sack, they emerge as fry from the substrate [19,20]. During this early ontogenetic development, they are very susceptible to pulsed-flow operations as they utilize high-risk habitats in the ramping zone and have little swimming capacities, entailing drift and stranding of individuals [5,6,50–53]. In the Saltdalselv River, Norway, high flows during the alevin and fry stage significantly increased the mortality of Atlantic salmon and brown trout [54]. Similarly, fry recruitment was negatively related to the number of hydropeaks during the emergence period in the Lez River, France [55].

#### 2.3.1. Thresholds for Impact Mitigation

Rapid flow reductions due to downramping can increase the stranding probability of fish through quickly receding water levels, causing sub-lethal impacts or direct fish mortality [16,18]. Studies have shown that stranding is species- and size-selective, whereby recently emerged fry are the most vulnerable life-history stage [15,52,56,57]. This finding is supported by the analysis of ten-year flow downramping monitoring data of Canadian rivers, showing that the highest stranding probabilities occur from May to August when juveniles inhabit nearshore areas which are likely to be dewatered [50]. Field surveys at the Drava River, Austria, revealed 50–500 stranded larvae of European grayling (*Thymallus thymallus*) per 100 m shoreline after single hydropeaking events [58]. In general, a reduction of ramping rates to <0.17–0.25 cm·min$^{-1}$ is related to less stranding and, therefore, a greater probability of attaining a higher fish ecological status [4,59]. On a more detailed level, experimental studies indicate that, during fry emergence, mitigation thresholds on downramping velocity must be rather low to prevent stranding, e.g., 0.23–0.31, 0.2, or 0.1 cm·min$^{-1}$ for larvae of Atlantic salmon, European grayling, or brown trout, respectively [6,52,56,60].

Early juvenile life stages are also susceptible to downstream displacement [52], especially during nighttime hydropeaking [5]. However, little is known about the long-term population effects caused by drifted fish. Nevertheless, a reduction of drift is advisable. Therefore, a lowered upramping rate and a reduced peak amplitude are recommended [5,29].

#### 2.3.2. Emergence Window Establishment

Generally, it is advisable to stabilize the flow as much as possible in the early growing season [59]. Alfredsen et al. [33] suggest introducing a cap flow and restricting rapid flow changes during swim-up. Since fry are especially vulnerable to sub-daily flow fluctuations and are present only at specific periods of the year [16,52], a feasible management approach is to define temporal 'emergence windows' where stringent thresholds, e.g., regarding downramping to prevent stranding, are enforced [8,15,52,61]. These emergence windows should start with the highly sensitive alevin phase [44–46] just before fry emerge from the gravel, whereas the length of the window depends on the growth, which is mainly related to temperature. Stranding experiments [52,56] indicate that the temporal duration must be around two weeks for European grayling and four weeks for brown trout, as grayling improves its reactivity to drawdown events quicker than brown trout. Even though these two species spawn at different periods (brown trout in fall/winter and European grayling in late winter/spring), their larvae occur in the same season (mid-April to early August) due to their temperature-dependent egg development [62], underlining the feasibility of the emergence window approach. Approximate start and end dates can be calculated with temperature data if the time of spawning is known [19,33,62]. For example, Figure 2 depicts a first river-specific assessment of emergence windows for brown trout and European grayling for selected Austrian hydropeaking rivers based on information from anglers. Due to the imprecise data situation, the proposed time frames are still rather long and represent only a rough estimate. It would be possible to confine the time period of the emergence window by assessing the exact emergence time through electrofishing surveys and modeling based on spawning time and day degrees of egg development. The results would allow the deduction of a river-specific period of early fish life cycle development by analyzing the water temperature of the sampling year. This information could then possibly be transferred to the following years. Another feasible approach would be to visually observe (stranding of) larvae from the end of April onwards to determine a river reach-specific emergence window. In rivers with different species of Pacific salmon, the implementation might be more difficult due to varying emergence periods [48].

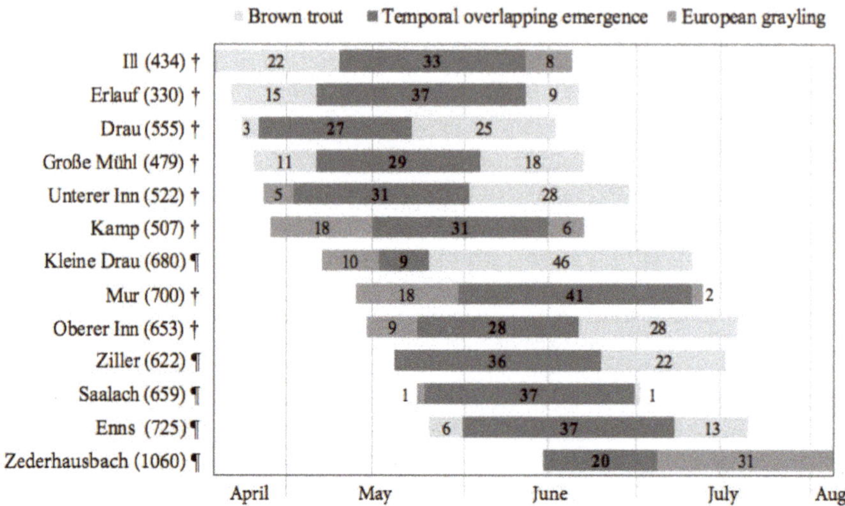


**Figure 2.** Emergence window proposal for brown trout (*Salmo trutta*) and European grayling (*Thymallus thymallus*) in selected Austrian hydropeaking rivers. The depicted time frame is based on temperature-dependent models for predicting the time of median hatch as reported in the literature. For the brown trout, we used the formula from Crisp [19], and for the grayling, the one published by Jungwirth & Winkler [62]. The begin and end of the spawning time for each species was reported by anglers and served as the starting point for the calculations, which were then based on daily mean temperature data of ten years (gauging stations of Austrian Hydrographic Service; meters above the Adriatic in brackets; † = metarhithral (i.e., lower trout) fish region; ¶ = hyporhitrhral (i.e., grayling) fish region). Emergence of brown trout and European grayling takes place between mid-April and early August, whereby the emergence windows of both species also partially overlap. The duration of the emergence periods is displayed in the bars as a mean number of days.

*2.4. Parr to Adults*

Parr are also vulnerable to stranding and drift, whereby the risk is reduced as they grow in size and increase their swimming performance, and additionally shift to deeper habitats away from the dewatered ramping zone [53]. Therefore, less restrictive ramping rates—in comparison to fry—can be sufficient [5,15,16,52,56,57]. However, since stranding probability is also determined by other factors aside from downramping velocity (e.g., wetted history, baseflow conditions, time of day), these must be considered in the establishment of mitigation rules as well [15]. For example, before a large flow reduction, lower reductions are recommended prior to higher ones to shorten the wetted history [50]. Furthermore, the time of day can play a significant role. Some studies report that, during summer, young-of-year brown trout and grayling are less vulnerable during the day and more susceptible to stranding during the night [5,56,60]. In contrast, during colder water temperatures in fall or winter, higher stranding is reported for daytime than for nighttime hours for brown trout, rainbow trout (*O. mykiss*), Atlantic salmon, and Coho salmon (*O. kisutch*) [6,63,64]. Both can be related to diel behavior changes [65] as, for example, in winter, salmonids are passive during the day and active in the night [64,66]. Therefore, ramping rate restrictions should be more stringent during darkness in summer and during daylight hours in winter. It must be noted, however, that Connor & Pflug [48] have reported exactly the opposite as the above cited studies. They recommend limiting downramping to nighttime hours between the emergence and outmigration period. However, this focus on daytime mitigation might be because, in the Skagit River case study, daytime flow reduction represented 89% of all events during the peak stranding period [48]. Differences in stranding or drift might also be triggered by other factors, such as water turbidity or predation. In this regard, more research is necessary. In the meantime, however, case-specific solutions are required.

Regarding fish movement patterns during hydropeaking, Robertson et al. [67] found that, in winter, flow neither affected fish activity or habitat use, nor displaced Atlantic salmon parr. Only in late winter was fish activity reduced during high flows in the night. Similarly, Stickler et al. [68] did not detect differences in fish activity between high and low flows, which was also confirmed by Berland et al. [69], who analyzed parr movement in September related to river flow and ramping rate. Therefore, it can be concluded that, for the most part, hydropeaking does not affect salmonid fish movements in winter. In contrast, in summer, juvenile Atlantic salmon show higher movement rates in hydropeaking channels than in control channels [70], a pattern which was also confirmed for 1+ salmon in a telemetry study during spring [71]. Considering the increase in juvenile fish movement during summer in combination with inhibited feeding during peak flows [72], it is not surprising that, at the end of the growing season, fish that were subjected to fluctuating flows had a lower body fat and body mass than fish subjected to stable flows. Although the effects were small [70] or, in some cases, not detectable [58], the long-term impact on the population can be potentially many times higher. Simulating the effects of stranding on the salmon population in the Dale River, Norway, Sauterleute et al. [73] conclude that the most substantial adverse effect on the population abundance in hydropeaking rivers is related to the stranding of older juvenile fish during winter daylight conditions, suggesting that the stranding of salmon at this life cycle stage is likely to have greater population impacts than that of earlier life cycle stages. Furthermore, the stored energy reserves from the summer may be a critical factor in determining overwinter survival [74,75].

## 3. Discussion

Future sustainable hydropower management relies on the development of well-informed and targeted mitigation strategies [76]. Here, we propose a management framework to mitigate adverse impacts of hydropeaking operations on salmonid fish, whereby we advocate that, in each season, the most sensitive life cycle stage should be the decisive element regarding peaking restrictions. Figure 1 constitutes a graphical depiction of this mitigation framework approach by illustrating a hydropeaking hydrograph scheme (two peaks and a baseflow phase), as well as the above-described flow restrictions necessary for aiding the environmental enhancement of hydropeaking rivers (i.e., depending on season: implementing flow caps, allocating minimum environmental flows, decreasing the flow ratio, minimizing dewatering durations or prohibiting dewatering, or lowering ramping rates). Alfredsen et al. [33] used a similar approach by defining flow blocks for environmental flow allocation to meet the need of Atlantic salmon life stages in the Daleelva River, Norway. Due to sub-daily flow fluctuations, hydropeaking represents a specific sub-group within environmental flow in rivers where hydrological stress on aquatic ecosystems is intensified. Therefore, there is a need to develop qualitative seasonal flow rules for hydropeaking rivers as well. Here, we focused solely on salmonid species as most available literature deals with this fish family [15,17,18]. Many aspects of the presented framework can potentially be transferred to other taxa as well, but further research must be conducted on other fish families, such as cyprinids, to validate these suggestions.

Literature indicates that different elements of the hydrograph can be tweaked to improve the survival of the respective life stages (Figure 1), whereby the temporal windows must be adapted to the local river conditions. However, mitigation thresholds (e.g., for baseflow, peak flow, ramping rate) depend not only on the life cycle stage, but also on the respective species present, as well as the time of day. Nevertheless, only a few quantitative thresholds have been proposed so far, highlighting the need for further research [15]. Therefore, for the most part, we were only able to extract qualitative/conceptual mitigation recommendations. Related to that is the question of which life cycle stage is the most sensitive to sub-daily flow fluctuations regarding having the greatest adverse effect for the fish population.

The field of environmental flow has progressed towards advocating function- and process-oriented flows [14,77], thereby moving away from static water allocations and towards dynamic environmental flows to sustain ecological communities [14,78]. Considering that hydropeaking rivers are essentially residual flow stretches—just that, in addition to water abstraction, they are also highly impacted by sub-daily flow fluctuations [79]—mitigation measures for hydropeaking must be incorporated into the seasonal and inter-annual environmental flow requirements. Therefore, scientific advancement must merge the concepts of dynamic environmental flow and hydropeaking mitigation to propose sustainable and holistic management recommendations for flow-altered watercourses.

### 3.1. The Effects of River Hydromorphology

Many studies indicate that the ecological effects related to hydropeaking, e.g., stranding or drift, are also dependent on river morphology, including bank slope, grain size distribution, or cover [5,12,64,80]. Higher stranding, for example, occurs more frequently on lower gradient bars than on steeper banks [64], but steeper banks provide less larval and juvenile habitats. Coarse substrate will trap fish more during dewatering than small grain sizes [80]. Due to water retention, vertical dewatering speed reduces with distance to the hydropower outlet, therefore lowering the stranding risk along the river's course [3,59]. In morphologically unfavorable river sections (e.g., in channelized rivers), spawning, larval, and juvenile habitats are often lacking, whereas in nature-like channels, ramping rate reductions may improve the fish ecological status [4]. Hauer et al. [3,80] highlight that the changes in wetted width between baseflow and peak flow and, thereto related, changes in the lateral ramping velocity depend on the river's channel bar form, as well as the baseflow magnitude (see also [81]). Depending on this combination, a <1:2 peaking ratio can have a greater impact on cross-sectional wetted width than a 1:5 ratio [80]. Therefore, Halleraker et al. [59] recommend different dewatering thresholds for distinct flow ranges in the Surna River, Norway, where more stringent flow limits are needed for lower discharges than for high ones [50,59].

Self-forming gravel bars (e.g., point bars or alternating gravel bars) have been identified as both suitable structures for young-of-year trout, as well as areas of reduced stranding risk due to self-forming backwater habitats [80,82]. These findings underline the need to combine hydrological rehabilitation with morphological restructuring measures to minimize the ecological impact of hydropeaking [8,12,80,83]. Next to river restoration measures, connectivity to tributaries may also play an essential role in supporting fish populations in hydropeaking rivers. Tributaries often exhibit more stable hydrological conditions and less risk of erosion and, therefore, may provide suitable spawning and rearing habitats for fish [3].

### 3.2. The Effects of Sub-Daily Temperature Changes

Another abiotic factor that can influence the effects of hydropeaking on fish is water temperature, as the release of pulsed flows is often coupled with thermal changes (i.e., thermopeaking) [84]. The ecological effects of long-term thermal alterations below dams have received some attention [85], whereas less is known about the reaction of eggs or fish to sub-daily thermopeaking. Therefore, further analyses have to be performed to study potential effects in detail [4].

Research has shown that both hydropeaking and thermopeaking can influence macroinvertebrate communities [86]. Although we did not integrate macroinvertebrates into our conceptual mitigation framework, this group should be included in more holistic approaches in the future, not only because benthic communities are an essential food source for fish, but also because they are an indispensable aspect of functioning river systems [87]. Also, as benthic dwellers, macroinvertebrates are particularly sensitive to sediment composition and habitat conditions in the river bed [34].

### 3.3. Other Hydropeaking-Related Impacts

Aside from thermopeaking, two other hydropeaking-related impacts have recently been described. 'Saturopeaking' refers to fluctuations of gas saturation which follow the rapid, periodic, and frequent pattern of hydropeaking operations [88], and 'soundpeaking' to hydropeaking-induced changes in river soundscapes, whereby sound pressure levels can be strongly correlated with turbine discharge [89]. Although some guidelines exist for supersaturation (i.e., when total dissolved gases saturation exceeds 100%) [88], possible ecological effects of saturopeaking and soundpeaking in hydropeaking rivers still remain to be studied [15], as well as the combined effects of these pressures.

### 3.4. Achieving Hydrological Mitigation Measures and Their Economic Implications

To achieve mitigation, either structural or operational measures can be utilized [10,12], whereby similar positive hydrological changes in the tailrace can be obtained, e.g., by changes in the power plant operation scheme, as well as through the construction of retention basins [2,90]. Considering the economic implications, however, these two approaches show quite different outcomes [2]. Hydropeaking power plants operate competitively according to immediately changing market prices, which means that the quicker their turbines can be turned on and off, the higher the economic benefits are. Therefore, operational restrictions affect the ability to produce highly valuable peak energy [2,12,91], especially if less favorable turbines types are installed [92] or the water availability imposes constraints. In contrast, peak retention basins might initially require significant investment costs, but, according to Person et al. [12], they show high beneficial ecological effects by reducing sub-daily flow fluctuations at reasonable costs. Since retention basins allow ramping rate reductions, they may be especially useful for applying seasonal flow rules during ecologically sensitive periods [90]. Only limited space availability may be the major problem for the construction of such basins. In contrast, instream velocity refuges such as deflectors require less lateral space [93]. In the very critical larval phase, a combination of compensation reservoirs and altered operational management might be most effective by avoiding over-sized reservoirs. Currently, the feasibility of the air cushion underground reservoir (ACUR) technology [94] to mitigate environmental hydropeaking effects is being tested in the European project "HydroFlex" [95].

### 3.5. Limitations of This Study and Research Needs

In recent years, researchers have established a firm knowledge basis regarding general fish ecological topics, such as the response of different salmonid life-history stages to different environmental parameters, including water flow, temperature, or substrate conditions [96,97] (Table 1). Of course, there are questions which still remain to be answered [96,98]. In the last years, hydropeaking research has significantly advanced in terms of scientific output [99], and this increasing amount of information has allowed us to formulate the presented mitigation framework (Figure 1). Nevertheless, although such a life cycle stage approach constitutes the most up-to-date framework on hydropeaking mitigation, it is to be expected that future studies will significantly expand the present knowledge base regarding the effects of hydropeaking on various life cycle stages, and proposed flow management recommendations may have to be adapted. Therefore, Table 1 presents a knowledge matrix which highlights crucial research areas in the field of hydropeaking impact and mitigation. Future research will, among other goals, have to better quantify the effects of hydropeaking on spawning activities and egg incubation phases (especially with regard to scour or siltation, and sub-daily temperature changes), and investigate impacts on the fish population and community level, including studying the effects of hydropeaking on the food web (e.g., between nutrients, periphyton, macroinvertebrates, and fish).

**Table 1.** Knowledge matrix on the general fish ecological and hydropeaking research conducted on salmonids. We categorized life-history stages (and their activity) into three classes: "−" = no or hardly any studies conducted; "±" = some research is published, but knowledge gaps remain; "+" = extensive studies have been conducted. Literature examples of each research field are given.

| Life-History Stage (and Activity) | | Ecological Research | Literature Example(s) [1] | Hydropeaking Research [2] | Literature Example(s) | Particularities on Hydropeaking Studies |
|---|---|---|---|---|---|---|
| Spawning | Migration | + | [20,98,100–102] | − | [22,23] | |
| | Behavior [3] | + | [27,102,103] | − | [24–26] | |
| Intragravel life stages | Egg incubation | + | [19,20,101] | ± | [31,43–46] | Aside from studies on short- or long-term desiccation, information on repeated wetting and drying is largely missing for different species. |
| | Alevin | + | [19,20,98] | − | [44] | |
| Young-of-the-year (0+) | Fry | + | [97,98] | ± | [52,55–57] | Only studies regarding few selected species, topics mostly restricted to stranding and drift. |
| | Parr | + | [19,20,104] | ± | [5,6,70,71] | Only studies regarding few selected species, topics mostly restricted to drift, stranding, movements, habitat use, and growth. |
| Juvenile | 1+ fish (smolt) | + | [19,102,104,105] | − | [61,106] | |
| Adult | 2+ fish | + | [105] | − | [107] | |

[1] We focused on reviews or books, as they provide the best overviews of the field of research. [2] Mostly regarding the impacts of hydropeaking. For a recent review on hydropeaking mitigation, see Moreira et al. [15], who summarized the current status (research and legislation) and presented research needs. [3] Common conditions contributing to spawning behavior include: Nest selection, building, probing, completion and oviposition, and covering [27].

## 4. Conclusions

Research shows that fish are sensitive to hydrological modifications, especially sub-daily flow fluctuations, which can influence each life cycle stage through various components of the hydropeaking hydrograph. We reviewed the literature to understand how hydropeaking influences each life cycle stage of salmonids. This approach allowed us to conceptualize a qualitative mitigation framework which is based on seasonal flow regulations to protect ecologically sensitive life cycle stages (Figure 1), whereby the following flow rules are recommended: During migration and spawning, flows should be kept relatively stable, and a flow cap should be implemented to prevent the dewatering of spawning grounds during intragravel life stages. While eggs may be comparably tolerant to dewatering, post-hatch stages are very vulnerable, which calls for minimizing or eliminating the duration of drawdown situations. Especially emerging fry are extremely sensitive to sub-daily flow fluctuations. Therefore, a temporally-limited 'emergence window', where stringent thresholds on ramping rates are enforced, is proposed. As fish grow in size, they become less vulnerable. Therefore, less restrictive ramping rates (in comparison to fry) can be acceptable. In all seasons, adequate environmental flows shall be allocated. Furthermore, when setting mitigation thresholds, interacting effects of daytime, river morphology, and water temperature also have to be considered.

The implementation of these seasonal restriction guidelines will not only counter possible hydropower-induced population bottlenecks but has the potential to entail less significant reductions in energy yield compared to all-year round hydrological limits [76]. Nevertheless, further research is necessary to evaluate the ecological effectiveness of the proposed concept and to quantify exact thresholds for different species, life cycle stages, seasons, and time of day in distinct river types [15] while minimizing flexible energy yield reductions in the implementation thereof.

**Author Contributions:** Conceptualization, D.S.H.; methodology, D.S.H., M.M., I.B., G.U., and B.Z.; investigation: D.S.H., M.M., and M.H.; writing—original draft preparation, D.S.H.; writing—review and editing, D.S.H., M.M., I.B., M.H., G.U., B.Z., F.G., S.A., T.F., and S.S.; visualization, D.S.H. and M.H.; supervision, S.S. and T.F.; project administration, F.G.; funding acquisition, T.F., F.G., and S.S.

**Funding:** D.S.H. and M.M. benefited from a Ph.D. grant from Fundação para a Ciência e a Tecnologia, Portugal (FCT) (PD/BD/114440/2016 and PD/BD/114336/2016). CEF is a research unit funded by FCT (UID/AGR/00239/2013). Some of the proposed life cycle measures were developed within the SuREmMa project (Sustainable River Management—Energiewirtschaftliche und umweltrelevante Bewertung möglicher schwalldämpfender Maßnahmen), which has been sponsored by the Austrian hydropower companies and the Federal Ministry of Agriculture, Forestry, Environment and Water Management via the COMET research program (alpS).

**Acknowledgments:** Supported by BOKU Vienna Open Access Publishing Fund.

**Conflicts of Interest:** The authors declare no conflict of interest. The funders had no role in the design of the study; in the collection, analyses, or interpretation of data; in the writing of the manuscript, or in the decision to publish the results.

## References

1. Ashraf, F.B.; Haghighi, A.T.; Riml, J.; Alfredsen, K.; Koskela, J.J.; Kløve, B.; Marttila, H. Changes in short term river flow regulation and hydropeaking in Nordic rivers. *Sci. Rep.* **2018**, *8*, 17232. [CrossRef] [PubMed]
2. Greimel, F.; Neubarth, J.; Zeiringer, B.; Hayes, D.S.; Haslauer, M.; Führer, S.; Auer, S.; Höller, N.; Hauer, C.; Holzapfel, P.; et al. Sustainable River Management in Austria. In Proceedings of the 12th International Symposium on Ecohydraulics 2018, Tokyo, Japan, 19–24 August 2018; p. 4.
3. Hauer, C.; Holzapfel, P.; Leitner, P.; Graf, W. Longitudinal assessment of hydropeaking impacts on various scales for an improved process understanding and the design of mitigation measures. *Sci. Total Environ.* **2017**, *575*, 1503–1514. [CrossRef] [PubMed]
4. Schmutz, S.; Bakken, T.H.; Friedrich, T.; Greimel, F.; Harby, A.; Jungwirth, M.; Melcher, A.; Unfer, G.; Zeiringer, B. Response of Fish Communities to Hydrological and Morphological Alterations in Hydropeaking Rivers of Austria. *River Res. Appl.* **2015**, *31*, 919–930. [CrossRef]

5.  Auer, S.; Zeiringer, B.; Führer, S.; Tonolla, D.; Schmutz, S. Effects of river bank heterogeneity and time of day on drift and stranding of juvenile European grayling (*Thymallus thymallus* L.) caused by hydropeaking. *Sci. Total Environ.* **2017**, *575*, 1515–1521. [CrossRef] [PubMed]

6.  Saltveit, S.J.; Halleraker, J.H.; Arnekleiv, J.V.; Harby, A. Field experiments on stranding in juvenile Atlantic salmon (*Salmo salar*) and brown trout (*Salmo trutta*) during rapid flow decreases caused by hydropeaking. *Regul. Rivers Res. Manag.* **2001**, *17*, 609–622. [CrossRef]

7.  Casas-Mulet, R.; Alfredsen, K.; Brabrand, A.; Saltveit, S.J. Hydropower operations in groundwater-influenced rivers: Implications for Atlantic salmon, *Salmo salar*, early life stage development and survival. *Fish. Manag. Ecol.* **2016**, *23*, 144–151. [CrossRef]

8.  Greimel, F.; Neubarth, J.; Fuhrmann, M.; Führer, S.; Habersack, H.; Haslauer, M.; Hauer, C.; Holzapfel, P.; Auer, S.; Pfleger, M.; et al. *SuREmMa, Sustainable River Management—Energiewirtschaftliche und umweltrelevante Bewertung möglicher schwalldämpfender Maßnahmen*; Bundesministerium für Land- und Forstwirtschaft, Umwelt und Wasserwirtschaft: Vienna, Austria, 2017.

9.  IHA. *Hydropower Status Report 2017*; International Hydropower Association: London, UK, 2017.

10. Bruder, A.; Tonolla, D.; Schweizer, S.P.; Vollenweider, S.; Langhans, S.D.; Wüest, A. A conceptual framework for hydropeaking mitigation. *Sci. Total Environ.* **2016**, *568*, 1204–1212. [CrossRef]

11. Greimel, F.; Schülting, L.; Wolfram, G.; Bondar-Kunze, E.; Auer, S.; Zeiringer, B.; Hauer, C. Hydropeaking Impacts and Mitigation. In *Riverine Ecosystem Management*; Schmutz, S., Sendzimir, J., Eds.; Springer: Berlin, Germany, 2018; pp. 91–110, ISBN 978-3-319-73250-3.

12. Person, E.; Bieri, M.; Peter, A.; Schleiss, A.J. Mitigation measures for fish habitat improvement in Alpine rivers affected by hydropower operations. *Ecohydrology* **2014**, *7*, 580–599. [CrossRef]

13. Freeman, M.C.; Bowen, Z.H.; Bovee, K.D.; Irwin, E.R. Flow and habitat effects on juvenile fish abundance in natural and altered flow regimes. *Ecol. Appl.* **2001**, *11*, 179–190. [CrossRef]

14. Hayes, D.S.; Brändle, J.M.; Seliger, C.; Zeiringer, B.; Ferreira, T.; Schmutz, S. Advancing towards functional environmental flows for temperate floodplain rivers. *Sci. Total Environ.* **2018**, *633*, 1089–1104. [CrossRef]

15. Moreira, M.; Hayes, D.S.; Boavida, I.; Schletterer, M.; Schmutz, S.; Pinheiro, A. Ecologically-based criteria for hydropeaking mitigation: A review. *Sci. Total Environ.* **2019**, *657*, 1508–1522. [CrossRef]

16. Hunter, M.A. *Hydropower Flow Fluctuations and Salmonids: A Review of the Biological Effects, Mechanical Causes and Options for Mitigation*; Department of Fish and Wildlife: Olympia, WA, USA, 1992.

17. Melcher, A.H.; Bakken, T.H.; Friedrich, T.; Greimel, F.; Humer, N.; Schmutz, S.; Zeiringer, B.; Webb, J.A. Drawing together multiple lines of evidence from assessment studies of hydropeaking pressures in impacted rivers. *Freshw. Sci.* **2017**, *36*, 220–230. [CrossRef]

18. Nagrodski, A.; Raby, G.D.; Hasler, C.T.; Taylor, M.K.; Cooke, S.J. Fish stranding in freshwater systems: Sources, consequences, and mitigation. *J. Environ. Manag.* **2012**, *103*, 133–141. [CrossRef]

19. Crisp, D.T. *Trout and Salmon: Ecology, Conservation and Rehabilitation*; Fishing News Books, Blackwell Science: Oxford, UK, 2000.

20. Quinn, T.P. *The Behavior and Ecology of Pacific Salmon and Trout*; University of Washington Press: Seattle, WA, USA, 2005.

21. Malcolm, I.A.; Gibbins, C.N.; Soulsby, C.; Tetzlaff, D.; Moir, H.J. The influence of hydrology and hydraulics on salmonids between spawning and emergence: Implications for the management of flows in regulated rivers. *Fish. Manag. Ecol.* **2012**, *19*, 464–474. [CrossRef]

22. Mendez, R. *Laichwanderung der Seeforelle im Alpenrhein*; Swiss Federal Institute of Aquatic Science and Technology: Dübendorf, Switzerland, 2007.

23. Jones, N.E.; Petreman, I.C. Environmental Influences on Fish Migration in a Hydropeaking River. *River Res. Appl.* **2015**, *31*, 1109–1118. [CrossRef]

24. Haas, C.; Zinke, P.; Vollset, K.W.; Sauterleute, J.; Skoglund, H. Behaviour of spawning Atlantic salmon and brown trout during ramping events. *J. Appl. Water Eng. Res.* **2016**, *4*, 25–30. [CrossRef]

25. Moir, H.J.; Gibbins, C.N.; Soulsby, C.; Webb, J.H. Discharge and hydraulic interactions in contrasting channel morphologies and their influence on site utilization by spawning Atlantic salmon (*Salmo salar*). *Can. J. Fish. Aquat. Sci.* **2006**, *63*, 2567–2585. [CrossRef]

26. Vollset, K.W.; Skoglund, H.; Wiers, T.; Barlaup, B.T. Effects of hydropeaking on the spawning behaviour of Atlantic salmon *Salmo salar* and brown trout *Salmo trutta*. *J. Fish Biol.* **2016**, *88*, 2236–2250. [CrossRef]

27. Esteve, M. Observations of spawning behaviour in Salmoninae: Salmo, Oncorhynchus and Salvelinus. *Rev. Fish Biol. Fish.* **2005**, *15*, 1–21. [CrossRef]

28. Chapman, D.W.; Weitkamp, D.E.; Welsh, T.L.; Dell, M.B.; Schadt, T.H. Effects of river flow on the distribution of Chinook salmon redds. *Trans. Am. Fish. Soc.* **1986**, *115*, 537–547. [CrossRef]

29. IRKA. *Alpenrhein: Quantitative Analyse von Schwall/Sunk-Ganglinien für unterschiedliche Anforderungsprofile*; IRKA: Vaduz, Liechtenstein, 2012.

30. Moir, H.J.; Gibbins, C.N.; Soulsby, C.; Webb, J. Linking channel geomorphic characteristics to spatial patterns of spawning activity and discharge use by Atlantic salmon (*Salmo salar* L.). *Geomorphology* **2004**, *60*, 21–35. [CrossRef]

31. Becker, C.D.; Neitzel, D.A. Assessment of intergravel conditions influencing egg and alevin survival during salmonid redd dewatering. *Environ. Biol. Fishes* **1985**, *12*, 33–46. [CrossRef]

32. McMichael, G.A.; Rakowski, C.L.; James, B.B.; Lukas, J.A. Estimated Fall Chinook Salmon Survival to Emergence in Dewatered Redds in a Shallow Side Channel of the Columbia River. *N. Am. J. Fish. Manag.* **2005**, *25*, 876–884. [CrossRef]

33. Alfredsen, K.; Harby, A.; Linnansaari, T.; Ugedal, O. Development of an inflow-controlled environmental flow regime for a Norwegian river. *River Res. Appl.* **2012**, *28*, 731–739. [CrossRef]

34. Hauer, C.; Holzapfel, P.; Tonolla, D.; Habersack, H.; Zolezzi, G. In situ measurements of fine sediment infiltration (FSI) in gravel-bed rivers with a hydropeaking flow regime. *Earth Surf. Process. Landf.* **2018**. [CrossRef]

35. Eberstaller, J.; Pinka, P. *Trübung und Schwall Alpenrhein. Einfluss auf Substrat, Benthos, Fische. Teilbericht Fischökologie*; BOKU: Wien, Austria, 2001.

36. Meile, T.; Fette, M.; Baumann, P. *Synthesebericht Schwall/Sunk*; Rhone-Thur Project, Swiss Federal Institute of Aquatic Science and Technology (EAWAG): Dübendorf, Switzerland, 2005.

37. Casas-Mulet, R.; Alfredsen, K.; Hamududu, B.; Timalsina, N.P. The effects of hydropeaking on hyporheic interactions based on field experiments. *Hydrol. Process.* **2015**, *29*, 1370–1384. [CrossRef]

38. Sternecker, K.; Geist, J. The effects of stream substratum composition on the emergence of salmonid fry. *Ecol. Freshw. Fish* **2010**, *19*, 537–544. [CrossRef]

39. O'Connor, W.C.K.; Andrew, T.E. The effects of siltation on Atlantic salmon, *Salmo salar* L., embryos in the River Bush. *Fish. Manag. Ecol.* **1998**, *5*, 393–401. [CrossRef]

40. Schälchli, U.; Abegg, J.; Hunzinger, L. *Kolmation: Methoden zur Erkennung und Bewertung*; Swiss Federal Institute of Aquatic Science and Technology (EAWAG): Dübendorf, Switzerland, 2002.

41. Buxton, T.H.; Buffington, J.M.; Yager, E.M.; Hassan, M.A.; Fremier, A.K. The relative stability of salmon redds and unspawned streambeds. *Water Resour. Res.* **2015**, *51*, 6074–6092. [CrossRef]

42. Unfer, G.; Hauer, C.; Lautsch, E. The influence of hydrology on the recruitment of brown trout in an Alpine river, the Ybbs River, Austria. *Ecol. Freshw. Fish* **2011**, *20*, 438–448. [CrossRef]

43. Casas-Mulet, R.; Saltveit, S.J.; Alfredsen, K. The Survival of Atlantic Salmon (*Salmo salar*) Eggs During Dewatering in a River Subjected to Hydropeaking. *River Res. Appl.* **2015**, *31*, 433–446. [CrossRef]

44. Becker, C.D.; Neitzel, D.A.; Fickeisen, D.H. Effects of dewatering on Chinook salmon redds: Tolerance of four developmental phases to daily dewaterings. *Trans. Am. Fish. Soc.* **1982**, *111*, 624–637. [CrossRef]

45. Becker, C.D.; Neitzel, D.A.; Abernethy, C.S. Effects of Dewatering on Chinook Salmon Redds: Tolerance of Four Development Phases to One-Time Dewatering. *N. Am. J. Fish. Manag.* **1983**, *3*, 373–382. [CrossRef]

46. Becker, C.D.; Neitzel, D.A.; Carlile, D.W. Survival data for dewatered Rainbow Trout (Salmo gairdneri Rich.) eggs and alevins. *J. Appl. Ichthyol.* **1986**, *3*, 102–110. [CrossRef]

47. Saltveit, S.J.; Brabrand, Å. Incubation, hatching and survival of eggs of Atlantic salmon (*Salmo salar*) in spawning redds influenced by groundwater. *Limnologica* **2013**, *43*, 325–331. [CrossRef]

48. Connor, E.J.; Pflug, D.E. Changes in the Distribution and Density of Pink, Chum, and Chinook Salmon Spawning in the Upper Skagit River in Response to Flow Management Measures. *N. Am. J. Fish. Manag.* **2004**, *24*, 835–852. [CrossRef]

49. Harnish, R.A.; Sharma, R.; McMichael, G.A.; Langshaw, R.B.; Pearsons, T.N. Effect of hydroelectric dam operations on the freshwater productivity of a Columbia River fall Chinook salmon population. *Can. J. Fish. Aquat. Sci.* **2014**, *71*, 602–615. [CrossRef]

50. Irvine, R.L.; Thorley, J.L.; Westcott, R.; Schmidt, D.; Derosa, D. Why do fish strand? An analysis of ten years of flow reduction monitoring data from the Columbia and Kootenay rivers, Canada. *River Res. Appl.* **2015**, *31*, 1242–1250. [CrossRef]
51. Young, P.S.; Cech, J.J.; Thompson, L.C. Hydropower-related pulsed-flow impacts on stream fishes: A brief review, conceptual model, knowledge gaps, and research needs. *Rev. Fish Biol. Fish.* **2011**, *21*, 713–731. [CrossRef]
52. Schmutz, S.; Fohler, N.; Friedrich, T.; Fuhrmann, M.; Graf, W.; Greimel, F.; Höller, N.; Jungwirth, M.; Leitner, P.; Moog, O.; et al. *Schwallproblematik an Österreichs Fließgewässern—Ökologische Folgen und Sanierungsmöglichkeiten*; Bundesministerium für Land- und Forstwirtschaft, Umwelt und Wasserwirtschaft: Vienna, Austria, 2013.
53. Korman, J.; Walters, C.; Martell, S.J.D.; Pine Ill, W.E.; Dutterer, A. Effects of flow fluctuations on habitat use and survival of age-0 rainbow trout (Oncorhynchus mykiss) in a large, regulated river. *Can. J. Fish. Aquat. Sci.* **2011**, *68*, 1097–1109. [CrossRef]
54. Jensen, A.J.; Johnsen, B.O. The functional relationship between peak spring floods and survival and growth of juvenile Atlantic Salmon (*Salmo salar*) and Brown Trout (*Salmo trutta*). *Funct. Ecol.* **1999**, *13*, 778–785. [CrossRef]
55. Hurel, G. *Impacts du fonctionnement par éclusées d'une usine hydroélectrique sur une population de truites communes (Salmo trutta L.) dans les Pyrénées ariégeoises (09)*; L'Ecole Nationale Supérieure Agronomique de Toulouse: Toulouse, France, 2010.
56. Auer, S.; Fohler, N.; Zeiringer, B.; Führer, S. *Experimentelle Untersuchungen zur Schwallproblematik. Drift und Stranden von Äschen und Bachforellen während der ersten Lebensstadien*; Institute of Hydrobiology and Aquatic Ecosystem Management, University of Natural Resources and Life Sciences, Vienna: Vienna, Austria, 2014.
57. Bauersfeld, K. *Stranding of juvenile salmon by flow reductions at Mayfield Dam on the Cowlitz River, 1976*; Washington Department of Fisheries: Tacoma, WA, USA, 1978.
58. Unfer, G.; Leitner, P.; Graf, W.; Auer, S. *Der Einfluss von Schwallbetrieb auf den Fischbestand der Oberen Drau*; VERBUND—Austrian Hydro Power AG and Bundesministerium für Land- und Forstwirtschaft, Umwelt und Wasserwirtschaft: Vienna, Austria, 2011.
59. Halleraker, J.H.; Sundt, H.; Alfredsen, K.T.; Dangelmaier, G. Application of multiscale environmental flow methodologies as tools for optimized management of a Norwegian regulated national salmon watercourse. *River Res. Appl.* **2007**, *23*, 493–510. [CrossRef]
60. Halleraker, J.H.; Saltveit, S.J.; Harby, A.; Arnekleiv, J.V.; Fjeldstad, H.-P.; Kohler, B. Factors influencing stranding of wild juvenile brown trout (*Salmo trutta*) during rapid and frequent flow decreases in an artificial stream. *River Res. Appl.* **2003**, *19*, 589–603. [CrossRef]
61. Liebig, H.; Cereghino, R.; Lim, P.; Belaud, A.; Lek, S. Impact of hydropeaking on the abundance of juvenile brown trout in a Pyrenean stream. *Arch. für Hydrobiol.* **1999**, *4*, 439–454. [CrossRef]
62. Jungwirth, M.; Winkler, H. The temperature dependence of embryonic development of grayling (*Thymallus thymallus*), Danube salmon (*Hucho hucho*), Arctic char (*Salvelinus alpinus*) and brown trout (*Salmo trutta* fario). *Aquaculture* **1984**, *38*, 315–327. [CrossRef]
63. Bradford, M.J. An experimental study of stranding of juvenile salmonids on gravel bars and in sidechannels during rapid flow decreases. *Regul. Rivers Res. Manag.* **1997**, *13*, 395–401. [CrossRef]
64. Bradford, M.J.; Taylor, G.C.; Allan, J.A.; Higgins, P.S. An experimental study of the stranding of juvenile coho salmon and rainbow trout during rapid flow decreases under winter conditions. *N. Am. J. Fish. Manag.* **1995**, *15*, 473–479. [CrossRef]
65. Heggenes, J.; Saltveit, S.J. Summer stream habitat partitioning by sympatric Arctic charr, Atlantic salmon and brown trout in two sub-arctic rivers. *J. Fish Biol.* **2007**, *71*, 1069–1081. [CrossRef]
66. Heggenes, J.; Krog, O.M.W.; Lindås, O.R.; Dokk, J.G.; Bremnes, T. Homeostatic behavioural responses in a changing environment: Brown trout (*Salmo trutta*) become nocturnal during winter. *J. Anim. Ecol.* **1993**, *62*, 295–308. [CrossRef]
67. Robertson, M.J.; Pennell, C.J.; Scruton, D.A.; Robertson, G.J.; Brown, J.A. Effect of increased flow on the behaviour of Atlantic salmon parr in winter. *J. Fish Biol.* **2004**, *65*, 1070–1079. [CrossRef]
68. Stickler, M.; Alfredsen, K.; Scruton, D.A.; Pennell, C.; Harby, A.; Økland, F. Mid-winter activity and movement of Atlantic salmon parr during ice formation events in a Norwegian regulated river. *Hydrobiologia* **2007**, *582*, 81–89. [CrossRef]

69. Berland, G.; Nickelsen, T.; Heggenes, J.; Okland, F.; Thorstad, E.B.; Halleraker, J. Movements of wild Atlantic salmon parr in relation to peaking flows below a hydropower station. *River Res. Appl.* **2004**, *20*, 957–966. [CrossRef]

70. Puffer, M.; Berg, O.K.; Huusko, A.; Vehanen, T.; Forseth, T.; Einum, S. Seasonal Effects of Hydropeaking on Growth, Energetics and Movement of Juvenile Atlantic Salmon (*Salmo salar*). *River Res. Appl.* **2015**, *31*, 1101–1108. [CrossRef]

71. Boavida, I.; Harby, A.; Clarke, K.D.; Heggenes, J. Move or stay: Habitat use and movements by Atlantic salmon parr (*Salmo salar*) during induced rapid flow variations. *Hydrobiologia* **2017**, *785*, 261–275. [CrossRef]

72. Holzapfel, P.; Leitner, P.; Habersack, H.; Graf, W.; Hauer, C. Evaluation of hydropeaking impacts on the food web in alpine streams based on modelling of fish- and macroinvertebrate habitats. *Sci. Total Environ.* **2017**, *575*, 1489–1502. [CrossRef] [PubMed]

73. Sauterleute, J.F.; Hedger, R.D.; Hauer, C.; Pulg, U.; Skoglund, H.; Sundt-Hansen, L.E.; Bakken, T.H.; Ugedal, O. Modelling the effects of stranding on the Atlantic salmon population in the Dale River, Norway. *Sci. Total Environ.* **2016**, *573*, 574–584. [CrossRef] [PubMed]

74. Scruton, D.A.; Pennell, C.J.; Robertson, M.J.; Ollerhead, L.M.N.; Clarke, K.D.; Alfredsen, K.; Harby, A.; McKinsley, R.S. Seasonal Response of Juvenile Atlantic Salmon to Experimental Hydropeaking Power Generation in Newfoundland, Canada. *N. Am. J. Fish. Manag.* **2005**, *25*, 964–974. [CrossRef]

75. Scruton, D.A.; Pennell, C.; Ollerhead, L.M.N.; Alfredsen, K.; Stickler, M.; Harby, A.; Robertson, M.; Clarke, K.D.; LeDrew, L.J. A synopsis of "hydropeaking" studies on the response of juvenile Atlantic salmon to experimental flow alteration. *Hydrobiologia* **2008**, *609*, 263–275. [CrossRef]

76. Casas-Mulet, R.; Alfredsen, K.; Killingtveit, A. Modelling of environmental flow options for optimal Atlantic salmon, *Salmo salar*, embryo survival during hydropeaking. *Fish. Manag. Ecol.* **2014**, *21*, 480–490. [CrossRef]

77. Yarnell, S.M.; Petts, G.E.; Schmidt, J.C.; Whipple, A.A.; Beller, E.E.; Dahm, C.N.; Goodwin, P.; Viers, J.H. Functional Flows in Modified Riverscapes: Hydrographs, Habitats and Opportunities. *Bioscience* **2015**, *65*, 963–972. [CrossRef]

78. Naiman, R.J.; Latterell, J.J.; Pettit, N.E.; Olden, J.D. Flow variability and the biophysical vitality of river systems. *C. R. Geosci.* **2008**, *340*, 629–643. [CrossRef]

79. Greimel, F.; Zeiringer, B.; Höller, N.; Grün, B.; Godina, R.; Schmutz, S. A method to detect and characterize sub-daily flow fluctuations. *Hydrol. Process.* **2016**, *30*, 2063–2078. [CrossRef]

80. Hauer, C.; Unfer, G.; Holzapfel, P.; Haimann, M.; Habersack, H. Impact of channel bar form and grain size variability on estimated stranding risk of juvenile brown trout during hydropeaking. *Earth Surf. Process. Landf.* **2014**, *39*, 1622–1641. [CrossRef]

81. Tuhtan, J.A.; Noack, M.; Wieprecht, S. Estimating stranding risk due to hydropeaking for juvenile European grayling considering river morphology. *KSCE J. Civ. Eng.* **2012**, *16*, 197–206. [CrossRef]

82. Hauer, C.; Unfer, G.; Graf, W.; Holzapfel, P.; Leitner, P.; Habersack, H. Grundlagenuntersuchungen und Methodikentwicklung zur Bewertung des Wasserkraft-Schwalls bei unterschiedlichen Flusstypen. *Österreichische Wasser- und Abfallwirtschaft* **2013**, *65*, 324–338. [CrossRef]

83. Harby, A.; Noack, M. Rapid Flow Fluctuations and Impacts on Fish and the Aquatic Ecosystem. In *Ecohydraulics: An Integrated Approach*; Maddock, I., Harby, A., Kemp, P., Wood, P., Eds.; John Wiley & Sons, Ltd.: Hoboken, NJ, USA, 2013; pp. 323–335.

84. Zolezzi, G.; Siviglia, A.; Toffolon, M.; Maiolini, B. Thermopeaking in alpine streams: Event characterization and time scales. *Ecohydrology* **2011**, *4*, 564–576. [CrossRef]

85. Olden, J.D.; Naiman, R.J. Incorporating thermal regimes into environmental flows assessments: Modifying dam operations to restore freshwater ecosystem integrity. *Freshw. Biol.* **2010**, *55*, 86–107. [CrossRef]

86. Schülting, L.; Feld, C.K.; Graf, W. Effects of hydro- and thermopeaking on benthic macroinvertebrate drift. *Sci. Total Environ.* **2016**, *573*, 1472–1480. [CrossRef] [PubMed]

87. Wallace, J.B.; Webster, J.R. The role of macroinvertebrates in stream ecosystem function. *Annu. Rev. Entomol.* **1996**, *41*, 115–139. [CrossRef] [PubMed]

88. Pulg, U.; Vollset, K.W.; Velle, G.; Stranzl, S. First observations of saturopeaking: Characteristics and implications. *Sci. Total Environ.* **2016**, *573*, 1615–1621. [CrossRef] [PubMed]

89. Lumsdon, A.E.; Artamonov, I.; Bruno, M.C.; Righetti, M.; Tockner, K.; Tonolla, D.; Zarfl, C. Soundpeaking— Hydropeaking induced changes in river soundscapes. *River Res. Appl.* **2018**, *34*, 3–12. [CrossRef]

90. Tonolla, D.; Bruder, A.; Schweizer, S. Evaluation of mitigation measures to reduce hydropeaking impacts on river ecosystems—A case study from the Swiss Alps. *Sci. Total Environ.* **2017**, *574*, 594–604. [CrossRef] [PubMed]

91. Pragana, I.; Boavida, I.; Cortes, R.; Pinheiro, A. Hydropower Plant Operation Scenarios to Improve Brown Trout Habitat. *River Res. Appl.* **2017**, *33*, 364–376. [CrossRef]

92. L'Abée-Lund, J.H.; Otero, J. Hydropeaking in small hydropower in Norway—Compliance with license conditions? *River Res. Appl.* **2018**, *34*, 372–381. [CrossRef]

93. Costa, M.J.; Boavida, I.; Almeida, V.; Cooke, S.J.; Pinheiro, A.N. Do artificial velocity refuges mitigate the physiological and behavioural consequences of hydropeaking on a freshwater Iberian cyprinid? *Ecohydrology* **2018**, *11*, e1983. [CrossRef]

94. Storli, P.-T. Novel methods of increasing the storage volume at Pumped Storage Power plants. *Int. J. Fluid Mach. Syst.* **2017**, *10*, 209–2017. [CrossRef]

95. HydroFlex. Social Acceptance and Mitigation of Environmental Impact. Available online: https://h2020hydroflex.eu/work-packages/wp-5/ (accessed on 6 March 2019).

96. Northcote, T.G.; Lobón-Cerviá, J. Increasing experimental approaches in stream trout research—1987–2006. *Ecol. Freshw. Fish* **2008**, *17*, 349–361. [CrossRef]

97. Northcote, T.G. Comparative biology and management of Arctic and European grayling (Salmonidae, Thymallus). *Rev. Fish Biol. Fish.* **1995**, *5*, 141–194. [CrossRef]

98. Jonsson, B.; Jonsson, N. A review of the likely effects of climate change on anadromous Atlantic salmon *Salmo salar* and brown trout *Salmo trutta*, with particular reference to water temperature and flow. *J. Fish Biol.* **2009**, *75*, 2381–2447. [CrossRef] [PubMed]

99. Bejarano, M.D.; Jansson, R.; Nilsson, C. The effects of hydropeaking on riverine plants: A review. *Biol. Rev.* **2018**, *93*, 658–673. [CrossRef]

100. Lucas, M.C.; Baras, E. *Migration of Freshwater Fishes*; Blackwell Science Ltd.: Hoboken, NJ, USA, 2001.

101. Crisp, D.T. Environmental requirements of common riverine European salmonid fish species in fresh water with particular reference to physical and chemical aspects. *Hydrobiologia* **1996**, *323*, 201–221. [CrossRef]

102. Lobón-Cerviá, J.; Sanz, N. (Eds.) *Brown Trout: Biology, Ecology and Management*; John Wiley & Sons, Ltd.: Hoboken, NJ, USA, 2018.

103. Fabricius, E.; Gustafson, K.-J. Observations on the spawning behaviour of the grayling, *Thymallus thymallus* (L.). *Rep. Inst. Freshw. Res. Drottningholm* **1955**, *36*, 75–103.

104. Gibson, R.J. The Atlantic salmon in fresh water: Spawning, rearing and production. *Rev. Fish Biol. Fish.* **1993**, *3*, 39–73. [CrossRef]

105. Jonsson, B.; Jonsson, N. *Ecology of Atlantic Salmon and Brown Trout*; Springer Science & Business Media: Berlin, Germany, 2011.

106. Flodmark, L.E.W.; Vøllestad, L.A.; Forseth, T. Performance of juvenile brown trout exposed to fluctuating water level and temperature. *J. Fish Biol.* **2004**, *65*, 460–470. [CrossRef]

107. Vehanen, T.; Huusko, A.; Yrjänä, T.; Lahti, M.; Mäki-Petäys, A. Habitat preference by grayling (*Thymallus thymallus*) in an artificially modified, hydropeaking riverbed: A contribution to understand the effectiveness of habitat enhancement measures. *J. Appl. Ichthyol.* **2003**, *19*, 15–20. [CrossRef]

 *sustainability*

*Article*

# Effects of Substrate on Movement Patterns and Behavior of Stream Fish through Culverts: An Experimental Approach

**Kyla Johnson** [1,*], **Lindsay E. Wait** [2], **Suzanne K. Monk** [2], **Russell Rader** [1], **Rollin H. Hotchkiss** [2] **and Mark C. Belk** [1]

[1] Department of Biology, 4102 LSB, Brigham Young University, Provo, UT 84602, USA; russell_rader@byu.edu (R.R.); mark_belk@byu.edu (M.C.B.)

[2] Department of Civil and Environmental Engineering, Brigham Young University, Provo, UT 84602, USA; lindsayewait@gmail.com (L.E.W.); smonk@westconsultants.com (S.K.M.); rhh@byu.edu (R.H.H.)

* Correspondence: kyla.johnson1229@gmail.com; Tel.: +1-401-262-4907

Received: 25 November 2018; Accepted: 15 January 2019; Published: 17 January 2019

**Abstract:** Culverts can provide a significant barrier to fish passage by fragmenting fish habitats and impeding the passage success of small-bodied fish. Geographical connectivity is critical to the maintenance of diverse fish assemblages. Culverts with high cross-sectional velocity can cause population fragmentation by impeding passage of small, freshwater fish. Behavioral responses of small fish to high velocities can differ among functional groups, and swimming behavior of many species is not well known. We tested effects of substrate type on swimming behavior in two small, freshwater fish species—southern leatherside chub (*Lepidomeda aliciae*, a midwater species), and longnose dace (*Rhinichthys cataractae*, a benthic species)—across three substrate treatments: (1) a bare flume, (2) large flow obstacles, and (3) a natural cobble substrate. Both longnose dace and southern leatherside chub used paths of low velocity and swam in the near-substrate boundary area. Fish in the bare flume and large obstacle treatments swam along the corners of the flume in a straight swim path, whereas fish in the natural substrate treatment used all parts of the flume bed. There was no relationship between passage success of fish and substrate type, fish species, or their interaction. In contrast, substrate type, fish species, and their interaction were significant predictors of passage time. Southern leatherside chub passed through the test section about two to four times faster than longnose dace. Both species took longer to pass through the large flow obstacle treatment compared to the bare flume or natural substrate. The natural substrate created a complex velocity profile with areas of low velocity throughout the entire flume, in contrast to the other two treatments. Our data suggest natural substrates can improve the passage of small fish in high-velocity culverts for both benthic and midwater functional groups.

**Keywords:** fish passage; culvert hydraulics; native fishes; velocity refuge; substrate size

## 1. Introduction

Geographical connectivity to source populations through a network of rivers and streams is critical to the development of diverse fish assemblages. Manmade barriers to fish migration or dispersal have been linked to the decline of freshwater fish assemblages throughout the world [1]. Culverts in streams are ubiquitous in the United States [2,3] and frequently act as barriers to fish passage [2–7]. Recent studies estimate that 30% to 53% of culverts are barriers [2,8]. These structures adversely affect the survival of freshwater fish species [3,8,9], by preventing migration and fragmenting the habitats of freshwater fish, which leads to increased risk of extinction [10,11]. Culverts can also prevent dispersal [6,11], which is critical for population maintenance and access of spawning habitats [3].

Culverts with high velocity profiles fail to accommodate smaller fish, because they do not account for differences in swimming behavior and size-dependent swimming abilities [1,12]. However, correctly retrofitting culverts can effectively restore fish passage [13]. Research on how the morphological and behavioral adaptations of fish affect passage behaviors through culverts will provide valuable information that could inform future culvert design, particularly for small fish [1,9,14,15].

Salmonids are disproportionally represented in studies of fish passage through high-velocity environments [16,17]. In a recent review of fish passage, 77% of studies examining the ecological impact of culverts included salmonids, whereas 45% of all fish passage studies concerned salmonids exclusively [11]. Although the passage success of salmonids has historically been prioritized due to commercial value, scientists are now recognizing the importance of passage for other fish species [18,19]. In recent years, studies on lampreys [20–22], perch [1,12,23,24], sculpins [25], sturgeon [26–29], cyprinids [30,31], and galaxiids [32] have increased our knowledge on non-salmonid fish passage. Some species of small-bodied fish exhibit behavioral adaptations that facilitate upstream passage through high-velocity zones by using low-velocity paths. In non-uniform flows, small, weak-swimming fish select paths of low velocity to swim upstream [14,15,33]. For example, flume studies document small-bodied fishes consistently swimming in the reduced velocity zone (RVZ) produced by the culvert wall [33,34]. Similarly, many studies have indicated that substrate type may influence passage success in culverts by providing areas of reduced velocity [35–39]. However, given the diversity of behavior and functional form of small fish compared to salmonids, there still is relatively little known regarding the effect of substrate type on navigation of small, freshwater fish through culverts [3,11].

Culverts can become a barrier to fish passage for any species of fish if the design fails to account for a given species' behavioral adaptations [12]. Several studies call for research on adaptive swimming behaviors that might inform culvert design [11,15,40,41]. These adaptations include the selection of energy-efficient swim paths and using different swimming strategies. Small fish may use sprint or burst swimming speeds to cross velocity barriers [1,9], or swim at high speeds to minimize passage time [42,43]. Small fish may also conserve energy by taking refuge in areas of reduced velocity [40,44]. Morphological adaptations between functional groups can also influence swimming strategy [28]. Fish in the midwater functional group may take advantage of low-velocity refuges in the wake of obstacles by using "burst-and-coast" swimming [41,44]. Benthic fishes are morphologically adapted to hold position on the substratum by bracing and sheltering [25,45–47]. Behavioral and morphological responses of small fish to high velocities may differ between functional groups, suggesting that culverts designed for the passage of small fish may need to accommodate a range of swimming abilities and strategies, instead of treating small, non-salmonid fish as one group. However, if small fish from different functional groups employ similar swimming strategies for passage, one culvert design may be sufficient to allow for the passage of diverse assemblages of small, non-salmonid stream fish.

To inform culvert design in the context of stream restoration for small-bodied stream fish, species from two different functional groups were selected to model how morphological and behavioral differences between functional groups influences passage behavior and swim paths on different substrates. We tested effects of substrate type on swimming behavior in southern leatherside chub (*Lepidomeda aliciae*, a midwater species) and longnose dace (*Rhinichthys cataractae*, a benthic species) across three substrate treatments: (1) a bare flume with no substrate, (2) large obstacles (15 cm diameter), and (3) a natural cobble substrate (see Section 2.2 below). Because longnose dace is a benthic species, we expected it to use bracing behaviors and do best in the natural substrate [25,45]. Conversely, we expected southern leatherside chub to use "burst-and-coast" swimming and "refuging" to perform best in the "large obstacle" treatment. We compared the passage time and behavior, and swim paths of southern leatherside chub and longnose dace in each treatment to determine if functional groups (midwater and benthic) react similarly to different velocity profiles. We then quantified fine-scale patterns of velocity generated by the three treatments and mapped the velocity profile of swim paths in each treatment.

## 2. Materials and Methods

### 2.1. Study Species

Longnose dace are widely distributed across North America [48]. Like other benthic omnivores [28,29,45,47], they exhibit bracing behavior in high velocities, allowing them to endure powerful flows without large increases in energy expenditure [25,45,49]. Southern leatherside chub co-occur with longnose dace in streams in central Utah, USA, as midwater carnivores [50]. These species are not known to undergo spawning migrations, but are most active during the warmest part of the year, i.e., June through August. In Soldier Creek, UT, USA, where specimens were collected for our experiments (39°59′37.5″ N, 111°29′42″ W, elevation = 1554 m, gradient = 22.5 m/km), southern leatherside chub and longnose dace are two of the most abundant fish species, and they co-occur with mottled sculpin (*Cottus bairdi*), redside shiner (*Richardsonius balteatus*), mountain sucker (*Catostomus platyrhynchus*), cutthroat trout (*Oncorhynchus clarki*), and introduced brown trout (*Salmo trutta*) and rainbow trout (*Oncorhynchus mykiss*) [51]. We captured adult longnose dace (mean total length = 82.5 mm $\pm$ 3.75 mm; range = 75 mm–90 mm) and southern leatherside chub (mean TL = 72.5 mm $\pm$ 3.75 mm; range = 65 mm–80 mm) with handheld seine nets (mesh size = 6.5 mm) in Soldier Creek during June and July of 2010. We pulled seine nets through pool and riffle environments and collected all fish for a given week over the course of one to two hours.

### 2.2. Experimental Design

We conducted six passage tests each week: three on longnose dace and three on southern leatherside chub over the course of 4 weeks for a total of twelve replicates or trials for both species. This was four trials for each species in each of the three flow treatments described below. We randomly assigned 5 fish from one of the two species to each trial, for a total of 60 fish of both species used in this study. Thus, we collected at least 15 fish of each species at the beginning of each week.

Trials were conducted in a laboratory flume designed to simulate flow conditions in a culvert with a similar diameter (Figure 1). That is, high velocity conditions towards the central portion of the barrel, with reduced velocities near the edges. The flume was made of Plexiglas with an adjustable flow rate, depth, and slope. The recirculating supply of water was equipped with an inline Venturi meter to accurately monitor flow rate, and water velocities were set by manipulating the slope of the flume.

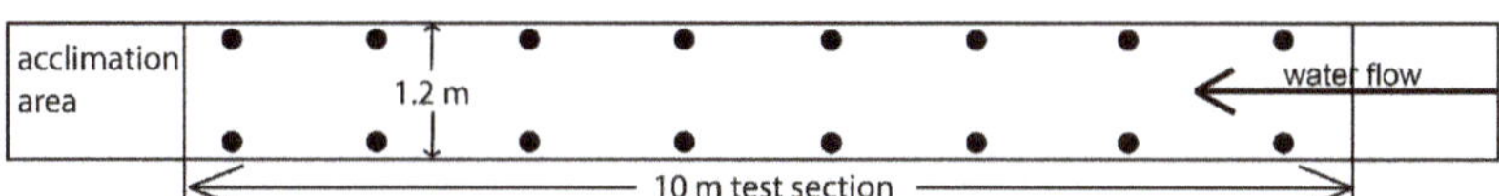


**Figure 1.** Plan view of the experimental flume representing the bare flume, large obstacle, and natural substrate treatments. Cylinders were placed in the large obstacle treatment only. In the natural substrate and bare flume treatments, the entirety of the experimental flume was covered with cobble substrate or left bare, respectively.

Trials consisted of three treatments: (1) the bare flume representative of bare culverts [32], (2) linearly arrayed obstacles made of concrete cylinders 15 cm in diameter (see Figure 1), representative of baffles, and (3) natural cobble substrate from the stream where the fish were collected. For the natural substrate trial, we positioned the substrate along the entire length of the flume upstream from the acclimation area, and spanning the entire width of the flume (Figure 1). A standard sieve analyses of four samples showed that the substrate median diameter ($d_{50}$) was 46 mm with a standard deviation of 6.7 mm.

Fish were allowed to acclimate for one hour in a holding area at a water velocity of 0.20 m/s (Figure 1). After the acclimation period, we increased the flow rate until the flume reached the desired cross-sectional average velocity for each test (see below). Once the flow rate was steady, we raised the containment barrier (a mesh panel) to begin a trial. Each trial lasted 60 minutes from the end of the acclimation period. In trials with natural substrate, cobbles were extended beyond the test section, such that fish were not suddenly confronted with an abrupt difference in bottom material at the end of the test section. Passage success was the number of fish per trial that traversed the test section of the flume. We also recorded the time for each fish to swim the length of the test section (Figure 1). If fish did not complete the test within 60 minutes, we recorded them as having failed. If a fish did not change position after 15 minutes, we lightly prodded their caudal fin to motivate movement [52]. If fish were impinged against the downstream gate or otherwise exhibited extreme exhaustion with no reaction to prodding, we removed them from the test [1,52].

We mapped fish swim paths by recording the position of each fish at five minute intervals as they moved upstream. Based on visual observations, positions were marked as the distance upstream from the beginning of the test section and the distance from the right edge of the flume. Swim paths were mapped as a sequence of positions with a straight line drawn between positions. Fish were visually monitored during the time between recording intervals (i.e., 5 minutes), but position was not recorded.

The average cross-sectional velocity for each trial was set at 25% greater than the prolonged swimming speeds for both species, which were 0.54 m/s for southern leatherside chub and 0.73 m/s for longnose [49]). Prolonged swimming speeds underestimate fish passage [53,54], since fish are often forced to swim above their prolonged swim speeds [1,49,55]. For longnose dace, we ran tests at a flow rate of 0.20 m$^3$/s, a velocity of 0.91 m/s, a depth of 0.2 m, and a Froude number of 0.66, whereas flow rate was 0.15 m$^3$/s, with a velocity of 0.68 m/s, a depth of 0.16 m, and a Froude number of 0.60 for southern leatherside chub. At these velocities the approach flow was fully developed according to the Blasius equation for flat planes [56], and Reynolds numbers were >100,000 for all experiments. To control turbulent surface waves that form at high velocities, which can impact the swimming ability of the fish [9], we tethered a 1 m long plywood board to the water surface at the inlet section of the flume. This board did not overlap with the test section.

We ran all trials within 4 days of collecting fish [57]. We transported fish to holding tanks (volume = 1100 liters) in aerated, insulated coolers filled with river water. The holding tanks were kept under 12 h light and dark cycles at about 18 °C, and fish were fed frozen brine shrimp twice daily. Mean water temperature was also kept at 18 °C $\pm$ 4 °C in the experimental flume. This temperature was in the mid-range of that experienced by fish in Soldier Creek during the growing season [51].

*2.3. Velocity Profiles and Swim Paths*

We used a SonTek 16-MHz Micro Acoustic Doppler Velocimeter (ADV) [58] to generate detailed maps of current velocity for each of our substrate treatments. The ADV is a three-pronged sensor that takes 3D velocity readings in a $\approx$0.3 cm$^3$ sampling volume 5 cm below the probe tip. ADV output data includes signal to noise ratio (SNR) and correlation (COR) values that filter out noise in the acoustic reflections. COR values are ideally greater than 70%, but COR values as low as 30% can be used for mean velocity measurements over variable terrain, and SNR values for mean current measurements can be as low as 5 dB. [59]. To reduce the effect of high turbulence near the substrate, only SNR values greater than 5 dB and COR above 30% that resulted in at least 70% good points were include in our analysis. We used the mean water-column velocity to characterize the velocity profile [46]. Velocity measurements were taken along a grid 1 cm above the substrate on 5 $\times$ 5 cm planform [14]. We used AutoCad software [60] with the velocity data to map isovels (contours of equal velocity) for a 3.5 m length of the flume located near the center of the 10 m test section.

To determine the lowest velocity swim path available to fishes in each of the substrate treatments, we plotted velocity along a simulated continuous swim path. We modeled this minimum velocity swim path after the observed swim paths for each substrate treatment. In the natural substrate treatment, where swim paths were highly variable, we simulated the path of minimum velocity through the measured velocity grid and plotted the resulting velocity profile. To simulate the swim paths, we used an algorithm that assumed fish could move any direction from a given point to follow the path of minimum velocity. We mapped minimum velocity profiles of a simulated fish swim path using AutoCad software [60].

*2.4. Statistical Analyses*

We used a logistic analysis of variance (Proc Logistic, SAS 9.3, SAS Institute, Cary, North Carolina) to examine the relationship between the categorical response variable (if individuals successfully traversed the 10 m test section; yes or no) and the categorical independent variables of fish species (2 levels), substrate treatments (3 levels), and their interaction [61]. For individuals that successfully traversed the test section (95 out of 120 individuals), we examined the relationship between passage time (a continuous dependent variable) and fish species (2 levels), substrate treatments (3 levels), and their interaction with a mixed model ANOVA in SAS 9.3 (Proc Mixed) [62]. We used a natural log-transform to meet the assumptions of the parametric model. A replicate trial was a random effect in both analyses.

## 3. Results

*3.1. Swim Paths*

In all substrate treatments, both longnose dace and southern leatherside chub passed through the flume within 1 cm above the substrate. Swim paths in the control (i.e., bare flume) and large flow obstacle treatments were completely uniform: fish swam in a straight swim path as close as physically possible to the wall of the flume, where velocity was lowest. They entered and exited the test section of the flume in the same position near the walls of the flume. None of the fish in these two treatments moved laterally from one side of the flume to the other. In the large flow obstacles treatment, none of the fish swam along the edge of the cylinders opposite the wall of the flume (i.e., interior edge). Because position of fish was recorded at 5 min intervals, swim paths for the natural substrate treatment appear as straight lines between recorded points. However, in the natural substrate treatment, in contrast to the uniform swim paths in the other two treatments, we observed fish swimming throughout the flume near the substrate with no clear preference for flume walls. There appeared to be no directional pattern to swim paths in the natural substrate treatment other than upstream, and fish made lateral movements across the flume, reversals in direction (including downstream), and entered and exited the test section in variable locations across the width of the flume (Figure 2).

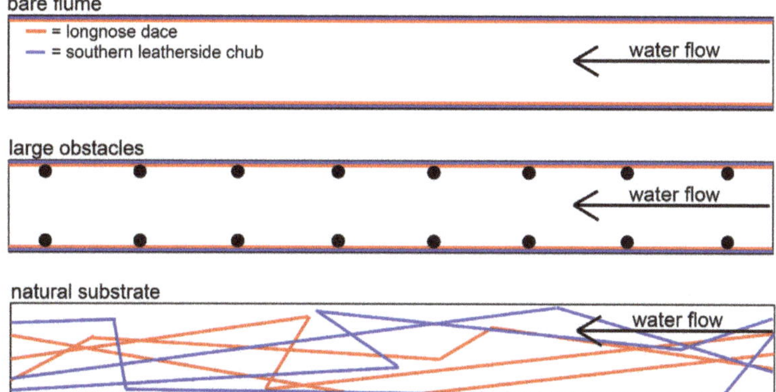


**Figure 2.** Plan view of sample swim paths of southern leatherside chub and longnose dace in bare flume (**top**), large flow obstacles (**middle**), and natural substrate (**bottom**) treatments in the 10 m test section of the flume. The flume test section is 10 m in length and 1.2 m in width, and the position and relative size of the large obstacles are indicated by the black circles. In these two treatments, the red line indicating longnose dace and the blue line indicating southern leatherside chub are slightly offset to show color differences, but do not represent differences in swim path.

### 3.2. Passage Success and Time

At least 70% of individuals in both species among all three treatments successfully passed through the test flume (Figure 3a). There was no relationship between passage success of fish and substrate type, fish species, or their interaction (Table 1). In contrast, substrate type, fish species, and their interaction were significant predictors of passage time (Table 2). Southern leatherside chub passed through the flume about two to four times faster than longnose dace. Both species took longer to pass through the large flow obstacles treatment compared to the bare flume or natural substrate. This effect was more pronounced in southern leatherside chub (Figure 3b).

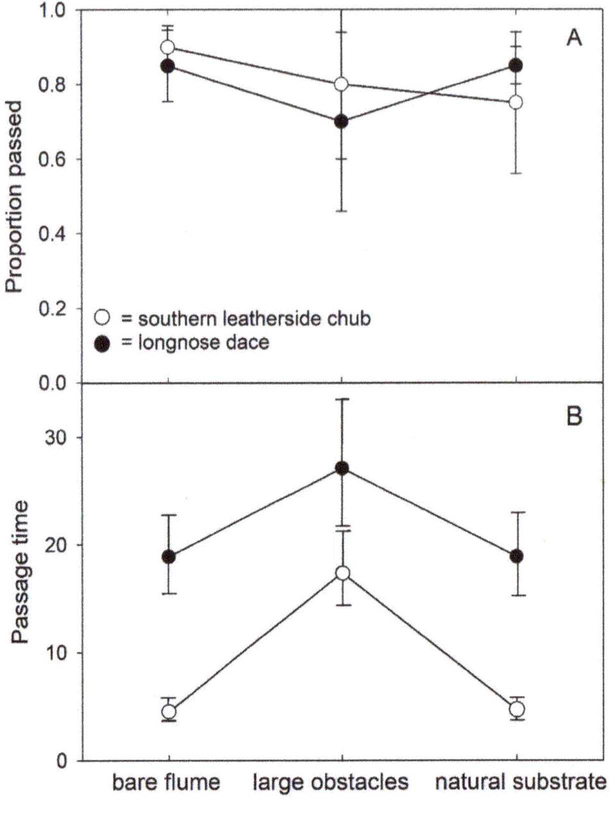


**Figure 3.** Mean (±1 SE) of (**A**) passage rate, and (**B**) passage time (in min) through the flume based on substrate type and fish species.

**Table 1.** Logistic regression analysis for fish passage success through the flume comparing the three substrate treatments (bare flume, large obstacles, and natural substrate) and two fish species (southern leatherside chub and longnose dace), and their interaction.

| Source of Variation | df | Chi-Square | P-Value |
|---|---|---|---|
| species | 1 | 0.09 | 0.76 |
| substrate | 2 | 2.09 | 0.35 |
| species × substrate | 2 | 1.43 | 0.49 |

**Table 2.** Analysis of variance for time of fish passage through the flume comparing the three substrate treatments (bare flume, large obstacles, and natural substrate) and two fish species (southern leatherside chub and longnose dace), and their interaction.

| Source of Variation | df | F Value | P-Value |
| --- | --- | --- | --- |
| species | 1/89 | 46.31 | <0.0001 |
| substrate | 2/89 | 11.94 | <0.0001 |
| species × substrate | 2/89 | 3.96 | 0.0224 |

### 3.3. Velocity Profiles and Swim Paths

The three substrate treatments created dramatically different velocity patterns in the test field. The lowest velocities in the bare flume were near the flume walls. Velocity increased uniformly from the walls to the center of the flume, where velocity was highest (Figure 4). In the large flow obstacle treatment, the areas of lowest velocity were located downstream from the obstacles and between the obstacles and the wall of the flume. Like the bare flume, the highest velocity in the large flow obstacles treatment was located in the center of the flume. (Figure 4). In contrast to both of these treatments, the natural cobble substrate produced a complex, nonuniform velocity pattern with low and high velocities distributed throughout the entire flume area (Figure 4).

bare flume

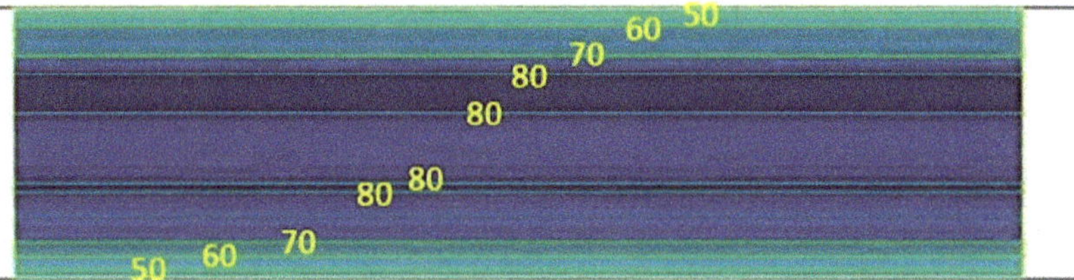


large obstacles

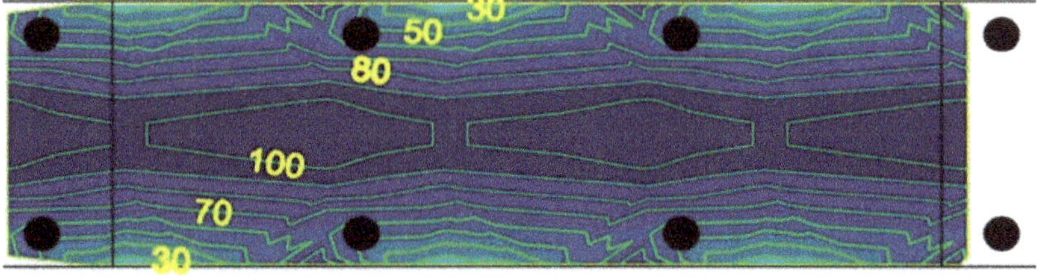


natural substrate

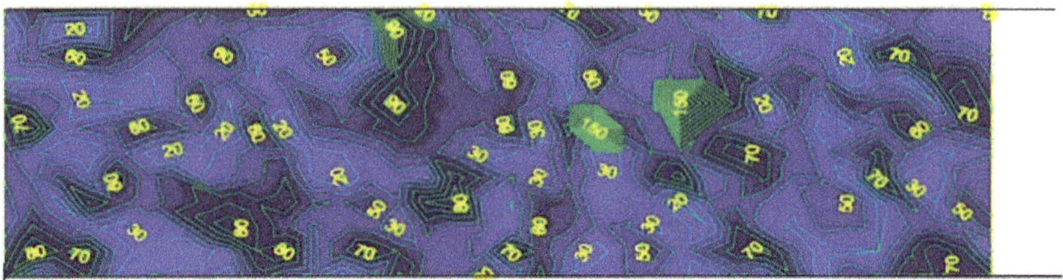

**Figure 4.** Plan view of velocity field in experimental flume for substrate treatments of bare flume (**top**), large flow obstacles (**middle**), and natural substrate (**bottom**). Length of the section shown above was 3.5 m. Water flow is from right to left. Velocity contours are in units of cm/s, and are shown at a flow rate of 0.91 m/s.

The velocity experienced by fish for simulated swim paths in the bare flume was constant at about 47 cm/s throughout the length of the flume near the flume walls (Figure 5a). In the large obstacle treatment, the velocity profile of simulated swim paths fluctuated regularly between velocities of 28 cm/s and 80 cm/s as the fish moved along the flume wall past the obstacles (Figure 5a). In the natural substrate treatment, the velocity profile of the simulated swim path fluctuated irregularly between about 15 cm/s and 49 cm/s as the fish moved in irregular paths across the flume following a low-velocity path (Figure 5b).

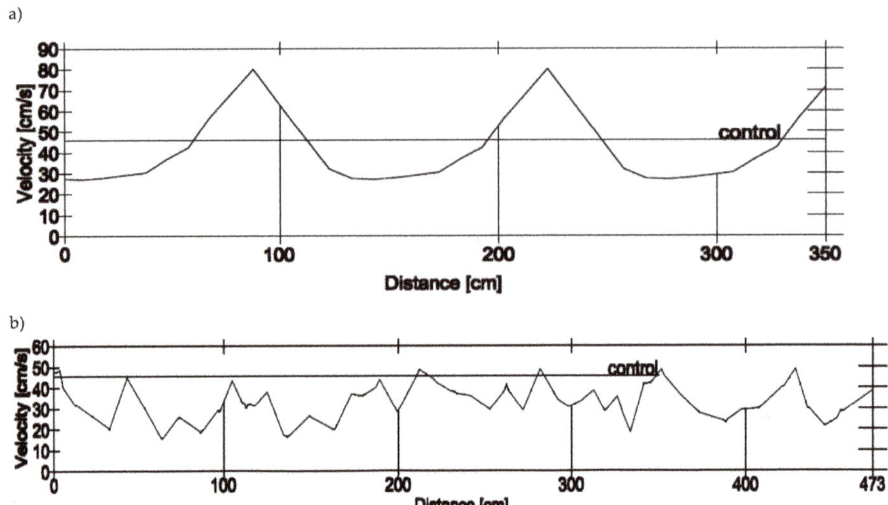


**Figure 5.** Velocity profiles for longnose dace flow conditions with velocity measurements taken 1 cm above the substrate along simulated typical swim paths; (**a**) bare flume (designated as control) and large flow obstacles treatment, and (**b**) bare flume (designated as control) and natural substrate treatment. The distance is longer in the natural substrate treatment because the swim path was not straight.

## 4. Discussion

Successful restoration of stream habitats depends on connectivity of fish assemblages within the stream system. Culverts can become barriers to fish movement and lead to fragmentation of otherwise suitable habitat [63]. The results of this study indicate that small-bodied stream fish, such as longnose dace and southern leatherside chub, select swim paths of low velocity to pass through culverts. It also suggests that natural substrate positively affects the ability of small fish to swim upstream by providing areas of reduced velocity. Fish consistently utilized paths of low velocity when navigating upstream through the experimental flume. In all treatments, fish swam close to the substrate, where velocity is lowest. Neither species used the midwater area, even though southern leatherside chub typically frequent midwater habitats [64]. The swim paths of fish also mirrored the path of lowest velocity. In the bare flume trial, fish swam up the side of the flume, where velocity was lowest [65]. Conversely, the natural substrate treatment produced a highly variable velocity profile [66]. The swim paths of fish in the natural substrate treatment reflected this variability. Swim paths lacked a clear pattern and covered the entirety of the flume bed, mirroring the non-uniformity of the velocity profile.

The difference in passage time between the two species is likely due to differences in swimming behavior, which is influenced by morphological features, such as rostrum shape, size of pectoral fins, and body shape [28]. Behavioral and morphological responses to high velocities differ between functional groups [23,25]. In high velocities, longnose dace exhibit a bracing behavior that allows them to maintain position without expending much energy [45,49,67]. Southern leatherside chub, on the other hand, must swim continuously to maintain position, even in lower velocity zones [49]. These behavioral adaptations minimize energy expenditure [25,40,45]; for southern leatherside chub, rapid passage through high velocity areas conserves energy, whereas longnose dace are able to brace without increasing energy expenditure [25,46]. Rough bottom substrates may provide an energetic advantage to both species if substratum dimensions are similar to the body size of the fish [27,46]. Fish seek refuge from the main current by "flow refuging" in areas of reduced flow velocity [44]. As the water velocity increases, the maximum distance that fish can travel without resting decreases [68,69].

The availability of low-velocity zones is an important factor to consider in high velocity flow fields [70]. Our observations support the hypothesis that small, weak-swimming fish navigate upstream by using paths of low velocity [15,33]. Our findings are also consistent with reports of small-bodied fish moving upstream along the side of the flume in the reduced velocity zone or skimming the substrate [15,28,33,34,71]. This study suggests that average cross-sectional velocities may not provide a good metric for predicting passage of small-bodied stream fish, because both midwater and benthic swimmers use the low-velocity area directly above the substrate for passage. This study also indicates that constricted structures that create high velocities are more likely to be a barrier to midwater swimmers, which don't have energy-saving bracing behaviors like benthic species [49].

The use of natural substrate and artificial or natural rugosities to create low velocity zones gives greater flexibility to culvert design in stream restoration plans. The average cross-sectional velocity in a culvert is used to accommodate the prolonged swimming speed of the weakest swimmer [72], which results in dramatically increased culvert sizes [15]. The size of the smallest fish that must transverse the culvert influences its design and cost [12]. These results demonstrate that for multiple species of small-bodied fish, upstream passage is greatly facilitated by lining the culvert bottom with the substrate found in the fish species' native environment. This is a simple and economical mitigation compared to the expense of replacing or otherwise modifying the culvert barrel.

Other studies on fish passage have shown that in addition to velocity, turbulence can have strong effects on swimming behavior and movement [9,12,73]. We did not measure turbulence in the bare flume or large obstacles treatments; however, we did observe that fish in the large obstacle treatment did not use the turbulent wake region downstream of the obstacles, suggesting that perhaps turbulence influenced swim path behavior in this treatment. In the natural substrate treatment, we did not observe any noticeable effects of turbulence. Turbulence was likely lower in the natural substrate treatment because the size of the substrate particles were smaller than in the large obstacle treatment. Effects of combined velocity and turbulence measurements would be an interesting area for future experimental research.

In summary, our data suggest that the type of substrate in a culvert can have a strong influence on movement and swimming behavior of both mid-water and benthic-oriented fish. Natural cobble substrate provides a much less restrictive environment for passage with multiple low velocity paths, compared to large obstacles or a bare culvert. These results add to our understanding of how small stream fish move through high flow environments created by culverts and other flow restrictions.

**Author Contributions:** M.C.B. and R.H.H. conceived and designed the experiments. L.E.W. and S.K.M. performed the experiments. M.C.B., L.E.W., R.R., and S.K.M. analyzed the data; R.R. and R.H.H. contributed reagents, materials, and analysis tools. K.J., M.C.B., R.R., and L.E.W. wrote the paper.

**Acknowledgments:** This work has been carried out with funding from the Utah Department of Transportation (UDOT). Thanks for knowledge and assistance in fish capture and care from Eric Billman and undergraduates in the Brigham Young University Department of Biology. Additional thanks to Karsten Busby, David Anderson, and Rodney Mayo for technical and logistical assistance in lab experiments. All fish capture, care, and testing was in accordance with the Institutional Animal Care and Use Committee (IACUC) protocol #10-0401.

**Conflicts of Interest:** The authors declare no conflict of interest. The funding sponsors had no role in the design of the study; in the collection, analyses, or interpretation of data; in the writing of the manuscript, and in the decision to publish the results.

## References


1. Starrs, D.; Ebner, B.; Lintermans, M.; Fulton, C. Using sprint swimming performance to predict upstream passage of the endangered Macquarie perch in a highly regulated river. *Fish. Manag. Ecol.* **2011**, *18*, 360–374. [CrossRef]
2. Price, D.M.; Quinn, T.; Barnard, R.J. Fish passage effectiveness of recently constructed road crossing culverts in the Puget Sound region of Washington State. *N. Am. J. Fish. Manag.* **2010**, *30*, 1110–1125. [CrossRef]
3. Warren Jr, M.L.; Pardew, M.G. Road crossings as barriers to small-stream fish movement. *Trans. Am. Fish. Soc.* **1998**, *127*, 637–644. [CrossRef]
4. Blank, M.; Cahoon, J.; Burford, D.; McMahon, T.; Stein, O. *Studies of Fish Passage through Culverts in Montana*; Road Ecology Center: Berkeley, CA, USA, 2005.
5. Clark, S.P.; Toews, J.S.; Tkach, R. Beyond average velocity: Modelling velocity distributions in partially filled culverts to support fish passage guidelines. *Int. J. River Basin Manag.* **2014**, *12*, 101–110. [CrossRef]
6. MacPherson, L.M.; Sullivan, M.G.; Foote, A.L.; Stevens, C.E. Effects of culverts on stream fish assemblages in the Alberta foothills. *N. Am. J. Fish. Manag.* **2012**, *32*, 480–490. [CrossRef]
7. Park, D.; Sullivan, M.; Bayne, E.; Scrimgeour, G. Landscape-level stream fragmentation caused by hanging culverts along roads in Alberta's boreal forest. *Can. J. For. Res.* **2008**, *38*, 566–575. [CrossRef]
8. Gibson, R.J.; Haedrich, R.L.; Wernerheim, C.M. Loss of fish habitat as a consequence of inappropriately constructed stream crossings. *Fisheries* **2005**, *30*, 10–17. [CrossRef]
9. Rodríguez, T.T.; Agudo, J.P.; Mosquera, L.P.; González, E.P. Evaluating vertical-slot fishway designs in terms of fish swimming capabilities. *Ecol. Eng.* **2006**, *27*, 37–48. [CrossRef]
10. Winston, M.R.; Taylor, C.M.; Pigg, J. Upstream extirpation of four minnow species due to damming of a prairie stream. *Trans. Am. Fish. Soc.* **1991**, *120*, 98–105. [CrossRef]
11. Roscoe, D.W.; Hinch, S.G. Effectiveness monitoring of fish passage facilities: Historical trends, geographic patterns and future directions. *Fish Fish.* **2010**, *11*, 12–33. [CrossRef]
12. Mallen-Cooper, M.; Brand, D. Non-salmonids in a salmonid fishway: What do 50 years of data tell us about past and future fish passage? *Fish. Manag. Ecol.* **2007**, *14*, 319–332. [CrossRef]
13. Franklin, P.A.; Bartels, B.; Ecosystems, F. Restoring connectivity for migratory native fish in a New Zealand stream: Effectiveness of retrofitting a pipe culvert. *Aquat. Conserv. Mar.* **2012**, *22*, 489–497. [CrossRef]
14. House, M.R.; Pyles, M.R.; White, D. Velocity distributions in streambed simulation culverts used for fish passage. *J. Am. Water Resour. Assoc.* **2005**, *41*, 209–217. [CrossRef]
15. Richmond, M.C.; Deng, Z.; Guensch, G.R.; Tritico, H.; Pearson, W.H. Mean flow and turbulence characteristics of a full-scale spiral corrugated culvert with implications for fish passage. *Ecol. Eng.* **2007**, *30*, 333–340. [CrossRef]
16. Belford, D.A.; Gould, W.R. An evaluation of trout passage through six highway culverts in Montana. *N. Am. J. Fish. Manag.* **1989**, *9*, 437–445. [CrossRef]
17. Fitch, G.M. *Nonanadromous Fish Passage in Highway Culverts. Final Report*; The National Academies of Sciences, Engineering, and Medicine: Washington, DC, USA, 1995.
18. Mallen-Cooper, M.; Stuart, I. Optimising Denil fishways for passage of small and large fishes. *Fish. Manag. Ecol.* **2007**, *14*, 61–71. [CrossRef]
19. Cooke, S.J.; Bunt, C.M.; Hamilton, S.J.; Jennings, C.A.; Pearson, M.P.; Cooperman, M.S.; Markle, D.F. Threats, conservation strategies, and prognosis for suckers (Catostomidae) in North America: Insights from regional case studies of a diverse family of non-game fishes. *Biol. Conserv.* **2005**, *121*, 317–331. [CrossRef]
20. Hard, A.; Kynard, B. Video evaluation of passage efficiency of American shad and sea lamprey in a modified Ice Harbor fishway. *N. Am. J. Fish. Manag.* **1997**, *17*, 981–987. [CrossRef]
21. Laine, A.; Kamula, R.; Hooli, J. Fish and lamprey passage in a combined Denil and vertical slot fishway. *Fish. Manag. Ecol.* **1998**, *5*, 31–44. [CrossRef]
22. Moser, M.L.; Matter, A.L.; Stuehrenberg, L.C.; Bjornn, T.C. Use of an extensive radio receiver network to document Pacific lamprey (*Lampetra tridentata*) entrance efficiency at fishways in the Lower Columbia River, USA. In *Aquatic Telemetry*; Springer: New York, NY, USA, 2002; pp. 45–53.
23. Tudorache, C.; Viaene, P.; Blust, R.; Vereecken, H.; De Boeck, G. A comparison of swimming capacity and energy use in seven European freshwater fish species. *Ecol. Freshw. Fish* **2008**, *17*, 284–291. [CrossRef]

24. Knaepkens, G.; Baekelandt, K.; Eens, M. Fish pass effectiveness for bullhead (*Cottus gobio*), perch (*Perca fluviatilis*) and roach (*Rutilus rutilus*) in a regulated lowland river. *Ecol. Freshw. Fish* **2006**, *15*, 20–29. [CrossRef]

25. Facey, D.; Grossman, G. The relationship between water velocity, energetic costs, and microhabitat use in four North American stream fishes. *Hydrobiologia* **1992**, *239*, 1–6. [CrossRef]

26. Peake, S.; Beamish, F.W.; McKinley, R.; Scruton, D.; Katopodis, C. Relating swimming performance of lake sturgeon, Acipenser fulvescens, to fishway design. *Can. J. Fish. Aquat. Sci.* **1997**, *54*, 1361–1366. [CrossRef]

27. Downie, A.T.; Kieffer, J.D. A split decision: The impact of substrate type on the swimming behaviour, substrate preference and UCrit of juvenile shortnose sturgeon (*Acipenser brevirostrum*). *Environ. Biol. Fishes* **2017**, *100*, 17–25. [CrossRef]

28. Kieffer, J.; Arsenault, L.; Litvak, M. Behaviour and performance of juvenile shortnose sturgeon Acipenser brevirostrum at different water velocities. *J. Fish Biol.* **2009**, *74*, 674–682. [CrossRef] [PubMed]

29. May, L.; Kieffer, J. The effect of substratum type on aspects of swimming performance and behaviour in shortnose sturgeon Acipenser brevirostrum. *J. Fish Biol.* **2017**, *90*, 185–200. [CrossRef] [PubMed]

30. Branco, P.; Santos, J.M.; Katopodis, C.; Pinheiro, A.; Ferreira, M.T. Pool-type fishways: Two different morpho-ecological cyprinid species facing plunging and streaming flows. *PLoS ONE* **2013**, *8*, e65089. [CrossRef]

31. Romão, F.; Quaresma, A.L.; Branco, P.; Santos, J.M.; Amaral, S.; Ferreira, M.T.; Katopodis, C.; Pinheiro, A.N. Passage performance of two cyprinids with different ecological traits in a fishway with distinct vertical slot configurations. *Ecol. Eng.* **2017**, *105*, 180–188. [CrossRef]

32. Doehring, K.; Young, R.G.; McIntosh, A.R. Factors affecting juvenile galaxiid fish passage at culverts. *Mar. Freshw. Res.* **2011**, *62*, 38–45. [CrossRef]

33. Johnson, G.E.; Pearson, W.H.; Southard, S.L.; Mueller, R.P. Upstream movement of juvenile coho salmon in relation to environmental conditions in a culvert test bed. *Trans. Am. Fish. Soc.* **2012**, *141*, 1520–1531. [CrossRef]

34. Pearson, W.; Richmond, M.; Johnson, G.; Sargeant, S.; Mueller, R.; Cullinan, V.; Deng, Z.; Dibrani, B.; Guensch, G.; May, C. *Protocols for Evaluation of Upstream Passage of Juvenile Salmonids in an Experimental Culvert Test Bed*; Report No. PNWD-3525; Battelle Memorial Institute: Richland, WA, USA, 2005.

35. Kristensen, E.A.; Baattrup-Pedersen, A.; Thodsen, H. An evaluation of restoration practises in lowland streams: Has the physical integrity been re-created? *Ecol. Eng.* **2011**, *37*, 1654–1660. [CrossRef]

36. Rodgers, E.M.; Heaslip, B.M.; Cramp, R.L.; Riches, M.; Gordos, M.A.; Franklin, C.E. Substrate roughening improves swimming performance in two small-bodied riverine fishes: Implications for culvert remediation and design. *Conserv. Physiol.* **2017**, *5*, cox034. [CrossRef] [PubMed]

37. Pedersen, M.L.; Kristensen, E.A.; Kronvang, B.; Thodsen, H. Ecological effects of re-introduction of salmonid spawning gravel in lowland Danish streams. *River Res. Appl.* **2009**, *25*, 626–638. [CrossRef]

38. Santos, J.M.; Branco, P.; Katopodis, C.; Ferreira, T.; Pinheiro, A. Retrofitting pool-and-weir fishways to improve passage performance of benthic fishes: Effect of boulder density and fishway discharge. *Ecol. Eng.* **2014**, *73*, 335–344. [CrossRef]

39. Heimerl, S.; Hagmeyer, M.; Echteler, C. Numerical flow simulation of pool-type fishways: New ways with well-known tools. *Hydrobiologia* **2008**, *609*, 189. [CrossRef]

40. Liao, J.C. A review of fish swimming mechanics and behaviour in altered flows. *Philos. Trans. R. Soc. Lond. B Biol. Sci.* **2007**, *362*, 1973–1993. [CrossRef] [PubMed]

41. Blake, R. The biomechanics of intermittent swimming behaviours in aquatic vertebrates. In *Biomechanics in Animal Behaviour*; CRC Press: Boca Raton, FL, USA, 2000; pp. 79–103.

42. Nursall, J. Some behavioral interactions of spottail shiners (*Notropis hudsonius*), yellow perch (*Perca flavescens*), and northern pike (*Esox lucius*). *J. Fish. Board Can.* **1973**, *30*, 1161–1178. [CrossRef]

43. Webb, P.W. Control of posture, depth, and swimming trajectories of fishes. *Integr. Comp. Biol.* **2002**, *42*, 94–101. [CrossRef]

44. Webb, P.W. Entrainment by river chub Nocomis micropogon and smallmouth bass Micropterus dolomieu on cylinders. *J. Exp. Biol.* **1998**, *201*, 2403–2412.

45. Ward, D.L.; Schultz, A.A.; Matson, P.G. Differences in swimming ability and behavior in response to high water velocities among native and nonnative fishes. *Environ. Biol. Fishes* **2003**, *68*, 87–92. [CrossRef]

46. Mullen, D.M.; Burton, T.M. Size-related habitat use by longnose dace (*Rhinichthys cataractae*). *Am. Midl. Nat.* **1995**, *133*, 177–183. [CrossRef]

47. Webb, P.W.; Gerstner, C.L.; Minton, S.T. Station-holding by the mottled sculpin, Cottus bairdi (Teleostei: Cottidae), and other fishes. *Copeia* **1996**, 488–493. [CrossRef]

48. Edwards, E.A.; Li, H.; Schreck, C.B. *Habitat Suitability Index Models: Longnose Dace*; Western Energy and Land Use Team: Fort Collins, CO, USA, 1983.

49. Aedo, J.; Belk, M.C.; Hotchkiss, R.H. *Swimming Performance and Morphology of Utah Fishes: Critical Information for Culvert Design in Utah Streams*; The National Academies of Sciences, Engineering, and Medicine: Washington, DC, USA, 2009.

50. Wilson, K.W.; Belk, M.C. Habitat characteristics of leatherside chub (*Gila copei*) at two spatial scales. *West. N. Am. Nat.* **2001**, *61*, 36–42.

51. Belk, M.; Johnson, J.; Wilson, K.; Smith, M.; Houston, D. Variation in intrinsic individual growth rate among populations of leatherside chub (Snyderichthys copei Jordan & Gilbert): Adaptation to temperature or length of growing season? *Ecol. Freshw. Fish* **2005**, *14*, 177–184.

52. Ward, D.; Maughan, O.; Bonar, S. A Variable-Speed Swim Tunnel for Testing the Swimming Ability of Age-0 Fish. *N. Am. J. Aquac.* **2002**, *64*, 228–231. [CrossRef]

53. Peake, S. An evaluation of the use of critical swimming speed for determination of culvert water velocity criteria for smallmouth bass. *Trans. Am. Fish. Soc.* **2004**, *133*, 1472–1479. [CrossRef]

54. Peake, S.J.; Farrell, A.P. Postexercise physiology and repeat performance behaviour of free-swimming smallmouth bass in an experimental raceway. *Physiol. Biochem. Zool.* **2005**, *78*, 801–807. [CrossRef]

55. Castro-Santos, T. Optimal swim speeds for traversing velocity barriers: An analysis of volitional high-speed swimming behavior of migratory fishes. *J. Exp. Biol.* **2005**, *208*, 421–432. [CrossRef]

56. Schlichting, H. *Boundary-Layer Theory*, 7th ed.; McGraw-Hill Book Co.: New York, NY, USA, 1979.

57. Belk, M.C.; Benson, L.J.; Rasmussen, J.; Peck, S.L. Hatchery-induced morphological variation in an endangered fish: A challenge for hatchery-based recovery efforts. *Can. J. Fish. Aquat. Sci.* **2008**, *65*, 401–408. [CrossRef]

58. Song, T.; Chiew, Y. Turbulence measurement in nonuniform open-channel flow using acoustic Doppler velocimeter (ADV). *J. Eng. Mech.* **2001**, *127*, 219–232. [CrossRef]

59. SonTek. *ADV Principles of Operation*; Sontek Inc.: San Diego, CA, USA, 2001.

60. Lockhart, S. *Tutorial Guide to AutoCad*; Schroff Development Corporation: Mission, KS, USA, 2012.

61. SAS Institute. *SAS-STAT Software: Changes and Enhancements through Release 6.12*; SAS Institute: Cary, NC, USA, 1997.

62. Littell, R.C.; Milliken, G.A.; Stroup, W.W.; Wolfinger, R.D.; Schabenberger, O. *SAS for Mixed Models*; SAS Institute: Cary, NC, USA, 2007.

63. Stoll, S.; Kail, J.; Lorenz, A.W.; Sundermann, A.; Haase, P. The importance of the regional species pool, ecological species traits and local habitat conditions for the colonization of restored river reaches by fish. *PLoS ONE* **2014**, *9*, e84741. [CrossRef] [PubMed]

64. Sigler, W.F.; Sigler, J.W. *Fishes of Utah: A Natural History*; University of Utah Press Salt: Lake City, UT, USA, 1996.

65. Ead, S.; Rajaratnam, N.; Katopodis, C.; Ade, F. Turbulent open-channel flow in circular corrugated culverts. *J. Hydraul. Eng.* **2000**, *126*, 750–757. [CrossRef]

66. Chin, C.-L.; Murray, D.W. Variation of velocity distribution along nonuniform open-channel flow. *J. Hydraul. Eng.* **1992**, *118*, 989–1001. [CrossRef]

67. Billman, E.J.; Pyron, M. Evolution of form and function: Morphology and swimming performance in North American minnows. *J. Freshw. Ecol.* **2005**, *20*, 221–232. [CrossRef]

68. Boubée, J.; Jowett, I.; Nichols, S.; Williams, E. *Fish Passage at Culverts: A Review, with Possible Solutions for New Zealand Indigenous Species*; Department of Conservation: Wellington, New Zealand, 1999.

69. Mitchell, C. Swimming performances of some native freshwater fishes. *N. Z. J. Mar. Freshw. Res.* **1989**, *23*, 181–187. [CrossRef]

70. MacDonald, J.; Davies, P. Improving the upstream passage of two galaxiid fish species through a pipe culvert. *Fish. Manag. Ecol.* **2007**, *14*, 221–230. [CrossRef]

71. Powers, P.; Bates, K. *Culvert Hydraulics Related to Upstream Juvenile Salmon Passage*; Washington Department of Fish and Wildlife, Land and Restoration Services Program, Environmental Engineering Services: Washington, DC, USA, 1997.

72. Hotchkiss, R.H.; Frei, C.M. *Design for Fish Passage at Roadway-Stream Crossings: Synthesis Report*; Federal Highway Administration: Washington, DC, USA, 2007.
73. Lacey, R.J.; Neary, V.S.; Liao, J.C.; Enders, E.C.; Tritico, H.M. The IPOS framework: Linking fish swimming performance in altered flows from laboratory experiments to rivers. *River Res. Appl.* **2012**, *28*, 429–443. [CrossRef]

 *sustainability*

*Article*

# Conversion of Vertical Slot Fishways to Deep Slot Fishways to Maintain Operation during Low Flows: Implications for Hydrodynamics

**Luís Pena, Jerónimo Puertas, María Bermúdez *, Luis Cea and Enrique Peña**

Water and Environmental Engineering Group, Department of Civil Engineering, University of A Coruña, 15071 A Coruña, Spain; luis.pena@udc.es (L.P.); jeronimo.puertas@udc.es (J.P.); luis.cea@udc.es (L.C.); enrique.penag@udc.es (E.P.)
* Correspondence: maria.bermudez@udc.es; Tel.: +34-981-167-000

Received: 28 May 2018; Accepted: 6 July 2018; Published: 10 July 2018

**Abstract:** Deep slot fishways (DSF) are similar to vertical slot fishways (VSF) except that a sill has been placed at the base of the slot, and thus require a lower discharge to operate. The conversion of a VSF to a DSF, which requires minimal design modifications, can make for a more flexible design in inflow management, maintaining the correct operation of the fishway in periods of limited water availability. It is, however, crucial to understand the new flow conditions that will be created inside the fishway, and their implications for fish passage. In this paper, the hydrodynamics of DSF were studied for two different pool configurations and five sill heights. The investigation comprised the analysis of the water surface configuration, the velocity and turbulence fields, as well as the definition of the equations that related discharges to depths in the pools. The DSF designs compared well in terms of water surface patterns and maximum velocities with VSFs, but resulted in a more complex three-dimensional flow pattern and increased turbulence levels. Further testing with fish is needed to analyze whether the benefits of retrofitting a VSF by adding a sill during low flows are cancelled out by increased fish passage difficulty.

**Keywords:** fishway; upstream passage; hydraulic models; flow patterns; velocity distribution; turbulent flow

---

## 1. Introduction

Hydraulic works (dams, diversion dams, and dikes) can lead to major changes in the characteristics of river ecosystems. One of the most important negative impacts on the ichthyofauna is that they can create an insurmountable physical barrier impeding the natural movements of fish. Fish passage structures are designed to restore the longitudinal (downstream-upstream) connectivity of streams and rivers affected by such obstacles [1], facilitating the passage of fish.

Fishway designs have traditionally been developed targeting salmonid species, to facilitate their passage during the migration period. Low passage efficiencies have been found in such devices for potamodromous species and other non-salmonid diadromous species [2–4], with failure often attributed to the diversity in behaviours, morphology, physiological capacity, and swimming ability. Free movement of non-salmonid species is however crucial to sustain stocks in a moderately natural state and to maintain fish community structure and dynamics [5]. The biological objectives of building a fish pass are thus developing towards allowing permanent free movement of the complete fish community [6]. The introduction of temporary or permanent design modifications in existing fishways can be a solution to improve their efficiency for a wider range of species during the whole year, contributing to the achievement of this goal. A few studies have already shown the potential of retrofitting technical fishways as an economic solution for improving fish passage [7,8].

Pool-type fishways, which are the most common type worldwide [9–11], consist of a channel with a sloping bed divided by cross-walls into a series of pools. The total height of the obstacle to surmount is divided into a number of small drops and suitable hydrodynamic conditions are recreated in the pools to facilitate the passage of fish. Vertical slot fishways (VSF) and deep slot fishways (DSF) are particular types of pool-type fishways, which differ on the type of connection between pools. They are both modular systems in which the opening that joins two successive pools is known as a slot. The slot is vertical when the opening extends all the way down the transversal wall, whereas the deep slot is the design where the opening is limited by a sill at its base.

The most straightforward advantage of DSF with respect to VSF is that they require a lower discharge to operate. The conversion of a VSF to a DSF, which requires minimal design modifications, can thus make for a more flexible design in inflow management, maintaining the correct operation of the fishway in periods of limited water availability. It is however crucial to understand the flow conditions that will be created inside the fishway, and their implications for fish passage.

Previous studies in VSFs have shown that the flow field is mainly two-dimensional, with nearly uniform velocities along the water column and small vertical velocities in comparison to the horizontal ones [12–14]. The velocity fields in the pools are relatively insensitive to variations on the discharge, and the water depths are proportional to a dimensionless discharge with an almost linear relation. Two main flow regions can be distinguished: the recirculation regions characterized by low velocities, horizontal eddies and reversed flows; and the main flow region defined by a high velocity jet, where maximum velocities occur [15].

The introduction of a sill at the base of the slot entails a change in the hydraulic characteristics of the flow. The sill may improve the orientation of the jet diagonally over the pool, thus avoiding short-circuiting from one pool to the next without the dissipation of an adequate proportion of kinetic energy of the flow [10]. DSFs are associated with larger pool depths for a given discharge in the fishway [16]; the height at which the sill is positioned makes it possible to increase or decrease the pool depths. The velocity field is also expected to be significantly different, with velocities being clearly three-dimensional in DSFs. It is essential to study these differences in order to evaluate the possible impact on the fish species that use these structures to overcome barriers.

The hydraulic conditions in the fishway must be compatible with the fish swimming capabilities and behavior [17]. Extensive data on swimming performance are currently available for many species [18], which can be compared to the velocity field developed in the fishway. It seems obvious that maximum water velocities must be less than the burst speed of the ascending fish. The majority of the flow velocity within the pools must also be below the species' critical swimming speed, which has been often used to define the transition from the use of purely aerobic red muscle fibres to the recruitment of anaerobic white muscle fibres that result in muscle fatigue and oxygen debt [19]. High vertical velocity components are also likely to influence the behavior of the fish, and can force them to shift from one depth to another [6,8]. On the contrary, low-velocity zones seem to play an important role, allowing fish to rest during upstream passage [20].

Turbulence also affects fish swimming performance and behavior [21]. Turbulence criteria have been incorporated into the design of pool-type fishways, typically through simple indicators of the average turbulent kinetic energy in the pools such as the volumetric power dissipation [22,23]. Recent studies have analyzed in more detail the fish response to turbulence parameters in fishways, with the aim of assessing the effect of potential key-variables that should be considered for future fishway designs. Their findings suggest that high turbulent kinetic energy can confuse fish in their efforts to move through the fishway along energy efficient paths, increasing fish fatigue [24]. On the contrary, pool areas with low turbulent kinetic energy and Reynolds shear stress values can be used by fish for resting during the ascent though the fishway [25,26]. It should be noted, however, that fish response to variations of these parameters is still not well documented, and that other turbulence descriptors such as the eddy size and strength are also suspected to be important for effective fish passage [27].

In this paper, the hydrodynamics of DSFs were studied experimentally. The objective was to investigate the feasibility of DSF as a retrofitting option to make existing VSF more suitable for fish passage during low flows. Two different basic pool configurations of VSF, reported to be effective by Rajaratnam et al. [28], and previously studied by the authors [12] were considered. A sill was positioned at the base of the slot, and five different sill heights were evaluated. The performance of the fishway was evaluated only in terms of hydrodynamics, and no experiments with fish were conducted. The experimental model study of the hydrodynamics of DSFs allowed us to acquire fundamental knowledge on the operation of this type of fishways. This is intended to contribute to the development of more effective fish passage structures, capable of accommodating all movements of a wide range of species and sizes of fishes.

## 2. Materials and Methods

The experimental work was carried out at the CITEEC (Centro de Innovación Tecnolóxica en Edificación e Enxeñería Civil) at the University of A Coruña (Spain). The fishway scale model consisted of a metallic structure 12 m long, 10% slope and $1 \times 1$ m$^2$ rectangular section. The fishway was divided into eleven pools. The first four pools presented a T2 configuration, the next three were transition type pools and the last four had a T1 configuration (Figure 1). The two pool designs differ in the dimensions and shape of the cross walls. In design T1, the left side cross-wall is shorter and has a baffle fixed to its upstream face, and the right side cross-wall is longer. The flume bed, side walls and the cross-walls (vertical) separating the pools were made of transparent methacrylate sheets making it possible to observe the flow. The experimental measurements were recorded in pools number 3 and 7 (Figure 1a). Discharge was measured by means of an electromagnetic flowmeter. All the elements in the recirculating water circuit were automated and their operation was centralized in a control computer.

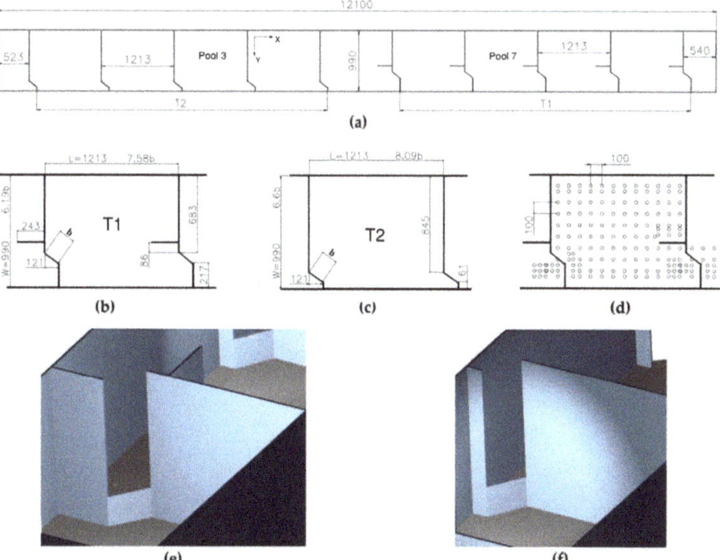

**Figure 1.** (a) Plan view of the laboratory model; (b) Details of design T1 pool, with slot width b = 160 mm; (c) Details of design T2 pool, with slot width b = 150 mm; (d) Data point mesh in a plane parallel to the bed for T1 design; (e) 3D-view of the slot in design T1, with a sill of a height of 200 mm; and (f) 3D-view of the slot in design T2, with a sill of a height of 200 mm. All dimensions are in millimeters.

Uniform flow conditions [13,28,29] were used in the tests, so that the mean depth measured at the middle transverse section ($y_0$) was the same in all the pools. At the lower end of the flume a tailgate causing overflow was used to reach the necessary boundary conditions for the uniform flow. For this reason, the tailwater levels were different for each discharge tested.

A traversing system was placed over the experimental pools to automate the positioning of the measurement instruments and could therefore be set automatically at any point in the pool. Two measurement devices—a depth probe and a velocimeter—were placed on the traversing system to perform the measurements. Velocities were measured by means of a Doppler Effect velocimeter (MicroAcoustic Doppler Velocimeter SonTek: San Diego, CA, USA). Through the MicroADV Data Acquisition System the velocity in the three Cartesian axes (Vx, Vy, and Vz) was obtained for each data point [30,31]. The water surface height in the pools was measured by means of a conductivity-based depth probe, DHI Wave Gauge Type 202. The instrument remained at each data point for 10 s to collect data on depth. The velocity data was measured at a frequency of 15 Hz during 15 s.

Velocity measurements were carried out in planes parallel to the flume bed with a separation of 10 cm between them, starting at 5 cm from the channel bed up to as close to the water surface as possible. In each plane, data points were distributed forming a $10 \times 10$ cm mesh, reduced to $5 \times 5$ cm in critical zones (Figure 1d). Therefore a three-dimensional mesh of $10 \times 10 \times 10$ cm was used to measure velocity. In each plane parallel to the flume bed, velocity was measured at 140 points in design T1 and at 132 points in design T2. The number of parallel planes ranged from 2–8 depending on the discharge and sill height. A summary of the experimental measurements obtained is presented in Table 1. Depth measurements in the pools were evaluated using a two-dimensional mesh with points at a $10 \times 10$ cm maximum separation in between. Depth was measured at 111 and 109 points in designs T1 and T2, respectively.

The experimental survey is summarized in Table 1. Five different sized sills were tested in each pool design (T1 and T2). The range of discharges used for each sill varied from the minimum discharge needed to use the measuring instruments up to the maximum discharge achievable in the laboratory model.

**Table 1.** Summary of the experiments: sill heights tested, flow discharges analyzed and number of measurement planes in each test.

| Design | Sill Height (cm) | Discharges Q (L/s) | # Measurement Planes | |
|---|---|---|---|---|
| | | | Lowest Q | Highest Q |
| T1 | 10 | 25, 35, 45, 55, 65, 75, 84, 94 | 2 | 7 |
| | 20 | 24, 35, 45, 55, 65, 75 | 3 | 7 |
| | 30 | 26, 35, 45, 55, 65, 75 | 3 | 7 |
| | 40 | 25, 34, 44, 55, 65 | 4 | 7 |
| | 50 | 25, 35, 45, 55 | 5 | 7 |
| T2 | 10 | 26, 35, 45, 55, 65, 75, 85, 95 | 2 | 7 |
| | 20 | 25, 35, 46, 55, 65, 75 | 3 | 8 |
| | 30 | 25, 35, 45, 55, 65 | 4 | 8 |
| | 40 | 25, 34, 44, 55 | 5 | 7 |
| | 50 | 25, 35, 44 | 6 | 7 |

The hydrodynamics of a DSF depend upon the discharge (Q), the geometric slope (S), the slot width (b) and the sill height (z). From the dimensional analysis, the following dimensionless variables were chosen for the representation of the experimental results: $Q^A = Q/\sqrt{gb^5}$, $y^A = (y - z)/b$. The variable $Q^A$ is the dimensionless discharge and $y^A$ is the relative flow depth. The variable y represents the depth measured at any given point in the pools. The following characteristic depths were defined: the mean depth at the transverse middle section of the pool ($y_0$), the depth at the slot measured from the lower sill base ($y_b$), the mean depth in the pool ($y_m$), and the maximum and minimum depths in the pool, ($y_{max}$ and $y_{min}$, respectively).

In order to quantify the turbulence generated in the DSFs, two key turbulent variables were selected: the turbulent kinetic energy k, calculated as $k = 0.5\left(\overline{v\prime_x^2} + \overline{v\prime_y^2} + \overline{v\prime_z^2}\right)$, where $\overline{v\prime_x^2}, \overline{v\prime_y^2}, \overline{v\prime_z^2}$ are the variance of the fluctuation velocity in each spatial direction; and the turbulence intensity $I_{kt}$, defined as $I_{kt} = k/\overline{V}^2$, where $\overline{V}$ is the mean velocity. It should be noted that the temporal resolution of the velocity measurements would not allow us to perform other types of analysis such as the calculation of the power spectrum.

## 3. Results

### 3.1. Discharge Equations

Discharge equations that provide a dimensionless relationship between depth and discharge were calculated by means of the analysis proposed by Puertas et al. [12] for vertical slot fishways. In this case, an independent term has been included, since the addition of the sill leads to $y_o \neq 0$ when $Q = 0$. The equation has the following form:

$$Q^A = \alpha\left(\frac{y_o}{b}\right) + \mu. \tag{1}$$

The experimental values of the dimensionless discharge $Q^A$, as compared to $y_o/b$ values for both designs and the five sills used are shown in Table 2. Also shown are the linear relationships computed from the experimental data along with the correlation factor in parentheses. The values of the proportionality factors depend on the configuration of the transverse cross-walls and sill height. An increase in sill height implies a proportional increase in depth. This characteristic is of utmost importance since it will allow designers to choose the most appropriate design and sill in terms of the available discharges and required depths.

The experimental results were compacted to obtain a single discharge equation for each pool design. The experimental results and linear relationships are shown in Figure 2, which also include the experimental values observed in the vertical slot fishways ($z = 0$). The discharge equations obtained provide a good fit to the experimental data, with a coefficient of correlation above 0.90 (Table 3). Design T2 has a higher proportionality coefficient in the equations than design T1 (0.90 versus 0.83).

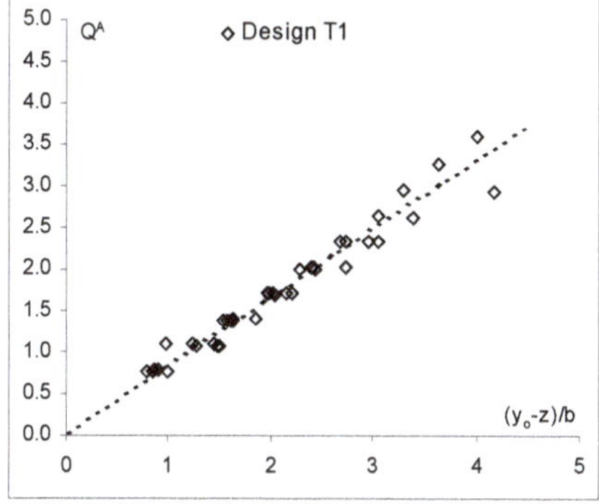
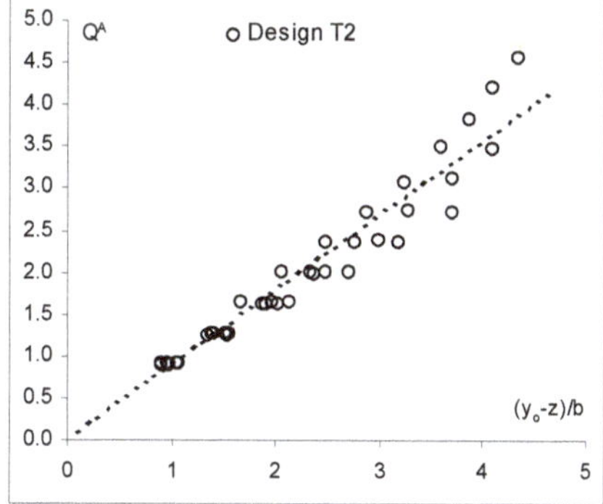

**Figure 2.** Trend line and experimental values of the relationship $(y_o - z)/b$ versus the dimensionless discharge $Q^A$ for the two pool designs.

**Table 2.** Summary of experimental results used to calculate the discharge equations. Discharge equations $Q^A - y_o$ (dimensionless discharge-dimensionless depth) for the two designs and the five sill heights used. Shown in parenthesis is the correlation coefficient $r^2$.

| Sill | Design T1 | | | Design T2 | | |
|---|---|---|---|---|---|---|
| (cm) | $Q^A$ | $y_o/b$ | Equation | $Q^A$ | $y_o/b$ | Equation |
| 10 | 0.777 | 1.483 | $Q^A = 0.794 y_o/b - 0.583$ | 0.935 | 1.731 | $Q^A = 0.844 y_o/b - 0.545$ |
|  | 1.078 | 2.115 | $(r^2 = 0.996)$ | 1.283 | 2.217 | $(r^2 = 0.999)$ |
|  | 1.403 | 2.476 |  | 1.658 | 2.620 |  |
|  | 1.715 | 2.841 |  | 2.008 | 2.997 |  |
|  | 2.021 | 3.361 |  | 2.376 | 3.421 |  |
|  | 2.332 | 3.679 |  | 2.743 | 3.926 |  |
|  | 2.628 | 4.010 |  | 3.119 | 4.240 |  |
|  | 2.937 | 4.809 |  | 3.478 | 4.756 |  |
| 20 | 0.757 | 2.241 | $Q^A = 0.818 y_o/b - 1.040$ | 0.921 | 2.369 | $Q^A = 0.677 y_o/b - 0.680$ |
|  | 1.092 | 2.688 | $(r^2 = 0.986)$ | 1.266 | 2.856 | $(r^2 = 0.999)$ |
|  | 1.402 | 2.885 |  | 1.668 | 3.453 |  |
|  | 1.701 | 3.295 |  | 2.019 | 4.040 |  |
|  | 2.016 | 3.663 |  | 2.373 | 4.497 |  |
|  | 2.326 | 4.205 |  | 2.733 | 5.026 |  |
| 30 | 0.796 | 2.769 | $Q^A = 0.857 y_o/b - 1.666$ | 0.916 | 2.961 | $Q^A = 0.731 y_o/b - 1.269$ |
|  | 1.071 | 3.348 | $(r^2 = 0.975)$ | 1.288 | 3.512 | $(r^2 = 0.999)$ |
|  | 1.383 | 3.507 |  | 1.643 | 4.007 |  |
|  | 1.725 | 4.034 |  | 2.010 | 4.488 |  |
|  | 1.997 | 4.313 |  | 2.388 | 4.977 |  |
|  | 2.343 | 4.560 |  |  |  |  |
| 40 | 0.773 | 3.345 | $Q^A = 0.817 y_o/b - 1.981$ | 0.900 | 3.631 | $Q^A = 0.772 y_o/b - 1.880$ |
|  | 1.068 | 3.783 | $(r^2 = 0.997)$ | 1.248 | 4.007 | $(r^2 = 0.997)$ |
|  | 1.383 | 4.111 |  | 1.625 | 4.534 |  |
|  | 1.707 | 4.479 |  | 2.000 | 5.038 |  |
|  | 2.017 | 4.905 |  |  |  |  |
| 50 | 0.769 | 3.915 | $Q^A = 0.774 y_o/b - 2.264$ | 0.903 | 4.233 | $Q^A = 0.720 y_o/b - 2.142$ |
|  | 1.094 | 4.362 | $(r^2 = 0.998)$ | 1.273 | 4.734 | $(r^2 = 0.999)$ |
|  | 1.392 | 4.691 |  | 1.630 | 5.243 |  |
|  | 1.707 | 5.142 |  |  |  |  |

**Table 3.** Discharge equations $Q^A - y_o$ (dimensionless discharge-dimensionless depth) and dimensionless relationships among the characteristic depths (maximum $y_{max}$, at the slot $y_b$, mean $y_m$, and minimum $y_{min}$) and the mean depth in the transverse middle section $y_o$, for both designs and all sill heights. Shown in parenthesis is the correlation coefficient $r^2$.

| | Design T1 | Design T2 |
|---|---|---|
| $Q^A$ | $0.827(y_o - z)/b$ <br> $(r^2 = 0.955)$ | $0.895(y_o - z)/b$ <br> $(r^2 = 0.933)$ |
| $(y_{max} - z)/b$ | $1.063(y_o - z)/b + 0.388$ <br> $(r^2 = 0.994)$ | $0.996(y_o - z)/b + 0.679$ <br> $(r^2 = 0.984)$ |
| $(y_b - z)/b$ | $0.981(y_o - z)/b + 0.242$ <br> $(r^2 = 0.986)$ | $1.053(y_o - z)/b + 0.172$ <br> $(r^2 = 0.954)$ |
| $(y_m - z)/b$ | $1.018(y_o - z)/b$ <br> $(r^2 = 0.998)$ | $1.015(y_o - z)/b$ <br> $(r^2 = 0.998)$ |
| $(y_{min} - z)/b$ | $0.982(y_o - z)/b - 0.433$ <br> $(r^2 = 0.997)$ | $0.973(y_o - z)/b + 0.496$ <br> $(r^2 = 0.992)$ |

### 3.2. Water Depth

The relationships obtained between $y_o$ and the other characteristic depths $y_b$, $y_{max}$, $y_{min}$, and $y_m$ are shown in Table 3. The slot is a section of unavoidable passage in the ascension of fish though the fishway and provides information on the maximum velocity required to overcome the barrier. Similarly to the other characteristic depths, $y_b/b$ increased linearly with the discharge Q. However, the water height over the sill in slot $(y_b - z)/b$ remained relatively constant regardless of the type of sill used (Figure 3). On the basis of these results, it is possible to foresee that the velocities at the slot reach a constant value that is independent of the discharge and the sill under consideration. The passage section through the slot is not a control section; the Froude number (F) was calculated for this section and the regime was found to be subcritical $F = u/\sqrt{g(y_b - z)} < 1$ (where u is the velocity at the slot).

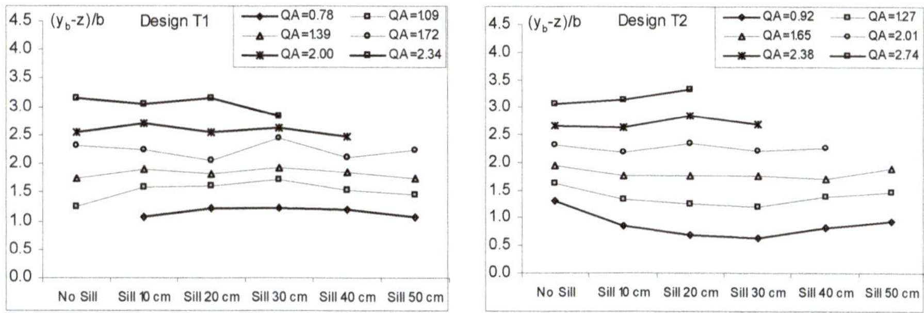

**Figure 3.** Depth at the slot ($y_b$) in terms of the dimensionless discharge ($Q^A$) and sill height (z).

## 3.3. Water Surface

The configuration of the water surface patterns in the DSFs was studied taking three factors into consideration: the basic design of the transverse cross-walls; the sill height; and the discharge. Designs T1 and T2 exhibited different water surface patterns as shown in Figures 4 and 5, although some similar features were detected: (1) a sharp drop in depth after the slot, continuing along in the direction of the flow; (2) the existence of a nearly circular area of minimum depths; and (3) the distribution of isodepth lines approximately perpendicular to the direction of the longitudinal axis of the fishway.

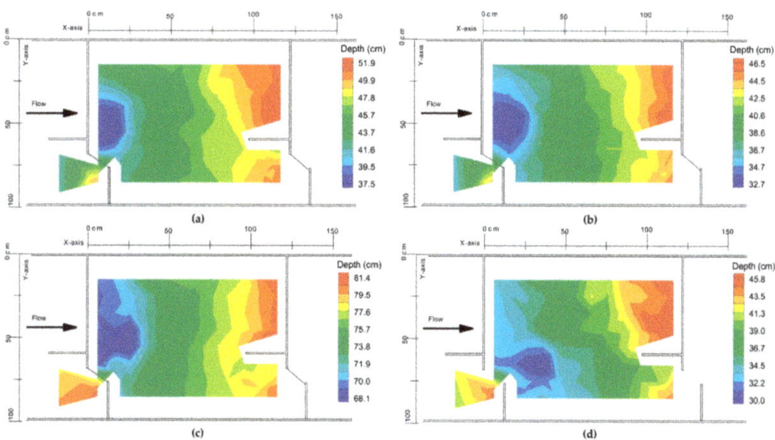

**Figure 4.** Water surface levels in designs T1: (**a**) Sill = 10 cm, Q = 55 L/s; (**b**) Sill = 10 cm, Q = 45 L/s; (**c**) Sill = 50 cm, Q = 45 L/s; and (**d**) VSF (no sill), Q = 65 L/s.

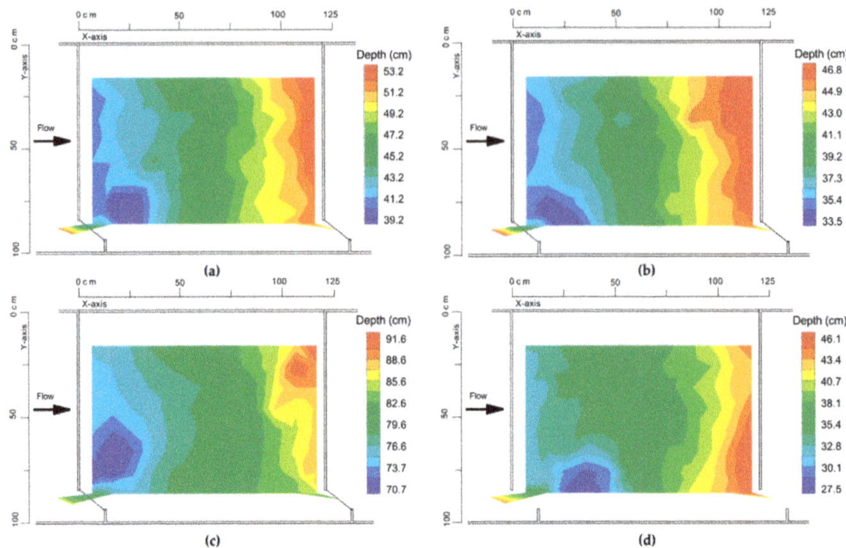

**Figure 5.** Water surface levels in designs T2: (**a**) Sill = 10 cm, Q = 55 L/s; (**b**) Sill = 10 cm, Q = 45 L/s, (**c**) Sill = 50 cm, Q = 45 L/s; (**d**) VSF (no sill), Q = 65 L/s.

The main differences with the pattern observed in the VSF designs (Figures 4d and 5d) were: (1) the location of the area of minimum depths inside the pool, which moved towards the interior of the pool, close to the crosswall, in both DSF designs and (2) the distribution of isodepth lines in the design T1, that changed from an oblique to a perpendicular orientation to the direction of the longitudinal axis of the fishway.

In the examples presented in Figures 4 and 5 we can see the effect of the sill height and flow discharge on the configuration of the water surface. The figures show the water surface for two different sill heights (10, 50 cm) and two discharges (45, 55 L/s). These representative examples illustrate the similarity of the water surface patterns within each pool type (T1 and T2), regardless of changes in these parameters.

### 3.4. Velocity Fields

The flow in the VSFs is nearly two-dimensional and the velocity vectors are parallel to the flume bed. Horizontal velocities are uniform along the vertical direction and vertical velocities are close to zero. On the contrary, the flow generated in the DSFs is very complex. In DSFs there is a wide variability in the circulation patterns which are dependent upon the cross-wall design, the sill height and the vertical position over the bed, these factors being invariable in terms of the circulating discharge. In Figure 6, it is possible to observe the differences between the velocity fields, particularly in the velocities occurring near the bed and the velocity fields above the greatest sill height. Also evident is the importance of the vertical velocities in some of the pool zones which favor the vertical circulation of the water.

As expected, the horizontal velocity fields in planes below the sill height are very different to those found in VSF (Figure 6e,f). In these planes, a jet flow region is not formed and maximum velocities are significantly lower. However, substantial differences can also be observed between the horizontal velocity fields above the sill height and those measured in VSFs. The addition of the sill, even if it has a low height, significantly modifies the overall flow pattern in the pools, which in turn results in changes in the velocity fields in planes closer to the surface, primarily: (1) the main flow follows a more curved path as it crosses from one slot to the next in the DSFs; (2) the size of the recirculation regions developed in the upstream part of the pools in design T1 of the VSF is reduced; and (3) the large recirculation eddy developed in design T2 of the VSF is not formed, and a smaller eddy is developed on the right side of the jet.

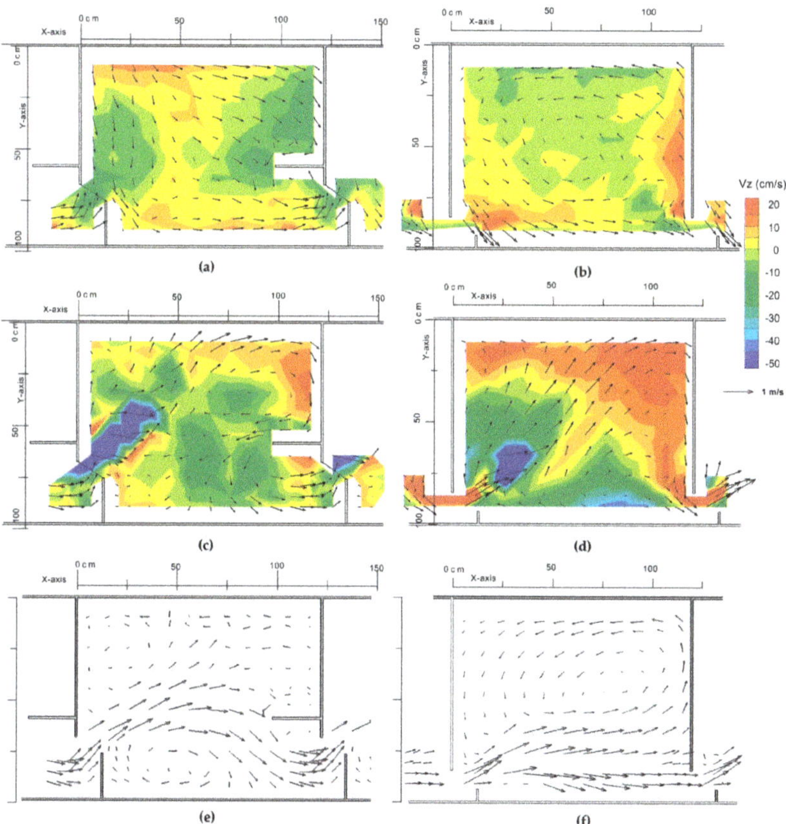

**Figure 6.** Velocity fields in planes parallel to the bed: (**a**) Design T1, plane lower than sill height; (**b**) Design T2, plane lower than sill height; (**c**) Design T1, plane above sill height; (**d**) Design T2, plane above sill height; (**e**) Design T1, VSF (no sill); and (**f**) Design T2, VSF (no sill). Vectors represent the mean horizontal velocity (Vx − Vy) and the color scale represents the vertical velocities (Vz). Please note that vertical velocities in the VSF are negligible except in the slot area, so they are not plotted in subfigures **e** and **f**.

As well as acquiring knowledge on the spatial distribution of the velocity, it is necessary to quantify these velocities and relate them to the discharges supplied. The spatial variability of the velocity fields in DSFs makes it very difficult to find these relationships. The slot velocity is chosen as a characteristic velocity, since this is an unavoidable passage section for fish. The velocity magnitude at the slot remains independent of the height over the bed, and it even proved to be stable when subjected to discharge variations. The velocity at the slot is indeed approximately constant, regardless of the discharge and the sill used, as shown in Figure 7. The mean velocity values at the slot in design T2 are slightly higher than those in design T1 (averaged values for all discharges and sill heights of 1.1 m/s vs. 0.9 m/s, respectively). As can be seen in Figure 7, these values are very similar to those measured in the VSF (averaged values for all discharges of 1.1 m/s in design T1 and 1.2 m/s in design T2).

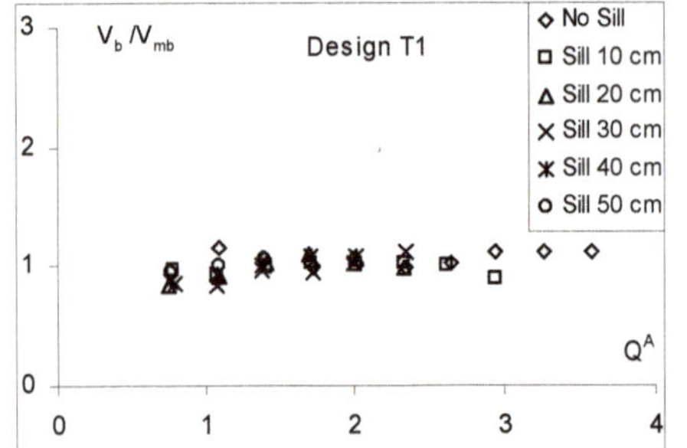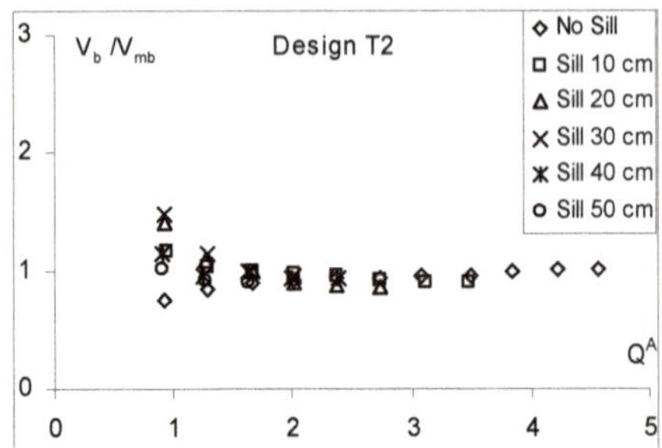

**Figure 7.** Experimental values of the velocities at the slot in both designs and different sill heights. Note: Vb = velocity at slot, Vmb = average velocity at the slot.

In addition to the slot velocity, it is useful to know the maximum velocities in the pools that the fish will potentially have to negotiate (Table 4). The maximum velocity values in DSFs are achieved at heights greater than the sill elevation, and are similar to those measured in the corresponding VSF design. These maximum velocities are slightly higher in design T2 than in T1.

**Table 4.** Maximum velocity Vmax/Vmb for both designs. Note: Vmb = average velocity at the slot; PA = planes parallel to the bed with an elevation greater than sill height; and PB = planes parallel to the bed with an elevation lower than sill height. In the case of DSF designs (with sill), a range of maximum velocities is indicated, considering all studied planes above or below the sill.

|  |  | Design T1 | Design T2 |
|---|---|---|---|
| VSF (no sill) |  | 1.1 | 1.3 |
| DSF (sill) | PA | 0.7–1.9 | 0.5–1.5 |
|  | PB | 0.4–1.1 | 0.3–1.3 |

*3.5. Turbulent Kinetic Energy*

The distribution of turbulent kinetic energy k and turbulence intensity $I_{kt}$ in the pools is extremely complex. The values of k and $I_{kt}$ for the three different zones in the pools defined in Figure 8 are summarized in Table 5. The high variability in the values of both designs is notable, and it is also clear that the values of the turbulent kinetic energy generated in planes greater than the sill height (in italicsin Table 5) are, in general, substantially higher.

The distribution of k values in the planes parallel to the flume bed for the different experimental situations is shown in Figure 8. In the lower planes (Figure 8a,b) it is possible to note small areas with high levels of turbulence, whereas these areas become increasingly larger in the higher planes. In the planes above the sill height, a very high turbulence zone is visible at the slot exit going in the direction of the side wall (Figure 8c,d). The area of high turbulence, with k greater than 1000 $cm^2/s^2$, occupies nearly the entire pool. On the contrary, the turbulent energy is localized in the main flow region in the corresponding VSFs designs, with a pronounced decay towards the walls (Figure 8e,f). Very low turbulence levels are observed in the recirculation regions.

**Table 5.** Values of the turbulent kinetic energy (k) and turbulent intensity ($I_{kt}$) at three different locations in the pools (1, 2, and 3 in Figure 8) and different heights (h) above the bed. The values of k and $I_{kt}$ corresponding to heights above the sill are indicated in italics.

| | | | T1 | | | | | |
| | | | k (cm$^2$/s$^2$) | | | $I_{kt}$ | | |
| Sill (cm) | Q (L/s) | h (cm) | k1 | k2 | k3 | $I_{kt}1$ | $I_{kt}2$ | $I_{kt}3$ |
|---|---|---|---|---|---|---|---|---|
| 10 | 75 | 5 | 428 | 773 | 473 | 2.880 | 0.915 | 0.157 |
| | | *25* | *6655* | *3254* | *482* | *0.365* | *0.701* | *0.563* |
| | | *45* | *5583* | *1514* | *556* | *1.325* | *0.420* | *0.993* |
| 20 | 65 | 5 | 285 | 511 | 251 | 0.831 | 0.766 | 0.115 |
| | | 15 | 498 | 361 | 371 | 3.142 | 1.170 | 0.765 |
| | | *35* | *5532* | *1874* | *252* | *0.273* | *0.314* | *0.967* |
| 30 | 65 | 5 | 274 | 349 | 325 | 0.189 | 0.277 | 0.253 |
| | | 15 | 462 | 331 | 193 | 1.237 | 3.770 | 0.389 |
| | | *45* | *10,164* | *5426* | *301* | *1.535* | *2.300* | *1.264* |
| 40 | 55 | 5 | 311 | 187 | 241 | 0.249 | 0.242 | 0.157 |
| | | 25 | 200 | 300 | 143 | 3.831 | 0.792 | 0.394 |
| | | *45* | *7189* | *3569* | *229* | *6.106* | *23.279* | *20.501* |
| 50 | 45 | 5 | 173 | 118 | 218 | 0.314 | 0.095 | 0.124 |
| | | 35 | 196 | 104 | 94 | 0.318 | 0.158 | 0.578 |
| | | *55* | *6988* | *4602* | *220* | *1.769* | *26.678* | *2.560* |

| | | | T2 | | | | | |
| | | | k (cm$^2$/s$^2$) | | | $I_{kt}$ | | |
| Sill (cm) | Q (L/s) | h (cm) | k1 | k2 | k3 | $I_{kt}1$ | $I_{kt}2$ | $I_{kt}3$ |
|---|---|---|---|---|---|---|---|---|
| 10 | 75 | 5 | 111 | 411 | 326 | 0.061 | 1.356 | 0.595 |
| | | *25* | *160* | *311* | *678* | *0.334* | *0.293* | *0.812* |
| | | *45* | *627* | *250* | *862* | *0.428* | *0.412* | *0.236* |
| 20 | 65 | 5 | 468 | 368 | 640 | 1.239 | 0.182 | 0.225 |
| | | 15 | 276 | 279 | 407 | 0.965 | 0.285 | 0.352 |
| | | *35* | *4107* | *4129* | *720* | *74.995* | *4.977* | *0.802* |
| 30 | 55 | 5 | 344 | 254 | 458 | 3.291 | 0.475 | 0.305 |
| | | 15 | 334 | 164 | 207 | 0.847 | 0.497 | 0.276 |
| | | *45* | *5767* | *4776* | *353* | *3.080* | *1.951* | *0.322* |
| 40 | 45 | 5 | 172 | 168 | 160 | 2.971 | 0.227 | 0.086 |
| | | 25 | 153 | 120 | 109 | 0.283 | 0.124 | 0.311 |
| | | *45* | *1049* | *1964* | *191* | *0.530* | *0.667* | *0.652* |
| 50 | 35 | 5 | 162 | 161 | 156 | 0.663 | 0.144 | 0.118 |
| | | 35 | 120 | 159 | 157 | 0.304 | 0.951 | 0.330 |
| | | *55* | *327* | *497* | *126* | *0.220* | *0.792* | *3.526* |

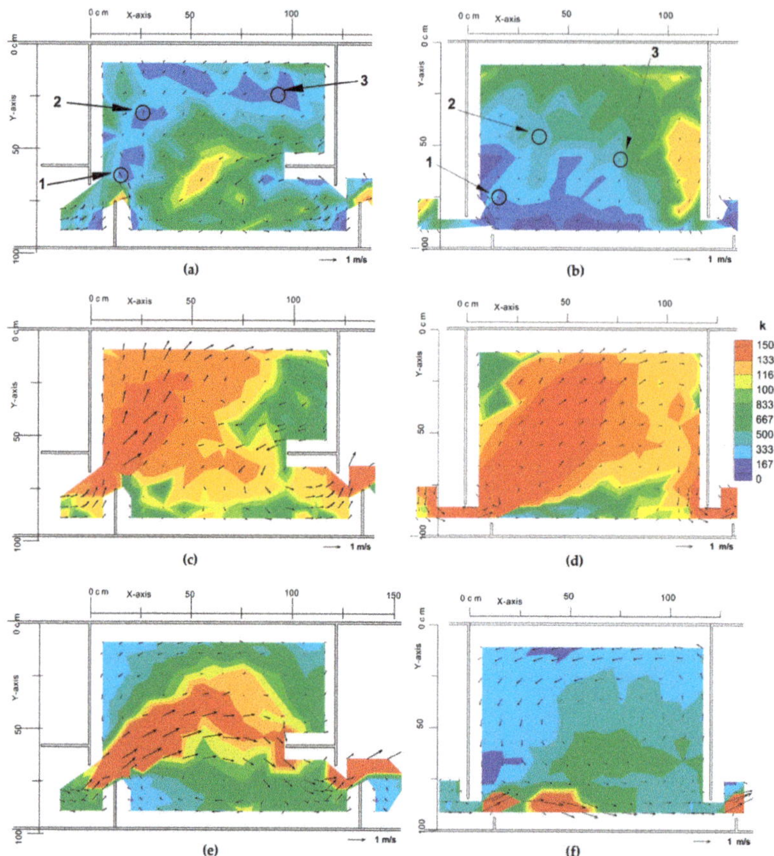

**Figure 8.** Turbulent kinetic energy level lines (cm$^2$/s$^2$) in planes parallel to the bed: (**a**) Design T1, plane lower than sill height; (**b**) Design T2, plane lower than sill height, (**c**) Design T1, plane above sill height; (**d**) Design T2, plane above sill height; (**e**) Design T1, VSF (no sill); and (**f**) Design T2, VSF (no sill). Vectors represent the mean horizontal velocity.

## 4. Discussion

This study analyzes the hydrodynamics of DSFs for two different pool configurations and five different sill heights, using a laboratory physical model. The experimental results included the data obtained in the VSFs, considering this type of fishway as a special case of DSFs in which sill height is zero. The inclusion of these data allowed us to compare the two types of fishways in addition to providing us with a more generalized view of these structures. Further details on the characteristics of the flow field developed in the two VSF designs can be found in the work of Puertas et al. [12].

Retrofitting of VSFs by adding a sill across the slot can be a solution to maintain adequate water depths for fish swimming during low flows. For the same discharges, the addition of a sill results in an increase in depth proportional to its height, while still allows fish to swim rather than leap over obstacles. Design T2 has a greater flow conveyance efficiency than design T1, which means that for the same discharge a lower depth is obtained (and consequently a higher velocity). In the extreme case in which the sill blocks most of the slot (very large sill height), the connection between pools is, effectively, a submerged notch operating in a streaming regimen [32]. In fact, the regimen in VSFs is sometimes referred to as streaming flow by analogy with these pool-type designs [33]. In laboratory experiments under controlled conditions, this flow regime has been found to enhance fish movements through surface notches, increasing the negotiation success of several species with different ecological characteristics, such as the Iberian chub [34,35] or the Iberian barbel [34,36]. Although further research is needed, these findings suggest that this flow regime favors multi-species fish passage, which can be considered an advantage of DSFs compared to other pool-type designs.

Unlike VSFs, the addition of the sill in DSFs prevents bottom dwelling species from swimming at their desired depth when passing through the slot. VSFs, which offer the full range of depth for passage, or pool-type fishways with submerged orifices would be preferable for these species, because of their preference for swimming close to the bottom. Moreover, the flow is no longer two-dimensional and flow conditions differ considerably along the vertical direction. This could also potentially contribute to different passage efficiencies of bottom-oriented and water-column species. In the lower planes (below the sill elevation), velocity and turbulence levels are low, and fish could find suitable resting areas to recover after negotiating the slot. On the contrary, fish face more challenging hydraulic conditions in the upper region of the pools (above the sill elevation). Maximum velocities are in the same order of magnitude as in the corresponding VSFs, but the three-dimensionality of the flow increases. High vertical velocity components, which only occurred in the slot region in VSF designs, are likely to influence the behavior of the fish [6]. Ascending velocities can disturb the fish behavior by generating a secondary flow that may force fish to shift from one depth to another [8]. Fish might move up in the water column, encountering less favorable hydraulic conditions.

Turbulence increases with respect to VSFs in this upper portion of the pools, which can also affect fish locomotion [21,37]. The highest turbulent kinetic energy values, both averaged over the horizontal plane and point values, are obtained in the planes higher than the sill level, showing a certain degree of coupling with the velocity fields. High turbulent kinetic energy can confuse fish in their efforts to move through the fishway along energy efficient paths, increasing fatigue [24]. In order to allow fish to rest, the pools should also provide large areas of low turbulent kinetic energy values ($<0.05$ m$^2$/s$^2$) [25]. In the designs tested, such areas would only be available in the lower portion of the pools, which would force fish to move vertically to find them. It should be noted, however, that the effect of turbulence on fish passage is still in the early stages of investigation, and other variables such as the Reynolds shear stress and eddy size are also suspected to be important in explaining fish swimming behavior [27]. Their effect is however likely to differ widely among species, and even among individuals within a species [38].

Thus, for further research, it would be useful to deepen the characterization of turbulence in DSFs, calculating additional descriptors that might correlate with fish response. Due to the high number of velocity measurements performed in this study, a relatively short measurement period was used. The experimental survey could be extended considering a more limited number of representative discharges and DSFs configurations, identified based on the results of this work. For these representative cases, longer time series of instantaneous velocity could be measured in order to increase the accuracy of the calculated turbulence descriptors. It would be also interesting to increase the ADV sampling frequency, in order to allow the analysis of inertial and dissipation subranges of the power spectrum.

Considering the potential implications for fish passage, it is necessary to conduct experiments with fish in order to study how hydraulic conditions in DSFs can affect fish passage. Given the variability in swimming performance, behavior and niche occupancy between species, several species representative of different morpho-ecological groups should be tested, similarly to what has been recently done for other fishway designs [34]. The findings suggest that the DSFs designs could be more species selective than VSFs, and therefore not appropriate for facilitating passage for a wide range of fish species. This would require exploring alternatives to reduce fishway selection, such as introducing operational changes that take advantage of temporal differences in movement patterns between species [39]. A potentially higher passage difficulty in DSFs would lead to a compromise between improved functionality of VSFs during the low-flow season, and potentially lower passage rates outside this season. If this were the case, the use of temporary modifications (e.g., removable sills) could be the way forward to improve the efficiency of the fishway during the whole year.

## 5. Conclusions

A wide variability was observed in the water circulation patterns of DSFs, depending on the cross-wall design, the sill height and the circulating discharge. The flow established was clearly three-dimensional, unlike that developed in VSFs. An uneven distribution of turbulent kinetic energy in the pools at different heights was found. The complexity of the flow obliges fish to tackle a three-dimensional water circulation with highly turbulent areas.

The discharge relationships that relate discharge and depth were calculated for both pool designs. Design T2 was verified to have greater flow conveyance efficiency, so given the same design discharge, the depths would be lower in this design. The use of different sill heights at the base of the slot makes inflow management more flexible. An increase in the sill height translates to a proportional increase in depth. However, the water height above the sill at the slot remains relatively constant regardless of the sill height used. The invariance of the velocity at the slot against the discharge and height over the flume bed was demonstrated. The velocities in design T2 were higher than in design T1.

The results show that retrofitting a VSF by adding a sill at the base of the slot might improve its functionality during low flows, ensuring adequate water depths for fish swimming. However, given the complexity of the flow developed in DSFs, further research is needed to evaluate the fish response to these challenging hydrodynamic conditions.

**Author Contributions:** Conceptualization, L.P., J.P., and E.P.; Data curation, L.P.; Funding acquisition, J.P.; Investigation, L.P., M.B., and L.C.; Project administration, J.P. and E.P.; Supervision, J.P.; Writing—original draft, L.P. and M.B.

**Funding:** This research received no external funding.

**Acknowledgments:** The authors would like to thank the University of A Coruña and the CITEEC (Centro de Innovación Tecnolóxica en Edificación e Enxeñería Civil) for their collaboration. María Bermúdez gratefully acknowledges financial support from the Spanish Regional Government of Galicia (Postdoctoral grant reference ED481B 2014/156).

**Conflicts of Interest:** The authors declare no conflict of interest.

## Abbreviations

| | |
|---|---|
| $b$ | Slot width |
| $Ikt$ | Turbulence intensity |
| $k$ | Turbulent kinetic energy |
| $Q$ | Discharge |
| $Q^A$ | Dimensionless discharge |
| $S$ | Bed slope |
| $Vb$ | Velocity at the slot |
| $Vmb$ | Velocity at the slot, averaged for all discharges tested |
| $Vx, Vy, Vz$ | Velocity components in the three Cartesian axes |
| $y$ | Flow depth |
| $y^A$ | Dimensionless flow depth |
| $y_o$ | Mean depth at the transverse middle section of the pool |
| $y_b$ | Depth at the slot measured from the base of the sill |
| $y_m$ | Mean depth in the pool |
| $y_{min}$ | Minimum depth in the pool |
| $y_{max}$ | Maximum depth in the pool |
| $z$ | Sill height |

## References


1. Ward, J.V. The Four-Dimensional Nature of Lotic Ecosystems. *J. N. Am. Benthol. Soc.* **1989**, *8*, 2–8. [CrossRef]
2. Mallen-Cooper, M.; Brand, D.A. Non-salmonids in a salmonid fishway: What do 50 years of data tell us about past and future fish passage? *Fish. Manag. Ecol.* **2007**, *14*, 319–332. [CrossRef]
3. Noonan, M.J.; Grant, J.W.A.; Jackson, C.D. A quantitative assessment of fish passage efficiency. *Fish Fish.* **2012**, *13*, 450–464. [CrossRef]
4. Thiem, J.D.; Binder, T.R.; Dumont, P.; Hatin, D.; Hatry, C.; Katopodis, C.; Stamplecoskie, K.M.; Cooke, A.S.J. Multispecies Fish Passage Behaviour in a Vertical Slot Fishway on the Richelieu River, Quebec, Canada. *River Res. Appl.* **2013**, *29*, 582–592. [CrossRef]
5. Lucas, M.C.; Mercer, T.; Peirson, G.; Frear, P.A. Seasonal movements of coarse fish in lowland rivers and their relevance to fisheries management. In *Management and Ecology of River Fisheries*; John Wiley & Sons: Hoboken, NJ, USA, 2000; pp. 87–100.
6. Marriner, B.A.; Baki, A.B.M.; Zhu, D.Z.; Cooke, S.J.; Katopodis, C. The hydraulics of a vertical slot fishway: A case study on the multi-species Vianney-Legendre fishway in Quebec, Canada. *Ecol. Eng.* **2016**, *90*, 190–202. [CrossRef]
7. Mallen-Cooper, M.; Zampatti, B.; Stuart, I.; Baumgartner, L. *Innovative Fishways—Manipulating Turbulence in the Vertical Slot Design to Improve Performance and Reduce Cost*; A Report to the Murray Darling Basin Commission; Fishway Consulting Services; Murray Darling Basin Commission: Sydney, Australia, 2008.
8. Santos, J.M.; Branco, P.; Katopodis, C.; Ferreira, T.; Pinheiro, A. Retrofitting pool-and-weir fishways to improve passage performance of benthic fishes: Effect of boulder density and fishway discharge. *Ecol. Eng.* **2014**, *73*, 335–344. [CrossRef]
9. Larinier, M.; Marmulla, G. Fish passes: types, principles and geographical distribution an overview. In Proceedings of the Second International Symposium on the Management of Large Rivers for Fisheries, Phnom Penh, Kingdom of Cambodia, 11–14 February 2003; Volume II.
10. Clay, C.H. *Design of Fishways and Other Fish Facilities*; Lewis Publishers, CRC Press: Boca Raton, FL, USA, 1995.
11. Santos, J.M.; Silva, A.; Pinheiro, P.; Pinheiro, A.; Bochechas, J.; Ferreira, M.T.; Katopodis, C.; Pinheiro, P.; Pinheiro, A.; Bochechas, J.; et al. Ecohydraulics of pool-type fishways: Getting past the barriers. *Ecol. Eng.* **2012**, *48*, 38–50. [CrossRef]
12. Puertas, J.; Pena, L.; Teijeiro, T. Experimental approach to the hydraulics of vertical slot fishways. *J. Hydraul. Eng.* **2004**, *130*, 10–23. [CrossRef]
13. Wu, S.; Rajaratnam, N.; Katopodis, C. Structure of flow in vertical slot fishway. *J. Hydraul. Eng.* **1999**, *125*, 351–359. [CrossRef]
14. Liu, M.; Rajaratnam, N.; Zhu, D.Z. Mean flow and turbulence structure in vertical slot fishways. *J. Hydraul. Eng.* **2006**, *132*, 765–777. [CrossRef]
15. Bermúdez, M.; Puertas, J.; Cea, L.; Pena, L.; Balairón, L. Influence of pool geometry on the biological efficiency of vertical slot fishways. *Ecol. Eng.* **2010**, *36*, 1355–1364. [CrossRef]
16. Larinier, M.; Porcher, J.P.; Travade, F.; Gosset, C. *Passes à Poissons: Expertise, Conception des Ouvrages de Franchissement*; Conseil Supérieur de la Pêche: Paris, France, 1992.
17. Williams, J.G.; Armstrong, G.; Katopodis, C.; Larinier, M.; Travade, F. Thinking like a fish: A key ingredient for development of effective fish passage facilities at river obstructions. *River Res. Appl.* **2012**, *28*, 407–417. [CrossRef]
18. Katopodis, C.; Gervais, R. Ecohydraulic analysis of fish fatigue data. *River Res. Appl.* **2012**, *28*, 444–456. [CrossRef]
19. Thiem, J.D.; Dawson, J.W.; Hatin, D.; Danylchuk, A.J.; Dumont, P.; Gleiss, A.C.; Wilson, R.P.; Cooke, S.J. Swimming activity and energetic costs of adult lake sturgeon during fishway passage. *J. Exp. Biol.* **2016**, *219*, 2534–2544. [CrossRef] [PubMed]
20. Rodríguez, Á.; Bermúdez, M.; Rabuñal, J.R.; Puertas, J. Fish tracking in vertical slot fishways using computer vision techniques. *J. Hydroinform.* **2015**, *17*, 275–292. [CrossRef]
21. Pavlov, D.; Lupandin, A.; Skorobogatov, M. The Effects of Flow Turbulence on the Behavior and Distribution of Fish. *J. Ichthyol.* **2000**, *20*, S232–S261.

22. Katopodis, C. Developing a toolkit for fish passage, ecological flow management and fish habitat works. *J. Hydraul. Res.* **2005**, *43*, 451–467. [CrossRef]
23. Bermúdez, M.; Rico, Á.; Rodríguez, Á.; Pena, L.; Rabuñal, J.R.; Puertas, J.; Balairón, L.; Lara, Á.; Aramburu, E.; Morcillo, F.; et al. FishPath: aplicación informática de diseño de escalas de peces de hendidura vertical. *Ing. Agua* **2015**, *19*, 179. [CrossRef]
24. Quaranta, E.; Katopodis, C.; Revelli, R.; Comoglio, C. Turbulent flow field comparison and related suitability for fish passage of a standard and a simplified low-gradient vertical slot fishway. *River Res. Appl.* **2017**, *33*, 1295–1305. [CrossRef]
25. Silva, A.T.; Santos, J.M.; Ferreira, M.T.; Pinheiro, A.N.; Katopodis, C. Effects of water velocity and turbulence on the behaviour of Iberian barbel (*Luciobarbus bocagei*, Steindachner 1864) in an experimental pool-type fishway. *River Res. Appl.* **2011**, *27*, 360–373. [CrossRef]
26. De Duarte, B.A.F.; Ramos, I.C.R.; de Santos, H.A. e Reynolds shear-stress and velocity: Positive biological response of neotropical fishes to hydraulic parameters in a vertical slot fishway. *Neotrop. Ichthyol.* **2012**, *10*, 813–819. [CrossRef]
27. Silva, A.T.; Katopodis, C.; Santos, J.M.; Ferreira, M.T.; Pinheiro, A.N. Cyprinid swimming behaviour in response to turbulent flow. *Ecol. Eng.* **2012**, *44*, 314–328. [CrossRef]
28. Rajaratnam, N.; Katopodis, C.; Solanki, S. New designs for vertical slot fishways. *Can. J. Civ. Eng.* **1992**, *19*, 402–414. [CrossRef]
29. Rajaratnam, N.; Van der Vinne, G.; Katopodis, C. Hydraulics of Vertical Slot Fishways. *J. Hydraul. Eng.* **1986**, *112*, 909–927. [CrossRef]
30. Kraus, N.C.; Lohrmann, A.; Cabrera, R. New Acoustic Meter for Measuring 3D Laboratory Flows. *J. Hydraul. Eng.* **1994**, *120*, 406–412. [CrossRef]
31. Nikora, V.I.; Goring, D.G. ADV Measurements of Turbulence: Can We Improve Their Interpretation? *J. Hydraul. Eng.* **1998**, *124*, 630–634. [CrossRef]
32. Fuentes-Pérez, J.F.; Sanz-Ronda, F.J.; de Azagra, A.M.; García-Vega, A. Non-uniform hydraulic behavior of pool-weir fishways: A tool to optimize its design and performance. *Ecol. Eng.* **2016**, *86*, 5–12. [CrossRef]
33. Larinier, M. Pool fishways, pre-barrages and natural bypass channels. *Bull. Franç. Pêche Piscic.* **2002**, 54–82. [CrossRef]
34. Branco, P.; Santos, J.M.; Katopodis, C.; Pinheiro, A.; Ferreira, M.T. Pool-Type Fishways: Two Different Morpho-Ecological Cyprinid Species Facing Plunging and Streaming Flows. *PLoS ONE* **2013**, *8*, e65089. [CrossRef] [PubMed]
35. Branco, P.; Santos, J.M.; Katopodis, C.; Pinheiro, A.; Ferreira, M.T. Effect of flow regime hydraulics on passage performance of Iberian chub (*Squalius pyrenaicus*) (Günther, 1868) in an experimental pool-and-weir fishway. *Hydrobiologia* **2013**, *714*, 145–154. [CrossRef]
36. Silva, A.T.; Santos, J.M.; Franco, A.C.; Ferreira, M.T.; Pinheiro, A.N. Selection of Iberian barbel *Barbus bocagei* (Steindachner, 1864) for orifices and notches upon different hydraulic configurations in an experimental pool-type fishway. *J. Appl. Ichthyol.* **2009**, *25*, 173–177. [CrossRef]
37. Lupandin, A. Effect of Flow Turbulence on Swimming Speed of Fish. *Biol. Bull.* **2005**, *32*, 461–466. [CrossRef]
38. Pon, L.B.; Hinch, S.G.; Cooke, S.J.; Patterson, D.A.; Farrell, A.P. Physiological, energetic and behavioural correlates of successful fishway passage of adult sockeye salmon *Oncorhynchus nerka* in the Seton River, British Columbia. *J. Fish Biol.* **2009**, *74*, 1323–1336. [CrossRef] [PubMed]
39. Johnson, E.L.; Caudill, C.C.; Keefer, M.L.; Clabough, T.S.; Peery, C.A.; Jepson, M.A.; Moser, M.L. Movement of Radio-Tagged Adult Pacific Lampreys during a Large-Scale Fishway Velocity Experiment. *Trans. Am. Fish. Soc.* **2012**, *141*, 571–579. [CrossRef]

*Article*

# Passage Performance of Potamodromous Cyprinids over an Experimental Low-Head Ramped Weir: The Effect of Ramp Length and Slope

Susana Dias Amaral [1,*], Paulo Branco [1], Christos Katopodis [2], Maria Teresa Ferreira [1], António Nascimento Pinheiro [3] and José Maria Santos [1]

[1]   Forest Research Centre, School of Agriculture, University of Lisbon, Tapada da Ajuda, 1349-017 Lisboa, Portugal; pjbranco@isa.ulisboa.pt (P.B.); terferreira@isa.ulisboa.pt (M.T.F.); jmsantos@isa.ulisboa.pt (J.M.S.)

[2]   Katopodis Ecohydraulics Ltd., 122 Valence Avenue, Winnipeg, MB R3T 3W7, Canada; katopodisecohydraulics@live.ca

[3]   CERIS—Civil Engineering for Research and Innovation for Sustainability, Técnico, University of Lisbon, Avenida Rovisco Pais, 1049-001 Lisboa, Portugal; antonio.pinheiro@tecnico.ulisboa.pt

*   Correspondence: samaral@isa.ulisboa.pt; Tel.: +351-213-653-492

Received: 7 February 2019; Accepted: 5 March 2019; Published: 9 March 2019

**Abstract:** Low-head ramped weirs are a common instream obstacle to fish movements. Fish passability of these structures, where water passes over but does not generate a waterfall, is primarily related to ramp length and slope, but their relative contribution has seldom been considered. This study aims to assess the passage performance of a potamodromous cyprinid, the Iberian barbel (*Luciobarbus bocagei*), negotiating an experimental ramped weir with varying ramp length (L) and slope (S). Four configurations were tested, with a constant discharge of 110 L·s$^{-1}$. Results suggest that both factors influenced passage performance of fish. Attraction efficiency (AE) increased with increasing L and S, whereas the number of successes (N) and passage efficiency (PE) decreased upon increasing L. For S, it was found that both N and PE peaked at the intermediate level (20%). These results suggest that configurations with the lowest slopes may not necessarily be the best option because they may be less attractive for the fish and their demand for space is higher. Higher slopes (but not excessive) could be more attractive to fish, less space-demanding, and therefore, more cost-effective. Future studies should investigate how discharge and boulder placement influence fish passage across ramped weirs, to improve habitat connectivity.

**Keywords:** potamodromous cyprinid species; low-head ramped weirs; upstream migration; ecohydraulics

---

## 1. Introduction

River fragmentation by small engineered structures, far more numerous than dams, has led to severe declines or local extinctions of many fish populations by blocking upstream movements for reproduction, feeding, and refuge needs [1–3]. By identifying the importance of aquatic connectivity for good ecological quality in rivers, the European Water Framework Directive (WFD) emphasized the need to re-establish free movements for all fish species and size classes, regulating that member states should assess all instream obstacles, even small weirs, and minimize their barrier effect [4–6]. Since then, a few studies on small obstacles (considering assessment protocols, e.g., [7–9], or field assessments, e.g., [10–12]) and projects, such as the European project AMBER and other operational programs like the EU LIFE programs, have been developed, aiming to enhance the knowledge on permeability of small obstacles and fish passage, recommend strategies for action, and rehabilitate river habitats [3,13,14].

Portuguese rivers have more than 8000 small weirs [14] that are, in general, less than 5 m in height. Along with small broad-crested weirs (designed with a vertical downstream face, [15]), low-head ramped weirs, with inclined faces that fish may be able to overcome by swimming, are the most usual design [8,16]. In fact, some old broad-crested weirs that, after assessment, could not be removed have undergone rehabilitation works to include ramps in their designs, in order to enhance fish passability (e.g., [17]). However, the effectiveness and efficiency of these structures remains poorly understood, particularly for potamodromous cyprinids, which are an important component of Mediterranean European fish assemblages [18].

In low-head ramped weirs, water passes over the ramp and does not generate a waterfall [8,19]. The permeability of such structures to fish movements is usually site-, season-, and species-specific, depending on the effect of hydraulic boundary conditions (e.g., roughness of the ramp surface, conditions at the ramp toe related with erosion processes, and/or structure maintenance), hydrodynamics (e.g., water depth, discharge, and turbulence) present in the vicinity of the structure [19,20], and on fish swimming abilities, which are closely related to fish species groups and body size [21–23]. Nevertheless, in the physical design of a ramped weir, length and slope play an important role on the efficiency of these structures to successful upstream passage of fish [9,19,24]. As mentioned by Baker [24], although the effect of ramp length and slope is difficult to discriminate and their relative contribution has seldom been assessed, it is particularly important to study the interaction of these key factors in order to establish more appropriate design considerations for these types of obstacles.

The goal of this study was to assess the passage performance of a medium-size potamodromous cyprinid, the Iberian barbel, *Luciobarbus bocagei* (Steindachner, 1864), negotiating an experimental low-head ramped weir with varying ramp length (L) and slope (S). Iberian barbel was selected as the target species for being considered a representative of several species from the genera *Barbus* and *Luciobarbus*, commonly present in rivers from Mediterranean and Western Europe [25,26]. It is expected that (i) passage performance of fish, considering the attraction as well as upstream successful passages, will be influenced by the different combinations of L and S; (ii) attraction efficiency would increase with increasing L and S, due to increasing water velocity near the ramp that may act as an attraction factor for fish; and (iii) successful passages, and consequently passage efficiency, would decrease with increasing L and increasing S, hampered by the increasing water velocity present downstream and over the ramped weir.

## 2. Materials and Methods

To study the influence of L and S on the passage performance of Iberian barbel, four configurations encompassing two ramp lengths (L = 1.50 and 3.00 m) and three different slopes (S = 10%, 20%, 30%) were assessed (L150 S10; L150 S20; L150 S30; L300 S10). The experimental ramped weirs (Figure 1a), made of maritime plywood, were tested in an indoor ecohydraulic flume (a rectangular steel frame 10.00 m long × 1.20 m high × 0.60 m wide, with glass-viewing panels on sidewalls that allow direct observation of fish where, due to its dimensions and facilities, it is possible to preform ecohydraulic studies, assessing the influence of key hydraulic variables on the behavior of specimens) installed at the Hydraulics and Environment Department of the National Laboratory for Civil Engineering (LNEC), in Lisbon. The flume (Figure 1b) includes an upstream and a downstream tank, separated from the channel by mesh panels (from where the water enters the flume and is recirculated), and it was tilted at a 3% slope to represent the average slope of central and southern Iberian rivers (Catchment Characterisation and Modelling, version 2 [CCM2]; [27]). The experimental ramped weir (Figure 1b), spanning the entire channel width, was fixed in the flume at 2.50 m upstream of the acclimation area, a 0.60 m² area created by two mesh panels in the downstream zone of the flume. Immediately downstream of the ramp toe, a zone 0.50 m in length was established as the approach area. Discharge was measured by a flow meter installed in the supply pipe and maintained constant at 110 L·s⁻¹. Consequently, in all the configurations tested, the water depths at the weir crest and along the ramps, measured using rulers placed along the glass-viewing panels of the channel, were similar. Values registered at the weir crest varied from 0.19 to 0.20 m (observed in L150 S10 and L300 S10, respectively). Along the ramp, water depths decreased from 0.10–0.11 m (registered in L300 S10 and L150 S10, respectively), registered at the upper part of the ramp, to 0.06–0.08 m (observed in L150 S30 and L150 S10, respectively). A minimum water depth of 0.20 m, which was found to be the most suitable according to literature [17,19] and previous studies by Amaral et al. [15,28], was maintained in the approach area to standardize that condition throughout the experiments. Since the water column over the tested ramps was not deep enough (≈0.10 m) to use a 3D acoustic Doppler velocimeter and there was too much aeration and turbulence downstream of the ramp, especially at the ramp toe, the water velocity along the ramps, as well as upstream and downstream of the ramp, was instead measured with a flow probe (model FP 101, Global Water Instrumentation) in 21 and 27 sampling points for L = 1.50 m and L = 3.00 m, respectively. Sampling points were established along three longitudinal planes—a plane along the center of the ramp and two lateral planes spaced 0.05 m from the walls, and at intervals of 0.75 m along the ramp. Measurements were also taken in the middle of the weir crest, as well as 0.50 m upstream and downstream (0.50 and 1.00 m) of the ramped weir. These measurements ($Vx$) were represented graphically by contour maps, to illustrate water velocity variation along the tested combinations.

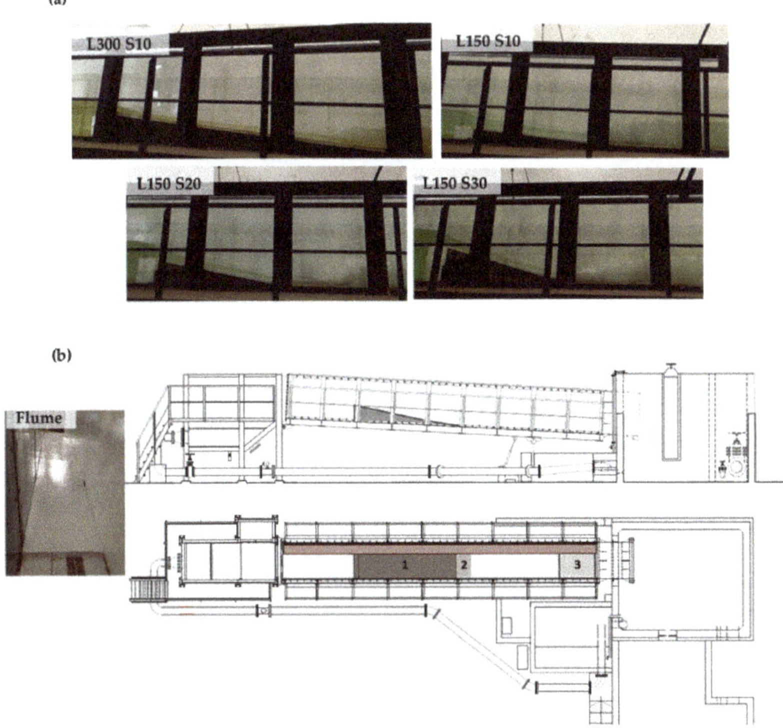



**Figure 1.** Images of (a) the four configurations tested (L represents the length (cm), while S the slope (%) of the ramp); (b) the experimental flume, representing a side view of the channel on a slope of 3% (scheme above), and a top view (scheme below) with the location of (1) the experimental low-head ramped weir (2.50 m upstream the acclimation area), (2) the approach area (the 0.30 m² shaded area immediately downstream of the ramp toe), and (3) the acclimation area (the 0.60 m² shaded area between the two removable fine mesh panels located downstream).

Adult Iberian barbel used in the experiments ($n$ = 80; mean total length (TL) ± standard deviation (SD) = 16.3 ± 2.1 cm) were captured by wadeable electrofishing (Hans Grassl IG-200) in the Lisandro

River, a small Atlantic coastal river near Lisbon. Fishing and handling permits for capture of wild fish (40/2017 and 222/2017/CAPT; 41/2017 and 223/2017/CAPT; 42/2017 and 224/2017/CAPT, respectively) were issued by the Portuguese Institute for Nature Conservation and Forests (ICNF, I.P.). A total of four electrofishing episodes were performed (two episodes per week during two consecutive weeks to not bias the fish motivation, collecting 20 fish per episode) according to the protocol adopted by the European Committee for Standardization (CEN 2003). To transport the fish to the laboratory facilities at LNEC, a fish transport box (Hans Grassl, 190 L) with external aeration was used. At LNEC, fish were maintained in filtered and aerated acclimation tanks (700 L tanks; Fluval Canister Filter FX5), where water quality was daily monitored (temperature = $23 \pm 1$ °C, pH = $7.7 \pm 0.1$, and conductivity = $174 \pm 14$ µs·cm$^{-1}$), using a multiparametric probe (HANNA, HI 9812-5), and high-quality levels (i.e., active fish, no mortality) were guaranteed by the mechanical and biological filtration system, with a turnover rate of 2300 L·h$^{-1}$. Fish were only tested after an acclimation period of 48 h from the holding conditions in the laboratory.

The study was conducted in agreement with national and international guidelines to maintain the welfare of the tested animals and minimise stress (J. M. Santos holds FELASA Level C certification (www.felasa.eu) to direct animal experiments). Fish experiments and maintenance in the laboratory and experimental facility were authorized (reference DGAV: 0420/000/000/2012) by the Department for Health and Animal Protection (Direcção de Serviços de Saúde e Protecção Animal) in accordance with the recommendations of the "Protection of animal use for experimental and scientific work". No fish were sacrificed during this study and, after finishing the experiments, all fish were taken back and released in their natural habitat.

Experiments were performed during late spring–early summer, reported by some authors as the main reproductive season for this species [19,29]. For each configuration tested (L150 S10; L150 S20; L150 S30; L300 S10), 4 replicates were carried out with schools of 5 fish ($n$ = 20 fish) that were haphazardly selected from the acclimation tanks and were used only once. The unit of analysis was therefore a school of five adult Iberian barbel with similar size, as this species tends to move in schools, rather than individually, as observed in other studies by Amaral et al. [15,28] and Romão et al. [26,30], to increase hydrodynamic efficiency [31]. For fish to adapt to the conditions in the flume, each replicate started with an acclimation period of 15 min (period previously tested by Amaral et al. [15,28,32] and considered to be appropriate for the acclimation of fish to the flume). After that time, the upstream mesh panel of the acclimation area was removed, and fish were able to volitionally explore the channel for a maximum of 60 min. Since both upstream and downstream passages were allowed, fish could approach, attempt to pass, and successfully negotiate the ramp multiple times. Fish movements were monitored by direct observation and recorded (top view) by a video camera (GoPro HERO5). The number of fish that entered the approach area (Ap), the number of fish that entered into the ramp and actively tried to negotiate it (At), and the number of fish that completely passed the ramp to upstream, i.e., completed successful passages (N), were registered. Metrics of passage performance, such as percentage of attraction efficiency (AE%) and percentage of passage efficiency (PE%), were then calculated from Equations (1) and (2), adapted from Amaral et al. [15]. For the statistical analysis, because this study did not have a full factorial design, and data were not homoscedastic nor normally distributed, a nonparametric Kruskal–Wallis $H$ test was performed to analyze the influence of L and S on the successful negotiation of the experimental ramps, pondering the results for N, AE%, and PE%. The *dunn.test* package [33], from the open-source software R [34], was used to compute the analysis.

$$AE\% = 100 \times At/Ap, \tag{1}$$

$$PE\% = 100 \times N/At, \tag{2}$$

## 3. Results

Upstream successful passages were registered in all the configurations tested. However, the number of N, Ap, and At, and consequently values of AE% and PE%, varied according to the tested configurations, highlighting the effect that factors L and S may have had on the passage performance of Iberian barbel. The total number of N, together with values of PE%, mainly decreased with the increase of tested L (Figure 2a) and S (Figure 2b). On the contrary, values of AE% registered an increase with the increasing values of both L and S (Figure 2a,b). Configuration L150 S20 recorded the highest number of Ap, At, and N (totals of 31, 21, and 17, respectively), being the configuration with higher PE% (81%). On the other hand, configuration L150 S30 registered the lowest numbers, with only Ap = 15, At = 11, and N = 4. However, it was the most attractive configuration for fish, with AE% = 73.3%, followed by L300 S10 (71.4%), which in turn was the least efficient configuration in terms of PE%, registering only 15% (Ap = 28, At = 20, N = 3). Configuration L150 S10 was the least attractive for fish (AE% = 53.6%), registering several approaches (Ap = 28) but few attempts (At = 15) to negotiate the experimental ramp.

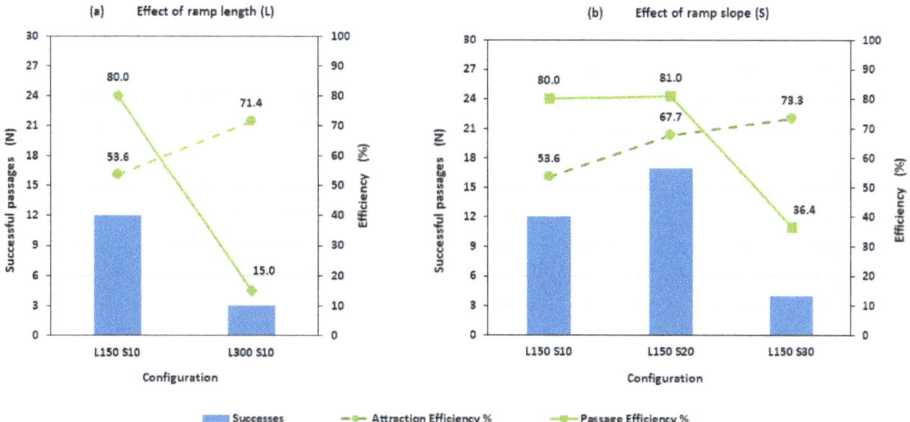

**Figure 2.** Results for the number of successful passages (N; bars), and attraction efficiency (AE%; dotted line) and passage efficiency (PE%; solid line) for the configurations tested, considering the variation of (**a**) ramp length (L); (**b**) ramp slope (S).

Results from the Kruskal–Wallis *H* test suggest a marginally significant influence (i.e., $P \leq 0.10$) of both factors L and S on the number of N (L: $H = 1.85$, 1 *d.f.*, $P = 0.10$; S: $H = 4.47$, 2 *d.f.*, $P = 0.10$), as well as on values of PE% (L: $H = 3.19$, 1 *d.f.*, $P = 0.07$; S: $H = 5.71$, 2 *d.f.*, $P = 0.05$). The ramp with L = 1.50 m achieved better results than the one with L = 3.00 m and, in terms of slope, S = 20% stood out from the other slopes tested as the most successful. As for AE%, however, results reveal no significant influence of factors L ($H = 0.004$, 1 *d.f.*, $P = 0.90$) and S ($H = 2.30$, 2 *d.f.*, $P = 0.31$).

Figure 3 displays the variation of water velocity (V$x$) for the different tested ramps. Contour maps revealed that water velocity values increased with L and S. This increase was particularly important in the case of L150 S30 and L300 S10, where values of water velocity above 3 m·s$^{-1}$ were registered close to the toe of the ramp. On the contrary, configuration L150 S10 was the one with the lowest water velocities (1.8 m·s$^{-1}$ close to the toe of the ramp, and a maximum of 2.3 m·s$^{-1}$ over the ramp).

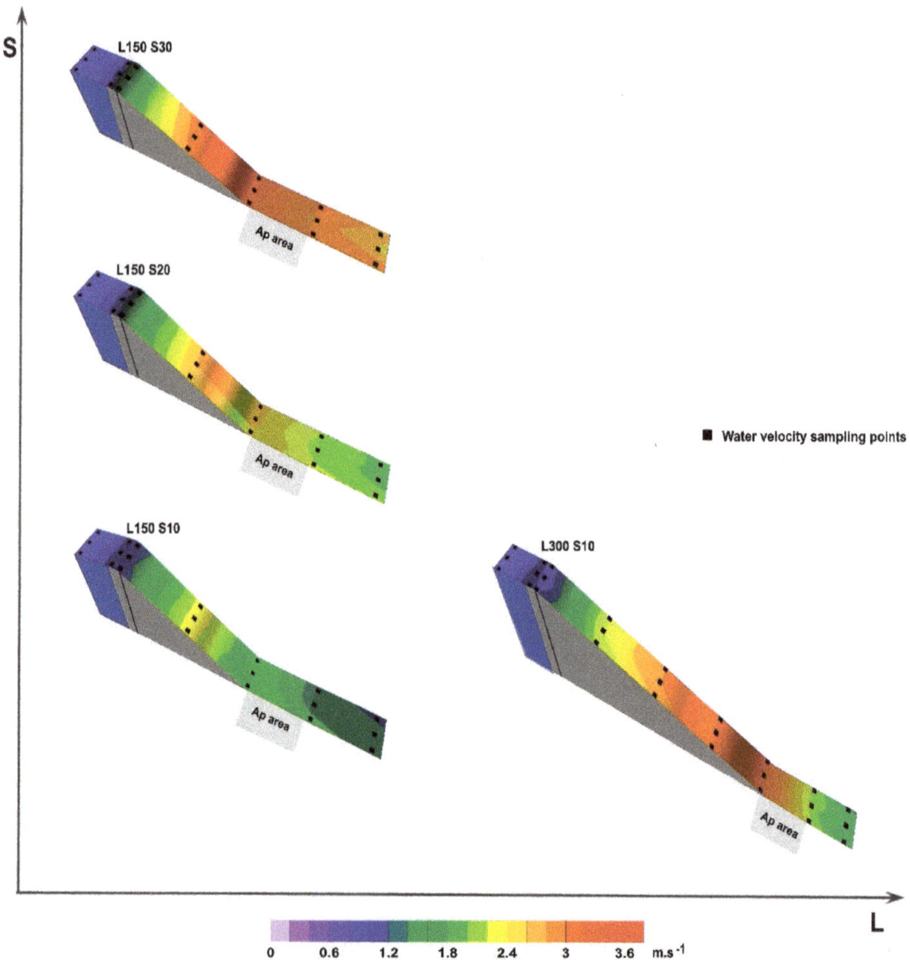



**Figure 3.** Contour maps of water velocity (V*x*) for the configurations tested, considering the variation of ramp length (L) and slope (S). Measurements were made with a flow probe (model FP 101, Global Water Instrumentation). Black dots represent water velocity sampling points. The approach area, located immediately downstream of the ramp toe between two lines of sampling points, is identified by the tag below (Ap area).

## 4. Discussion

In situ studies on the negotiation of small instream obstacles by fish—that must associate the assessment of fish movements and an extensive characterization of all the hydrodynamic conditions that fish need to overcome in order to successfully pass the obstacle—can be very complex and onerous [35–37]. All the requirements needed to carry out such studies, in terms of human resources and time, field equipment, and robust technology, may strongly constrain their developments [35,38], unless a long period of execution and provision of funding is ensured, conditions that most scientific field experiments often fail to achieve. Therefore, the use of full-scale or even scaled-down laboratory facilities, such as the ecohydraulic flume used in the present study, is presented as a more expeditious parallel approach to study fish behavior and negotiation of small instream obstacles [25,39,40]. Inherent to laboratory conditions, these ecohydraulic flumes provide the opportunity to easily manipulate important factors, control for confounding variables and effects that could bias the results, and observe responses that should improve the knowledge of events occurring in the wild [25,39,41].

In this study, the influence of L and S on the passage performance of the Iberian barbel negotiating an experimental low-head ramped weir was assessed, maintaining a constant discharge of 110 L·s$^{-1}$. Although experimental conditions tested in the flume were a simplification of what fish may encounter in nature, they allowed detailed observation of fish behavior (e.g., fish approaching the ramp, attempts to negotiate it, and successful passages) as well as the control and analysis of physical and hydraulic variables, such as ramp length and slope, discharge and consequently water velocity, and water depth at the toe of the ramp that, along with fish swimming abilities and other boundary conditions (e.g., roughness of the ramp surface, structural conditions of the ramp toe), are referred by some authors [8,9,19,20,24] as preponderant factors for the successful upstream passage of fish along ramped weirs.

Results of this experiment suggest that both factors L and S had a marginally significant influence on the number of N, and consequently on values of PE%, but their influence on AE% was not significantly determined. As in other experiments by Amaral et al. [15,32], and in Goering and Castro-Santos [42], the "fish passage paradox"—concerning the influence of water velocity and, consequently, of turbulence and energy dissipation present on these small barriers [9,19] in the attraction of fish and on the successful negotiation of the obstacle—was also observed in the present study. Fish were attracted to the ramped weir by high values of water velocity but, at the same time, it might have been a limiting factor for successful upstream passage—what attracts fish is what hampers movements.

Contrary to what was initially expected, configuration L150 S10, that combined the smallest L with the lowest S, and thus the one that registered the lowest values of water velocities, was not the configuration that recorded the highest values of N or PE%, and was also the least attractive (AE% ca. 50%). In this configuration, only half the fish that entered the approach area (Ap = 28) went into the ramp and actively tried to negotiate it, a fact that may suggest that the water velocity (Vx = 1.7 m·s$^{-1}$) was not the most appropriate to establish an attractive path for fish to proceed and successfully pass the obstacle [42–44]. On the other hand, configurations L150 S30 and L300 S10, which displayed high water velocity (registering values of 3.6 and 3.4 m·s$^{-1}$, respectively, at the ramp toe) due to the correspondingly steeper S and the longer L, achieved the highest values of AE% but registered a low number of N (only 4 and 3 successful passages, respectively) and, consequently, the lowest values of PE% (36% and 15%, respectively), suggesting that water velocity, and the potential turbulence associated to these type of obstacles [9,19], had a positive influence on the attraction of fish to the ramp but, at the same time, might have hampered their successful upstream passage possibly due to fish disorientation and fatigue [43–45]. This was especially observed in configuration L300 S10 where, in some attempts, fish were able to negotiate the ramp up to its half-length by sprinting (maximum-speed swimming), overcoming values of water velocity around 3 m.s$^{-1}$. However, most likely due to fatigue, fish stopped swimming and were dragged down to the end of the ramp.. Therefore, to enhance fish passage along long low-head ramped weirs, it would probably be important to retrofit these types of obstacles with substrates, such as different types of blocks or rocks for a more nature-like design, in order to create areas with diverse hydraulic conditions along the ramp [46,47], allowing fish to rest and to recover energy to continue successful negotiation of the ramp [22,47]. Since the swimming performance of the Iberian barbel is quite similar to the swimming performance of other rheophilic cyprinids and salmonids of the same length [22], these results may be more broadly applicable. Nevertheless, species swimming traits and the different strategies to negotiate obstacles should always be considered [30,48,49]. Finally, configuration L150 S20, which displayed intermediate values of water velocity when compared to the other configurations tested, was the combination that recorded the best results for N and PE%, and registered also nearly 70% of AE%, a value that may be considered as a reasonable percentage for attraction. Taken together, these results may suggest that, upon designing ramped-weirs, configurations with the lowest slopes may not necessarily be the best option, because they are less attractive for the fish and their demand for space is higher, thereby increasing construction costs. Conversely, as the present study shows, higher (but not excessive) slopes, though yielding a similar PE%, can be more attractive to fish, less expensive and, therefore, more cost-effective.

In conclusion, this study is in line with the outcomes of Baker [24] about the importance that L and S may have on the permeability of low-head ramped weirs for upstream movements of fish, both in terms of the attraction of fish to the ramp and especially regarding successful negotiation. However, the negotiation of ramped weirs by potamodromous fish species should be further investigated. Future studies should explore discharge variation and boulder placement, featuring different arrangements and geometries that influence fish passage across low-head ramped weirs, to further improve habitat connectivity. Thereby, the outcomes from the present work, complemented with future research pondering the above considerations, may significantly contribute to help engineers and biologists to design more appropriate passage structures for low-head instream obstacles.

**Author Contributions:** Conceptualization, S.D.A., P.B. and J.M.S.; methodology, S.D.A. and J.M.S.; formal analysis, S.D.A.; investigation, S.D.A., P.B. and J.M.S.; writing—original draft preparation, S.D.A., P.B. and J.M.S.; writing—review and editing, C.K., M.T.F. and A.N.P.; supervision, J.M.S.

**Funding:** Forest Research Centre (CEF) is a research unit funded by Fundação para a Ciência e a Tecnologia I.P. (FCT), Portugal (UID/AGR/00239/2013). Susana D. Amaral was funded by a PhD grant from University of Lisbon/Santander Totta (SantTotta/BD/RG2/SA/2011), and by FCT (SFRH/BD/110562/2015). Paulo Branco was financed by a post-doctoral grant from FCT (SFRH/BPD/94686/2013), and José M. Santos is presently the recipient of an FCT researcher contract (IF/00020/2015).

**Acknowledgments:** The authors want to thank the staff of the National Laboratory for Civil Engineering (LNEC), specially to João Manuel Pereira, for all the support during the experiments, and to the Portuguese Institute for Nature Conservation and Forests (ICNF, I.P.), for the fishing and handling permits for capture of wild fish. Thanks are also extended to two anonymous reviewers, for their helpful comments on an early draft of this manuscript.

**Conflicts of Interest:** The authors declare no conflict of interest.

## References

1. Aarts, B.G.; Van Den Brink, F.W.; Nienhuis, P.H. Habitat loss as the main cause of the slow recovery of fish faunas of regulated large rivers in Europe: The transversal floodplain gradient. *Regul. River* **2003**, *20*, 3–23. [CrossRef]

2. Nilsson, C.; Reidy, C.A.; Dynesius, M.; Revenga, C. Fragmentation and flow regulation of the world's large river systems. *Science* **2005**, *308*, 405–408. [CrossRef] [PubMed]

3. King, S.; O'Hanley, J.R.; Newbold, L.R.; Kemp, P.S.; Diebel, M.W. A toolkit for optimizing fish passage barrier mitigation actions. *J. Appl. Ecol.* **2017**, *54*, 599–611. [CrossRef]

4. European Commission. Directive 2000/60/EC of the European Parliament and of the Council of 23 October 2000 establishing a framework for the community action in the field of water policy. *Off. J. Eur. Comm.* **2000**, *22*, L327.

5. Reyjo, Y.; Argillier, C.; Bonne, W.; Borja, A.; Buijse, A.D.; Cardoso, A.C.; Daufresne, M.; Kernan, M.; Ferreira, M.T.; Poikane, S.; et al. Assessing the ecological status in the context of the European Water Framework Directive: Where do we go now? *Sci. Total Environ.* **2014**, *497–498*, 332–344. [CrossRef]

6. Barry, J.; Coghlan, B.; Cullagh, A.; Kerr, J.R.; King, J.J. Comparison of coarse-resolution rapid methods for assessing fish passage at riverine barriers: ICE and SNIFFER protocols. *River Res. Appl.* **2018**, *34*, 1168–1178. [CrossRef]

7. Ovidio, M.; Capra, H.; Philippart, J.C. Field protocol for assessing small obstacles to migration of brown trout Salmo trutta, and European grayling Thymallus thymallus: A contribution to the management of free movement in rivers. *Fish. Manag. Ecol.* **2007**, *14*, 41–50. [CrossRef]

8. Solà, C.; Ordeix, M.; Pou-Rovira, Q.; Sellarès, N.; Queralt, A.; Bardina, M.; Casamitjana, A.; Munné, A. Longitudinal connectivity in hydromorphological quality assessments of rivers. The ICF index: A river connectivity index and its application to Catalan rivers. *Limnetica* **2011**, *30*, 273–292.

9. Schmutz, S.; Mielach, C. *Measures for Ensuring Fish Migration at Transversal Structures—Technical Paper*; ICPDR—International Commission for the Protection of the Danube River: Vienna, Austria, 2013; p. 52.

10. Ovidio, M.; Philippart, J.C. The impact of small physical obstacles on upstream movements of six species of fish—Synthesis of a 5-year telemetry study in the River Meuse basin. *Hydrobiologia* **2002**, *483*, 55–69. [CrossRef]

11. Weibel, D.; Peter, A. Effectiveness of different types of block ramps for fish upstream movement. *Aquat. Sci.* **2013**, *75*, 251–260. [CrossRef]

12. Ordeix, M. Fish migration and fish ramp assessment at a gauging station on a Mediterranean river (Catalonia, NE Iberian Peninsula). *Limnetica* **2017**, *36*, 427–443. [CrossRef]

13. Birnie-Gauvin, K.; Candee, M.M.; Baktoft, H.; Larsen, M.H.; Koed, A.; Aarestrup, K. River connectivity reestablished: Effects and implications of six weir removals on brown trout smolt migration. *River Res. Appl.* **2018**, *34*, 548–554. [CrossRef]

14. Ordeix, M.; González, G.; Sanz-Ronda, F.J.; Santos, J.M. Restoring fish migration in the rivers of the Iberian Peninsula. In *From Sea to Source 2.0. Protection and Restoration of Fish Migration in Rivers Worldwide*; Brink, K., Gough, P., Royte, J., Schollema, P.P., Wanningen, H., Eds.; World Fish Migration Foundation: Groningen, The Netherlands, 2018; pp. 174–179.

15. Amaral, S.D.; Branco, P.; Silva, A.T.; Katopodis, C.; Viseu, T.; Ferreira, M.T.; Pinheiro, A.N.; Santos, J.M. Upstream passage of potamodromous cyprinids over small weirs: The influence of key-hydraulic parameters. *J. Ecohydraulics* **2016**, *1*, 79–89. [CrossRef]

16. Branco, P.; Amaral, S.D.; Ferreira, M.T.; Santos, J.M. Do small barriers affect the movement of freshwater fish by increasing residency? *Sci. Total Environ.* **2017**, *581–582*, 486–494. [CrossRef]

17. Food and Agriculture Organization (FAO)/DVWK. *Fish Passes—Design, Dimensions and Monitoring*; FAO: Rome, Italy, 2002; p. 119.

18. Ferreira, T.; Oliveira, J.; Caiola, N.; De Sostoa, A.; Casals, F.; Cortes, R.; Economou, A.; Zogaris, S.; Garcia-Jalon, D.; Ilhéu, M.; et al. Ecological traits of fish assemblages from Mediterranean Europe and their responses to human disturbance. *Fish. Manag. Ecol.* **2007**, *14*, 473–481. [CrossRef]

19. Baudoin, J.M.; Burgun, V.; Chanseau, M.; Larinier, M.; Ovidio, M.; Sremski, W.; Steinbach, P.; Voegtle, B. *Assessing the Passage of Obstacles by Fish. Concepts, Design and Application*; Onema: Paris, France, 2014; p. 200.

20. Harris, J.H.; Kingsford, R.T.; Peirson, W.; Baumgartner, L.J. Mitigating the effects of barriers to freshwater fish migrations: The Australian experience. *Mar. Freshw. Res.* **2016**, *68*, 614–628. [CrossRef]

21. Kemp, P.S.; O'Hanley, J.R. Procedures for evaluating and prioritising the removal of fish passage barriers: A synthesis. *Fish. Manag. Ecol.* **2010**, *17*, 297–322. [CrossRef]

22. Katopodis, C.; Gervais, R. *Fish Swimming Performance Database and Analyses*; Canadian Science Advisory Secretariat: Ottawa, Canada, 2016. Available online: http://www.dfo-mpo.gc.ca/csas-sccs/Publications/ResDocs-DocRech/2016/2016_002-eng.html (accessed on 30 November 2018).

23. Newton, M.; Dodd, J.A.; Barry, J.; Boylan, P.; Adams, C.E. The impact of a small-scale riverine obstacle on the upstream migration of Atlantic Salmon. *Hydrobiologia* **2018**, *806*, 251–264. [CrossRef]

24. Baker, C.F. Effect of ramp length and slope on the efficacy of a baffled fish pass. *J. Fish Biol.* **2014**, *84*, 491–502. [CrossRef]

25. Santos, J.M.; Branco, P.; Katopodis, C.; Ferreira, T.; Pinheiro, A. Retrofitting pool-and-weir fishways to improve passage performance of benthic fishes: Effect of boulder density and fishway discharge. *Ecol. Eng.* **2014**, *73*, 335–344. [CrossRef]

26. Romão, F.; Branco, P.; Quaresma, A.L.; Amaral, S.D.; Pinheiro, A.N. Effectiveness of a multi-slot vertical slot fishway versus a standard vertical slot fishway for potamodromous cyprinids. *Hydrobiologia* **2018**, *816*, 153–163. [CrossRef]

27. Vogt, J.; Soille, P.; De Jager, A.; Rimaviciute, E.; Mehl, W.; Foisneau, S.; Bodis, K.; Dusart, J.; Paracchini, M.L.; Haastrup, P.; et al. *A pan-European River and Catchment Database*; European Commission—Joint Research Centre—Institute for Environment and Sustainability: Luxembourg, 2007.

28. Amaral, S.D.; Branco, P.; Katopodis, C.; Ferreira, M.T.; Pinheiro, A.N.; Santos, J.M. To swim or to jump? Passage behaviour of a potamodromous cyprinid over an experimental broad-crested weir. *River Res. Appl.* **2018**, *34*, 1–9. [CrossRef]

29. Santos, J.M.; Ferreira, M.T.; Godinho, F.N.; Bochechas, J. Efficacy of a nature-like bypass channel in a Portuguese lowland river. *J. Appl. Ichthyol.* **2005**, *21*, 381–388. [CrossRef]

*Sustainability* **2019**, *11*, 1456

30. Romão, F.S.; Quaresma, A.; Branco, P.; Santos, J.M.; Amaral, S.D.; Ferreira, M.T.; Katopodis, C.; Pinheiro, A. Passage performance of two Cyprinids with different ecological traits in a fishway with distinct vertical slot configurations. *Ecol. Eng.* **2017**, *105*, 180–188. [CrossRef]

31. Pitcher, T.J.; Parrish, J.K. Functions of shoaling behaviour in teleosts. In *Behaviour of Teleost Fishes*; Pitcher, T.J., Ed.; Chapman and Hall: London, UK, 1993; pp. 363–439.

32. Amaral, S.D.; Branco, P.; Romão, F.; Viseu, T.; Ferreira, M.T.; Pinheiro, A.N.; Santos, J.M. The effect of weir crest width and discharge on passage performance of a potamodromous cyprinid. *Mar. Freshw. Res.* **2018**, *69*, 1795–1804. [CrossRef]

33. Dinno, A. dunn.test: Dunn's Test of Multiple Comparisons Using Rank Sums. R package version 1.2.3. 2015. Available online: http://CRAN.R-project.org/package=dunn.test (accessed on 30 November 2018).

34. R Core Team. *R: A Language and Environment for Statistical Computing*; R Foundation for Statistical Computing: Vienna, Austria, 2017. Available online: http://www.R-project.org/ (accessed on 30 November 2018).

35. Rice, S.P.; Lancaster, J.; Kemp, P. Experimentation at the interface of fluvial geomorphology, stream ecology and hydraulic engineering and the development of an effective, interdisciplinary river science. *Earth Surf. Process. Landf.* **2010**, *35*, 64–77. [CrossRef]

36. Ordeix, M.; Pou-Rovira, Q.; Sellarès, N.; Bardina, M.; Casamitjana, A.; Solà, C.; Munnè, A. Fish pass assessment in the rivers of Catalonia (NE Iberian Peninsula). A case study of weirs associated with hydropower plants and gauging stations. *Limnetica* **2011**, *30*, 405–426.

37. Lacey, R.W.J.; Neary, V.S.; Liao, J.C.; Enders, E.C.; Tritico, H.M. The IPOS Framework: Linking fish swimming performance in altered flows from laboratory experiments to rivers. *River Res. Appl.* **2012**, *28*, 429–443. [CrossRef]

38. Wang, R.W.; Hartlieb, A. Experimental and field approach to the hydraulics of nature-like pool-type fish migration facilities. *Knowl. Manag. Aquat. Ec.* **2011**, *400*, 5. [CrossRef]

39. Kemp, P.S.; Gessel, M.H.; Sandford, B.P.; Williams, J.G. The behaviour of Pacific salmonid smolts during passage over two experimental weirs under light and dark conditions. *River Res. Appl.* **2006**, *22*, 429–440. [CrossRef]

40. Martin, P.R.; Bateson, P.P.G. *Measuring Behaviour: An Introductory Guide*, 3rd ed.; Cambridge University Press: Cambridge, UK, 2007; p. 176.

41. Alexandre, C.M.; Quintella, B.R.; Silva, A.T.; Mateus, C.S.; Romao, F.; Branco, P.; Ferreira, M.T.; Almeida, P.R. Use of electromyogram telemetry to assess the behavior of the Iberian barbel (*Luciobarbus bocagei* Steindachner, 1864) in a pool-type fishway. *Ecol. Eng.* **2013**, *51*, 191–202. [CrossRef]

42. Goerig, E.; Castro-Santos, T. Is motivation important to brook trout passage through culverts? *Can. J. Fish. Aquat. Sci.* **2017**, *74*, 885–893. [CrossRef]

43. Pavlov, D.S.; Lupandin, A.I.; Skorobogatov, M.A. The effects of flow turbulence on the behavior and distribution of fish. *J. Ichthyol.* **2000**, *40*, S232–S261.

44. Elder, J.; Coombs, S. The influence of turbulence on the sensory basis of rheotaxis. *J. Com. Physiol. A* **2015**, *201*, 667–680. [CrossRef]

45. Liao, J.C. A review of fish swimming mechanics and behavior in altered flows. *Philos. T. Roy. Soc. B* **2007**, *362*, 1973–1993. [CrossRef]

46. Baki, A.; Zhu, D.; Rajaratnam, N. Flow Simulation in a Rock-Ramp Fish Pass. *J. Hydraul. Eng.* **2016**, *142*, 04016031. [CrossRef]

47. Muraoka, K.; Nakanishi, S.; Kayaba, Y. Boulder arrangement on a rocky ramp fishway based on the swimming behavior of fish. *Limnologica* **2017**, *62*, 188–193. [CrossRef]

48. Aramburu, E.; Lara, Á.; Morcillo, F.; Castillo, M.; Berges, J.A. *Escalas de Peces de Hendidura Vertical*; CEDEX: Madrid, Spain, 2016; p. 144.

49. Romão, F.; Quintella, B.R.; Pereira, T.J.; Almeida, P.R. Swimming performance of two Iberian cyprinids: The Tagus nase Pseudochondrostoma polylepis (Steindachner, 1864) and the bordallo Squalius carolitertii (Doadrio, 1988). *J. Appl. Ichthyol.* **2012**, *28*, 26–30. [CrossRef]

*Article*

# Passability of Potamodromous Species through a Fish Lift at a Large Hydropower Plant (Touvedo, Portugal)

Daniel Mameri [1], Rui Rivaes [1], João M. Oliveira [1,2], João Pádua [3], Maria T. Ferreira [1] and José M. Santos [1,*]

[1]  Forest Research Centre (CEF), School of Agriculture, University of Lisbon, Tapada da Ajuda, 1349-017 Lisboa, Portugal; dmameri@isa.ulisboa.pt (D.M.); ruirivaes@isa.ulisboa.pt (R.R.); joliveira@isa.ulisboa.pt (J.M.O.); terferreira@isa.ulisboa.pt (M.T.F.)
[2]  Centre for Ecology, Evolution and Environmental Changes (cE3c), Faculty of Sciences, University of Lisbon, 1749-016 Lisboa, Portugal
[3]  EDP Labelec—Estudos, Desenvolvimento e Actividades Laboratoriais, S.A., Rua Cidade de Goa, 4, 2685-038 Sacavém, Portugal; Joao.Padua@edp.com
*  Correspondence: jmsantos@isa.ulisboa.pt; Tel.: +351-213-653-489

Received: 11 October 2019; Accepted: 18 December 2019; Published: 24 December 2019

**Abstract:** River fragmentation by large hydropower plants (LHP) has been recognized as a major threat for potamodromous fish. Fishways have thus been built to partially restore connectivity, with fish lifts representing the most cost-effective type at high head obstacles. This study assessed the effectiveness with which a fish lift in a LHP on the River Lima (Touvedo, Portugal), allows potamodromous fish—Iberian barbel (*Luciobarbus bocagei*), Northern straight-mouth nase (*Pseudochondrostoma duriense*) and brown trout (*Salmo trutta fario*)- to migrate upstream. Most fish (79.5%) used the lift between summer and early-fall. Water temperature was the most significant predictor of both cyprinids' movements, whereas mean daily flow was more important for trout. Movements differed according to peak-flow magnitude: nase (67.8%) made broader use of the lift in the absence of turbined flow, whereas a relevant proportion of barbel (44.8%) and trout (44.2%) passed when the powerhouse was operating at half (50 $m^3s^{-1}$) and full-load (100 $m^3s^{-1}$), respectively. Size-selectivity found for barbel and trout could reflect electrofishing bias towards smaller sizes. The comparison of daily abundance patterns in the river with fish lift records allowed the assessment of the lift's efficacy, although biological requirements of target species must be considered. Results are discussed in the context of management strategies, with recommendations for future studies.

**Keywords:** potamodromous fish; migration; lift; hydropower; species management

## 1. Introduction

Rivers are currently one of the most threatened ecosystems in the world [1,2], with flow regulation and longitudinal fragmentation by dams and weirs being among the main causes of environmental degradation and reduction of available habitat for freshwater fauna [3–5]. Large hydropower plants (LHP) are particularly harmful for fish populations, not only by causing the blockage to their movements, but also by increasing the risk of fish stranding, drifting and dewatering of spawning grounds caused by flow variations, as results of peak-operations in response to energy demands [6–9]. In fact, a myriad of studies have reported significant declines or extinctions of many fish affected by LHP [10–13]. Particularly impacted are potamodromous species, i.e., freshwater species that seasonally undergo upstream migrations along the river, for the purpose of finding suitable habitats for reproduction, which are needed to complete their life-cycles [14–16]. A significant amount of research has therefore been carried out with the goal of restoring longitudinal connectivity in an upstream direction [15,17]. In this context, the development of fishways to transpose barriers stood up as a hydraulic structural solution which facilitates fish movements past the barriers, while partially restoring river connectivity [18–20].

From the different fishway types that have been constructed worldwide to address upstream migration of fish [18,21], fish lifts are the most used and cost-effective at high dams (>15 m; [22]), from the economic and biologic point of view [23,24]. A fish lift consists of a mechanical system which is located at the foot of an obstruction, which attracts (by a guiding flow) the fish into a cage with an inscale (non-return device), raising it and then emptying it in the reservoir upstream, transporting fish over the barrier (for schemes see [18,25]). Although much less studies on fish lifts are available when compared to other fishways, such as pool-type or nature-like facilities [17,26], these structures have nonetheless been monitored in different regions and targeted different fish species [27]. In Europe, fish lift studies have mostly addressed salmonids [23,24,28] and eels [29], but also cyprinids [30,31]. However, most assessments were performed without considering the abundance and size-structure of fish species downstream the dam (often costly due to the human resources and equipment involved) that potentially constitute the migrant population to use the fish lift (but see Discussion below). Such information, in addition to seasonal and daily patterns of fish migration and associated environmental triggers [32], is fundamental to address fish lift selectivity and efficacy, and may be useful to support management decisions.

Studies of the effectiveness of fishways on LHP often focus on high-value economic and recreational species, namely diadromous and salmonids [33,34], whereas studies on potamodromous fish have often been neglected [17,26]. These species, however, are well represented in riverine fish assemblages, particularly in Iberia [9,35,36] and free instream movement is crucial for their survival [37]. Moreover, this is particularly important as potamodromous fish are key components of the lower and middle reaches of temperate rivers [38] and sensitive to river regulation and longitudinal fragmentation [16]. Within the fish community present in the study area, the cyprinids Iberian barbel *Luciobarbus bocagei* (Steindachner, 1864) (hereafter barbel) and Northern straight-mouth nase *Pseudochondrostoma duriense* (Coelho, 1985) (hereafter nase), and the salmonid brown trout *Salmo trutta fario* (Linnaeus, 1758) (hereafter trout) are amongst the most abundant species in northern Iberian rivers [39] and were therefore the focus of this study.

The main goal of this study was therefore to assess the effectiveness with which a fish lift in a large hydropower plant on the River Lima (Portugal), allows potamodromous fish to migrate upstream. For this, we assessed the seasonal and daily use of the lift by the fish population and compared it with (1) the environmental factors that are known to be associated with the triggering of the fish upstream migration; (2) the peak-flow magnitudes at the power plant (0, 50 and 100 $m^3s^{-1}$, see Study Area); and (3) the size structure (to infer selectivity) and abundance (to calculate a ratio of effectiveness) of the fish population downstream.

We predict that (i) fish counts through the lift would vary between the different months for all three species following patterns outlined in the literature, i.e., cyprinid species movements should mainly occur during the reproductive season, i.e., summer [40,41], and further extending to early fall when species start to search for winter, feeding or thermal refuges [15,42] (with regard to the trout, movements are predicted as well to occur in the reproductive season, in this case, between late fall and early-winter [43]); (ii) daily activity of the studied species would not show marked diurnal/nocturnal preferences due to the absence of natural predators in the River Lima [16,44], which should not restrict movements to take place preferentially during the night, when survival is expected to be maximum [26]; (iii) water temperature and flow, two of the most important environmental factors responsible for triggering migration [35,45], would be the most significant ones for the target species; (iv) large-sized and faster-flowing species would be better able to cope with higher peak-flow magnitudes (50 and 100 $m^3s^{-1}$) than smaller ones, and hence expected to use the lift during such conditions; and (v) selectivity should be low, although the presence of larger individuals in the lift when compared to the downstream river segment, would be expected to occur as a result of upstream migration of adults to spawning sites [41].

## 2. Materials and Methods

### 2.1. Study Area

The River Lima runs for 135 km in the north-west part of Iberian Peninsula, being shared by Spain and Portugal (Figure 1). It runs on a NE-SW direction and is characterized by a relatively high run-off, as a result of a mean annual rainfall of about 2000 mm. Geology is mainly granitic and the topography consists of a series of steep and narrow valleys in the upper reaches, contrasting with the lower reaches, with milder gradients and wider valleys dominated by alluvial materials.

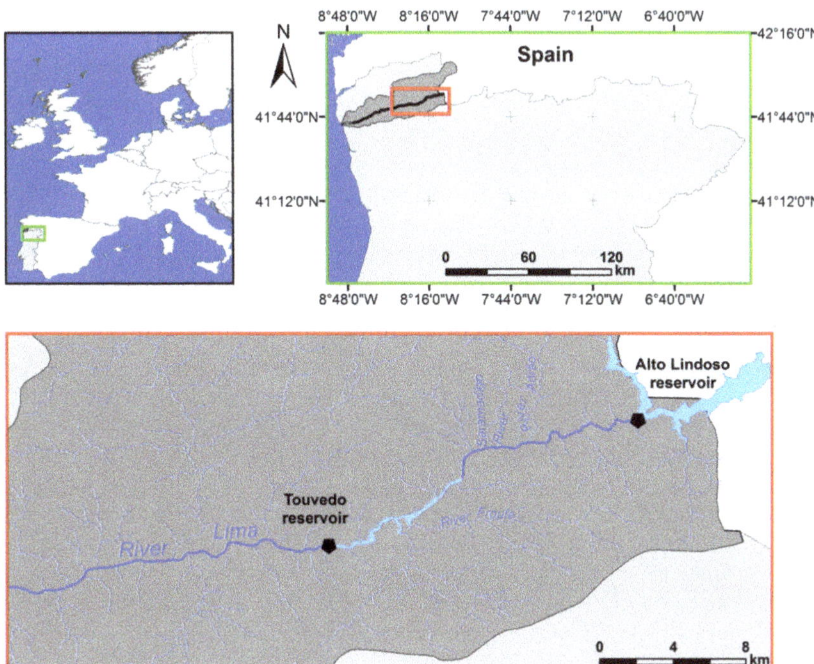


**Figure 1.** Map of the study area in the River Lima, North Portugal. The black pentagons refer to the dam locations.

The Touvedo LHP stands at 47 km from the river mouth and is the first large instream barrier to upstream fish migration. It is 42 m high and it serves as a tailwater reservoir for the high flows released by the Alto Lindoso Dam, located 16.5 km upstream (250 m$^3$s$^{-1}$ at full operation), by temporarily storing them, and then returning them to the river, with values not higher than 100 m$^3$s$^{-1}$. The dam is equipped with a 22-MW Kaplan turbine and three spillway gates, with a maximum discharge of 3200 m$^3$s$^{-1}$ when the reservoir reaches the run-off storage limit. Mean number of spillway discharge events is 32/year, which mainly occur (c. 80%) from mid-autumn to early spring [44]. The Touvedo LHP works under three peak-flow magnitudes: (i) 0 m$^3$s$^{-1}$, turbine shutdown (i.e., powerhouse off), which is compensated by a 5.5 m$^3$s$^{-1}$ constant ecological flow (a minimum of 4.0 m$^3$s$^{-1}$ + 1.5 m$^3$s$^{-1}$ from the fish lift) to ensure the connectivity of the different habitats and movements of species downstream; (ii) 50 m$^3$s$^{-1}$, half-load operation; and (iii) 100 m$^3$s$^{-1}$, full-load operation.

A network of spawning, feeding and refuge habitats is available for fish upstream the Touvedo dam. These are mainly located in the Rivers Adrão, Froufe and Saramadigo, which have no man-made obstacles (i.e., all free-flowing) and also no sources of pollution across their watersheds [9].

The dam features a fish lift (2.1 m long, 1.3 m wide and 2.9 m high) which is located on the left bank. It has 3 entrances (two placed above the turbine gates and another one displaced 20 m downstream, to take advantage of the turbined flow) and was initially designed to improve diadromous species movements, such as Atlantic salmon *Salmo salar* and sea trout *Salmo trutta trutta*. A maximum attraction flow of 4.5 $m^3s^{-1}$ is released to promote attraction towards the entrances, which have a mean water velocity varying between 0.21 and 0.55 m $s^{-1}$ (turbine shutdown, powerhouse off) and between 0.68 and 0.91 m $s^{-1}$ (turbine operating, powerhouse on), as previously measured with a SonTek FlowTracker Handheld ADV (model number P4267, Qualitas Instruments Ltd., Madrid, Spain, 2012) at nine points across their width [44]. Inside the attraction circuit, the fish move towards the trapping cage, which is set to raise and empty every 4-h.

## 2.2. Fish Passage through the Lift

To account for seasonal variations in migration patterns, monitoring of fish passage through the lift was made on a monthly basis, from March 2013 until February 2014. Continuous data was acquired through an automatic video-recording system, which included a video camera (Bosch, mod. MR700, Gerlingen, Germany) placed on the top of the fish lift (allowing the collection of trapping cage images during the final period of the cage ascent) and a video recorder (Bosch, mod. LTC455). Target species (barbel, nase and trout) were the most frequent and abundant potamodromous fish species previously recorded in the catchment [30,44]. Following this approach, no fish handling was required, as opposed to other monitoring techniques (e.g., mark-recapture or radio telemetry), thus avoiding causing injury or stress to the fish. The camera was installed on the upper part of the fish lift in order to acquire images of the lift cage during its final period of ascension. The trapping cage was sealed with 20 × 20 cm white quadrats to obtain clearer images for identification and estimates of fish lengths [29]. Collected data included: the timing of fish passage (day and hour), the number of fish per cycle, the identification of each fish to the species level and the estimated total length of individuals (TL, to the nearest cm). For further details on the video-recording system, see [29].

To determine the role of environmental variables on fish movements, six potential predictors were recorded: (1) water temperature, recorded on an hourly basis by a Vemco Minilog-II probe placed in the downstream river segment; (2) mean daily flow, defined as the amount of flow through the turbine, spillway or ecological flow provided by the dam reports on an hourly basis; (3) daily flow fluctuation, i.e., the standard deviation of hourly flows—turbined, spillway or ecological—provided by the dam reports; (4) mean daily rainfall, provided by a nearby weather station (code 03F/01G, managed by the Portuguese Institute for Sea and Atmosphere, I.P.), located 17 km downstream from the Touvedo dam; (5) accumulated rainfall, obtained by combination of the mean daily rainfall that occurred on the three preceding days (as we predicted that fish would move upstream a few days earlier in response to accumulated rainfall; (6) photoperiod, as the time of civil twilight, i.e., the length (in hours) of the daytime period, obtained at http://zenite.nu/ (accessed June 2018); and (7) the proportion of illumination of the moon, based on the ephemeris available at http://www.rodur.ago.net/en/ (accessed June 2018), obtained by dividing the lunar cycle into four phases.

## 2.3. Fish Catches Downstream

To obtain a measure of fish lift efficacy, surveys ($n = 9$) were performed once every month (unable to sample on March 2013 and January-February 2014, due to adverse weather conditions) in a river segment (total length: 340 m) located immediately downstream the dam, by using a combined wadable and boat electrofishing scheme (DC, 300–700 V, SAREL model WFC7-HV, Electracatch International, Wolverhampton, UK) to obtain the most reliable picture of fish abundance (unit effort = 1 fishing day—4h of effective sampling—along with ratios of fish-lift records to downstream catches; for further details on the sampling procedure, see [29]). Fish were then identified and measured for TL (nearest cm); native specimens were then returned to the river alive, whereas non-natives were sacrificed in accordance with Portuguese legislation. Fish surveys were not performed in March 2013 and January–February 2014 due to adverse weather conditions (high flow events) that prevented secure access to the river.

### 2.4. Data Analyses

Monthly fish counts recorded in the lift were initially plotted on a line chart to examine seasonal activity and search for migration periods. Next, to search for eventual daily patterns of passage through the lift, two periods were considered: 06:00–18:00 h (day) and 18:00–06:00 h (night) [41]. For both data, the chi-square test of goodness of fit was conducted to account for differences in the relative abundance of fish passing through the lift in each month and between day and night periods, respectively.

The relative influence of environmental variables on the fish counts in the lift was also evaluated through generalized linear models (GLM) following a Poisson distribution. For this purpose, a forward stepwise approach was conducted, based on the Akaike Information Criterion (AIC) for each fitted model, selecting only the variables leading to the most adequate model (i.e., lowest AIC). In each model, variable significance was set at $\alpha = 0.05$. To improve data distribution, we applied a log (x + 1) transformation to all environmental variables before fitting them into the GLM, with the exception of the proportion of illumination of the moon, which was arcsin-transformed. Durbin–Watson statistics for each model were also calculated to detect possible autocorrelation between residuals (values ranging from 1 to 2 are considered to be acceptable). To search for significant differences in species movements according to the different peak-flow magnitudes (0 $m^3s^{-1}$, powerhouse off; 50 $m^3s^{-1}$, powerhouse at half-load; 100 $m^3s^{-1}$, powerhouse at full-load), the chi-square test of goodness of fit was employed. Size selectiveness in the fish lift was assessed by comparing the population size structure of each species recorded in the lift with the one obtained downstream the dam, using Fisher's exact test. Size-classes were partitioned in 5-cm intervals, to allow a more detailed effect of selectivity [29].

Literature has outlined the absence of a standardized procedure to evaluate fish passage efficacy [17,46], a qualitative concept consisting of checking if the fishway is capable of allowing the target species to pass. This concept differs from efficiency, which focuses on its quantitative performance, defined as the percentage of marked fish that enter and successfully negotiate the fishway out of the total fish previously marked [47]. A ratio of fish lift efficacy was therefore calculated by dividing the number of fish observed ascending the lift (number $day^{-1}$) by the total number of fish captured below the dam (unit effort = number in 1 fishing day), being considered as a proxy of the lift's efficacy [48].

All analyses were conducted in R version 3.5.2, [49], using the packages *stats* (implemented in R) and *MASS* [50].

## 3. Results

### 3.1. Seasonal Fish Counts in the Lift

Fish counts through the lift varied significantly between the different months for all three species (barbel: $\chi^2 = 57.828$, df = 11, $p < 0.001$; nase: $\chi^2 = 232.440$, df = 11, $p < 0.001$; trout: $\chi^2 = 66.315$, df = 11, $p < 0.001$) (Figure 2). A total of 548 barbel, 1801 nase and 63 trout were recorded passing the fish lift within the study period, with most of the fish being observed between summer and early fall (79.5%), i.e., August and October. Among the three species, nase was the most abundant (74.7% of the total fish counts), with the highest counts being recorded in August (699) and October (491). Barbel (22.8%) was more abundant in October (125), while trout (2.6%), the least abundant species, peaked a maximum of 13 individuals in both April and October.

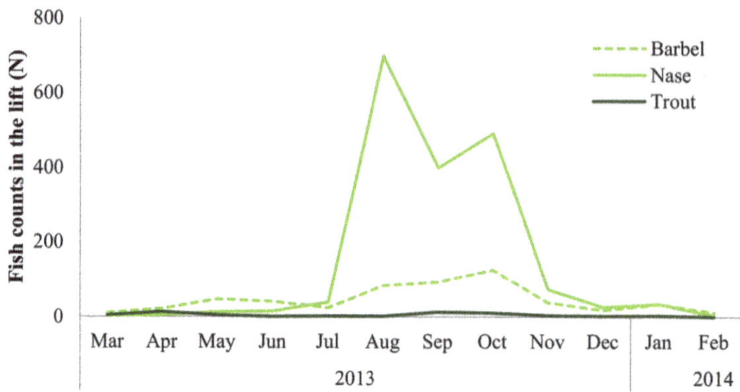


**Figure 2.** Fish counts for barbel (dotted green line), nase (green line) and trout (dark green line) in the fish lift, between March 2013 and February 2014.

### 3.2. Daily Patterns of Fish Passage

Fish recordings in the lift did not vary significantly with the time of day for barbel ($x^2 = 0.006$, df = 1, $p = 0.936$), nase ($x^2 = 0.810$, df = 1, $p = 0.368$); and trout ($x^2 = 3.028$, df = 1, $p = 0.082$). Among the three species, the highest difference in percentage of fish passing between two periods belonged to trout (58.7% of the fish counts recorded during the night period) (Figure 3).

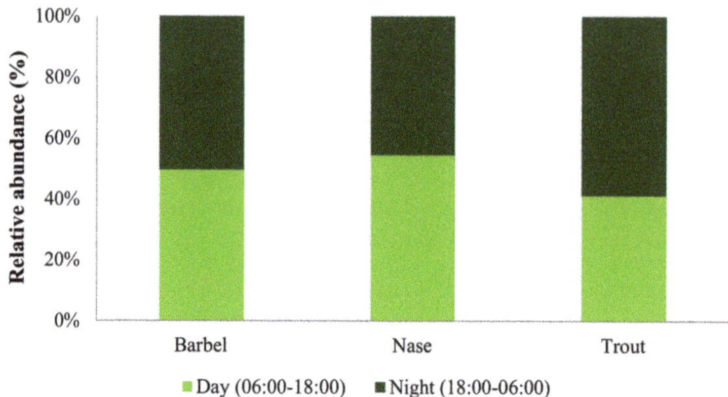


**Figure 3.** Abundance of barbel, nase and trout recorded in the fish lift during the day (06:00–18:00) and night (18:00–06:00).

### 3.3. Effects of Environmental Variables on Fish Passability

Water temperature (F = 22.425, $p < 0.001$) and daily flow fluctuation (F = 4.894, $p < 0.001$) were positively associated with nase passability. The fitted model for this species included a third significant variable positively associated—photoperiod (F = 3.955, $p = 0.049$) (Table 1). Water temperature was also included in the model for barbel passability (F = 7.138, $p = 0.008$), together with mean daily flow (F = 7.733, $p = 0.006$) and accumulated rainfall (F = 12.818, $p < 0.001$) as significant variables, all with positive associations (Table 1). For trout, only the mean daily flow (positively associated) was retained in the final model (F = 3.941, $p = 0.049$).

**Table 1.** Variables entered in the GLMs explaining species abundance in the Touvedo fish lift. Seven different factors were analyzed, but only those included in the final models are presented. For each species, significance of each variable in the final model was calculated through the F-test. Beta coefficients (ß) and Durbin–Watson statistics (D) for each model are also presented.

| Variable | ß | *F*-Test | *p*-Value | D |
|---|---|---|---|---|
| *P. duriense* | | | | 1.127 |
| Water temperature | 0.466 | 22.425 | <0.001 | |
| Flow variation | 0.228 | 4.894 | 0.029 | |
| Photoperiod | 0.198 | 3.955 | 0.049 | |
| *L. bocagei* | | | | 1.812 |
| Water temperature | 0.167 | 7.138 | 0.008 | |
| Mean daily flow | 0.155 | 7.733 | 0.006 | |
| Acumulated rainfall | 0.276 | 12.818 | <0.001 | |
| *S. trutta fario* | | | | 1.996 |
| Mean daily flow | 0.151 | 3.941 | 0.049 | |

*3.4. Fish Passage in Relation to Peak-Flow Magnitudes*

Fish lift use varied according to peak-flow magnitudes, with nase showing significant differences in passability ($\chi^2 = 55.460$, df = 2, $p < 0.001$). Accordingly, passability of this species was the highest (67.8%) when the powerhouse was off (0 m$^3$s$^{-1}$, Figure 4). Contrastingly, passability of larger species, i.e., barbel ($\chi^2 = 6.480$, df = 2, $p = 0.039$) and trout ($\chi^2 = 5.631$, df = 2, $p = 0.060$) occurred mainly when the powerhouse was operating, being the highest for the barbel (44.8%) upon operation at half-load (50 m$^3$s$^{-1}$), whereas for the trout, the largest portion of individuals (44.2%) migrating occurred when the powerhouse was operating at full-load, though differences in the last species were not significant (100 m$^3$s$^{-1}$; $\chi^2 = 5.631$, df = 2, $p = 0.060$).

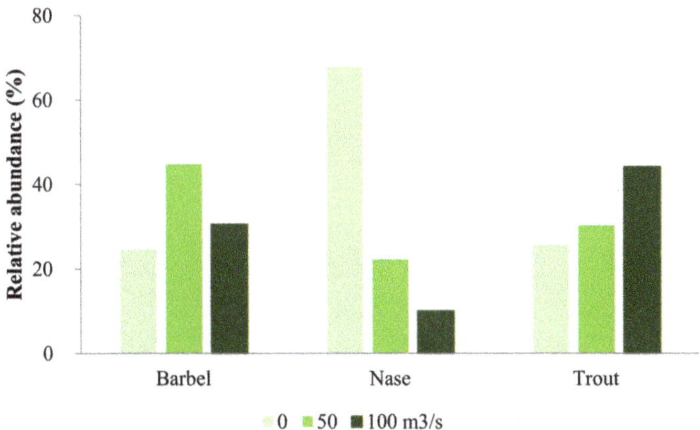


**Figure 4.** Relative abundance (%) of barbel, nase and trout recorded in the lift for different peak-flow magnitudes: 0 m$^3$s$^{-1}$ (pale green; powerhouse off), 50 m$^3$s$^{-1}$ (light green; powerhouse at half-load) and 100 m$^3$s$^{-1}$ (dark green; powerhouse at full-load).

*3.5. Fish Lift Selectivity*

Differences in population size structure were found when comparing the proportions of fish counts in the lift and captures downstream of the Touvedo dam for all three species (Fisher's exact test, $p < 0.05$), though these differences were more pronounced in barbel and trout (Fisher's exact test, $p < 0.001$), with some selectiveness being observed (Figure 5). For both species, the proportions of individuals observed in the lift (barbel: mean ± SD = 22.4 ± 6.9 cm; trout: 23.5 ± 4.2 cm) were generally larger than the ones captured in the river segment downstream (barbel: 18.1 ± 4.5; trout: 16.5 ± 4.5 cm).

Nonetheless, for barbel, both lift recordings and river surveys revealed 15–20 cm individuals as the most abundant size class (Figure 5). For nase, despite differences in size-class distributions were found (Fisher's exact test, $p = 0.016$; mean size ± SD = 13.0 ± 3.6 in the lift and 13.3 ± 2.9 in caught fish), the same size classes were represented in the lift and river surveys, but with a larger proportion of the smallest size individuals (TL ≤ 10 cm) occurring in the fish lift: 24.3% (10.1% in the river downstream) (Figure 5).

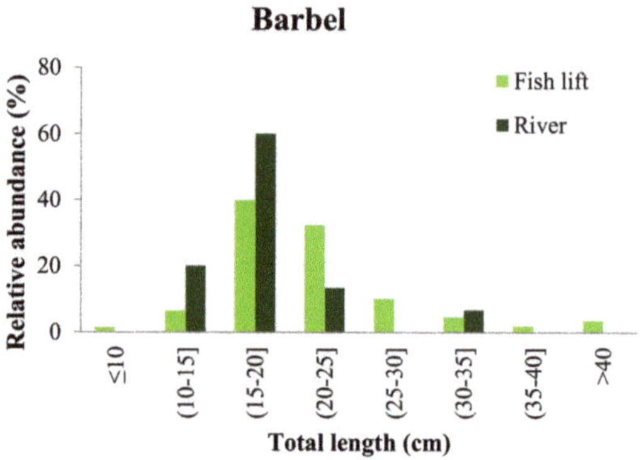


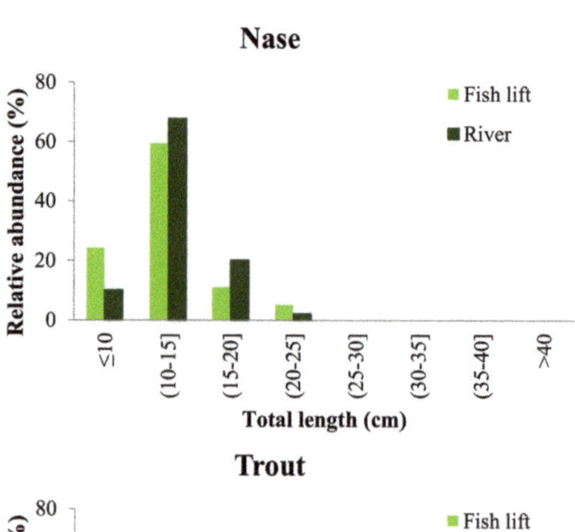


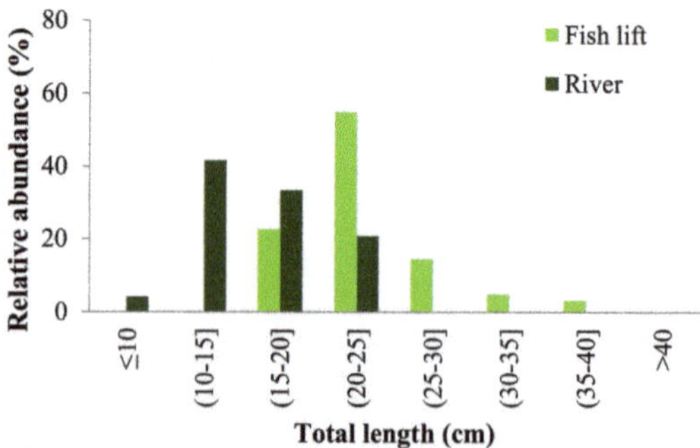


**Figure 5.** Size class (cm) distributions of barbel, nase and trout recorded in the lift (light green) and captured downstream of the Touvedo dam (dark green).

*3.6. Fish Lift Efficacy*

The passage-to-catch ratio was generally higher in barbel (mean ± SD = 1.36 ± 1.06) than in nase (mean ± SD = 0.26 ± 0.25) and trout (mean ± SD = 0.10 ± 0.10). Average monthly ratios for the barbel were superior to the unit in half of the study period, attaining a maximum value of 3.10 in September (Table 2), when this species recorded higher counts in the lift (Figure 2). Ratios for nase in its most active period (August-October; Figure 2) varied between 0.43 and 0.75, being considerably higher than the ratios obtained in the remaining months (≤0.10; Table 2). For trout, ratios were generally low throughout the year, despite an increase was observed in September–October, when mean values were superior to the overall mean obtained for this species (0.10).

**Table 2.** Mean daily number of barbel, nase and trout migrating through the Touvedo fish lift and captured downstream by electrofishing in 2013 (unit effort = 1 fishing day (4 h of effective sampling)), along with ratios of fish-lift records to downstream catches. (a) Undetermined ratio due to the absence of caught individuals by electrofishing, despite being observed in the fish lift.

| | Barbel | | | Nase | | | Trout | | |
|---|---|---|---|---|---|---|---|---|---|
| Month | Fish Lift $N\,Day^{-1}$ | Electr. Catch $N\,Unit\,Effort^{-1}$ | Ratio | Fish Lift $N\,Day^{-1}$ | Electr. Catch $N\,Unit\,Effort^{-1}$ | Ratio | Fish Lift $N\,Day^{-1}$ | Electr. Catch $N\,Unit\,Effort^{-1}$ | Ratio |
| Apr | 0.70 | 0 | (a) | 0.13 | 5 | 0.03 | 0.43 | 5 | 0.09 |
| May | 1.52 | 0 | (a) | 0.42 | 1 | 0.42 | 0.16 | 0 | (a) |
| Jun | 1.36 | 2 | 0.68 | 0.50 | 18 | 0.03 | 0.03 | 1 | 0.03 |
| Jul | 0.77 | 0 | (a) | 1.26 | 13 | 0.10 | 0.06 | 1 | 0.06 |
| Aug | 2.80 | 0 | (a) | 23.30 | 30 | 0.75 | 0.07 | 2 | 0.03 |
| Sep | 3.21 | 1 | 3.10 | 13.76 | 31 | 0.43 | 0.45 | 3 | 0.14 |
| Oct | 4.03 | 12 | 0.34 | 4.03 | 34 | 0.47 | 0.35 | 1 | 0.35 |
| Nov | 1.34 | 1 | 1.30 | 1.34 | 28 | 0.09 | 0.14 | 2 | 0.07 |
| Dec | 0.62 | 0 | (a) | 0.62 | 29 | 0.03 | 0.10 | 9 | 0.01 |
| Mean | 1.82 | 1.78 | 1.36 | 5.04 | 21.00 | 0.26 | 0.20 | 2.67 | 0.10 |

## 4. Discussion

This study assessed the effectiveness with which a fish lift in a LHP allows native potamodromous fish to migrate upstream. To accomplish such goal, we assessed the seasonal and daily use of the lift by fish and compared it with the environmental factors that are known to be associated with the triggering of the fish upstream migration, the peak-flow magnitudes at the power plant and the size structure and abundance of the fish population downstream. This continuous monitoring of fish stocks downstream the dam, though time-consuming and enclosing inherent technical difficulties of sampling a large river [51], has seldom been used in fish lift evaluation studies (e.g., [23,30]) and provided a useful proxy of the efficacy of the fish lift that, together with the seasonality of fish movements, can be used by managers to better plan fish lift operations and shutdowns.

As it was expected, fish counts through the lift varied significantly between the different months for all three species. Overall, the highest number of fish counts was attained in late summer–early fall, when almost 80% of the individuals used the fish lift, with nase being the most abundant species (74.7% of the total fish counts), followed by barbel (22.7%). The large proportions of both species observed in late summer is consistent with previous reports on these rheophilic cyprinids [52,53]. Similarly, in a work by De Leeuw and Winter [53] in the lowland rivers Meuse and Rhine in the Netherlands, the authors reported more movements of the common barbel *Barbus barbus* and common nase *Chondrostoma nasus* during both summer-early fall (July–October) and spring (spawning season for both species). It is highly likely that such activity observed in the lift is related to reproductive migrations, which takes place during these periods, particularly in summer [40,41], when these potamodromous species migrate upstream to seek areas for reproduction, typically in gravel and pebble beds located in upstream tributaries [39]. These species also showed movements outside their usual spawning period, displaying a second peak in early-fall (September–October), a result that is also consistent with other studies, in which "out of season" movements may reflect a search for winter, feeding or thermal refuges, as it has been observed in other potamodromous cyprinids [15,54].

The presence of trout in the fish lift was residual (only 2.6% of the total fish counts), with movements occurring throughout the year and not only restricted to the spawning period, which typically occurs during late fall and winter [43]. The similarity of observed seasonal fish counts in the lift, with the species migratory ecology, provides therefore an indication that the lift is not disrupting the seasonality of fish movements, serving therefore its purpose.

No significant differences were found in daily patterns of fish passage ascending the fish lift, which is in accordance with our expectations. Though some studies suggest that cyprinids are more active during the night to avoid predation (e.g., [37,41]), such patterns can be quite species-specific [16]. In a recent study conducted in the Meuse river basin (Belgium), Benitez et al. [16] found that the common barbel *Barbus barbus* did not show any differences in daily activity when passing through the existing fishways, contrarily to the trout, which was more active during the day, a result that was also supported by some authors (e.g., [43]), but not others (e.g., [42]). In a series of surveys conducted in the Zêzere River (Tagus river basin), Santos et al. [52] also did not find differences in daily activity for the Iberian straight-mouth nase (*Pseudochondrosoma polylepis*), a sister species of the present *P. duriense*. Such findings, as well those of the present study, reflect the absence of predators in the sampled river segment downstream of the Touvedo dam [44], which does not constrain the activity of native species to take place during night-time periods when survival would be expected to be maximum [26,41].

Water temperature was found to be the most significant predictor of the abundance of both cyprinids (barbel and nase), which is consistent with our expectations and with previous findings on the migratory ecology of these species, where increasing water temperature acts as an environmental trigger for the upstream movements of these species [30,55]. Flow variables (mean daily flow and daily flow fluctuations) were also important to explain upstream movements of both cyprinids, as it was previously expected: daily flow fluctuation, which can act as an environmental trigger for fish migration [45,56], also had a positive effect on the observed nase in the lift, whereas barbel abundance in the fish lift was positively associated with increasing mean daily flow. Moreover, accumulated rainfall, which also has a direct influence in river flow [57], was also present in the model for barbel passability. Taken together, these results corroborate the ones in the literature in which flow (in addition to water temperature) is also one of the most important variables in triggering fish migration [45,58], particularly under conditions of high-water level fluctuation [59] as in the Touvedo LHP. Photoperiod was also positively associated with barbel passability, which seems to indicate a higher activity during the day in the lack of predators [60], as previously mentioned. Mean daily flow was the only variable selected by the model to explain the number of trout individuals that migrated through the fish lift. This result is in agreement with some studies [61,62], but not with others [30,41] that pointed out water temperature as the most important factor in the upstream migration of this species. It should be noted, however, that the number of individuals recorded was considerably lower compared to the remaining potamodromous species, which may have reduced the statistical power of our analyses. On the other hand, it is possible that a different hierarchy of environmental factors stimulated the same behaviour in different years [45]. Hence, long-term studies could provide a broader understanding of the interaction between environmental variability and potamodromous fish movements, in order to clarify trends over long time series (>10 years), while also providing important data for scientists and ecosystem managers.

Fish passage in relation to the different peak-flow magnitudes differed in two of the three species, with the largest proportion of nase (67.8% of total abundance) using the fish lift in the absence of turbined flow (powerhouse off). Contrarily, barbel made broader use of the fish lift when the powerhouse was operating at half-load (50 m$^3$s$^{-1}$). As for trout, it should be noted that while differences were not significant, the largest proportion of movements occurred when the powerhouse was operating at full-load (100 m$^3$s$^{-1}$). It is our conviction that nase made larger use of the fish lift when the powerhouse was off (i.e., with turbines shutdown) due to the lower water velocities (0.21–0.55 m s$^{-1}$) that occur at the lift entrances upon this scenario [44]. Though nase is a medium-sized cyprinid [63], for which adults can cope critical swimming speeds up to 0.78 m s$^{-1}$ [64] and therefore theoretically being able to negotiate such a range of velocities, individuals found in both river segment downstream and lift were mainly juveniles, small-sized fish (mean: 13.0 cm TL), for which swimming performance is typically lower than larger conspecifics [65,66]. It is thus important to ensure that water velocities that nase will face within the entrances that lead to the lift are sufficiently attractive—not too low (<0.20 m s$^{-1}$) to hinder attraction [29], nor too high, above their critical swimming speed (>0.78 m s$^{-1}$, [64])—for appropriate entrance and passage, particularly during summer and early fall when most of the individuals (74.7%) used the fish lift. It is tempting to suggest that managers should try to implement management strategies, such as periodic turbine shutdown [67], that best balance the trade-off between energy production and the potential for upstream fish migration [68], at least during the critical migratory periods. However, this is often difficult to achieve and dependent on the characteristics of the national network of hydropower schemes as well as on the specificities of the energy market. Nevertheless, the specific requirements of the different species migrating, namely swimming performance, should be taken into account when planning for a mitigation flow scheme such as ecological flow releases. Both barbel (mean TL: 22.4 cm) and trout (mean TL: 23.5 cm), the two largest species that used the lift, have rheophilic habits during part (barbel) or the whole (trout) life-cycle [36], which, combined with their greater ability to withstand higher velocities during short time periods (adult barbel: $U_{crit} = 0.81$ m s$^{-1}$, [69]; trout: $U_{crit} = 0.65$ m s$^{-1}$ and $U_{max}$ varying between 0.94–1.26 m s$^{-1}$, depending on water temperature, [70]), may have determined the larger proportion of individuals of both species using the fish lift under the half- load (50 m$^3$s$^{-1}$) and full-load (100 m$^3$s$^{-1}$) conditions, respectively. It should be noted, however, that the willingness to enter and use the fish lift cannot be explained solely by the water velocity at the entrances nor the size of individuals. Such motivation can also be driven by other internal (such as the physiological condition or fatigue level) or external (such as turbidity or turbulence) factors not accounted for in the present study. For example, some recent studies on fish passage have pointed out the importance of turbulence in determining the success and timing of potamodromous fish migration upstream [71,72]. It is clear that future studies should focus on experimental controlled conditions, where the variables of interest (e.g., water velocity and associated turbulence parameters) can be manipulated while controlling for potential confounding factors (e.g., temperature), which provide an excellent opportunity to disentangle the effect that multiple factors have on fish attraction and passage through fish lifts.

The comparison of species size-structure between observed fish in the lift and those captured downstream, which gives an indicator of fishway selectivity, showed some differences for all three species, particularly for barbel and trout, with the occurrence of larger individuals in the fish lift relatively to the river downstream, as it was previously predicted (see Introduction). Such selectivity could have also arisen as a result of sampling the fish with electrofishing in specific habitats, such as deep pools, where typically the larger fish, like barbel and trout [44], dwell, and where capture efficiency is often lower than in shallower (up to 1.5 m) habitats [9,63]. On the other hand, the smaller individuals of these species may not display a marked migration stimulus, at least associated with reproduction (e.g., [15]), so their abundance in the fish lift should be lower than that of the larger ones. Another relevant aspect that could partially explain the lower abundance of smaller-sized individuals of these species is related to the potential effect of water velocity in the fish lift attraction circuit (up to 0.90 m s$^{-1}$ when the powerhouse is on), which may have limited the entrance of smaller individuals, for which swimming capacity is typically more limited comparatively to the larger ones [66]. Assessing their swimming capabilities would help clarify if the observed patterns are related with their lower swimming capabilities, or the lack of environmental cues for these smaller fish to perform upstream migrations.

As a proxy of the fish lift's efficacy, the standardized passage-to catch ratio was used, as there are presently, to the best of our knowledge, no standard methods nor metrics to evaluate efficacy, neither any defined thresholds (e.g., [46,73]). Our results showed that the mean value of this indicator was higher than 1 for the barbel, suggesting that more individuals were using the fish lift compared to those that were available downstream and captured by electrofishing. As outlined above, such results should be analyzed with caution as most barbel, particularly the larger individuals, often dwell in deep pool habitats [9] where electrofishing is clearly less effective [74], and thus their population downstream that is potentially available to migrate could have been under–evaluated. The use of other techniques, such as mark-recapture or passive integrated transponders (PIT) telemetry [75], can be useful to provide more accurate data on barbel stocks arriving at the foot of large–scale barriers. The mean ratio obtained for the nase (0.26), the most abundant species in the fish lift, was higher than those reported by Noonan et al. [26], who reviewed worldwide estimates of fish passage efficiency across all types of fishways. In the case of fish lifts and the presence of non–salmonid species, the mean value reported was only 0.10, which makes the present estimate (0.26) quite optimistic in the current context, although their work focuses on efficiency rather than efficacy as in the present study. However, since the concept of efficacy is not defined in terms of minimum standards (e.g., [41,46]), it should be specified with respect to the biological requirements of the species using the fishway, and not as an absolute value [47]. In the Lima basin, cyprinids are the most dominant and abundant species in the main river [9,41]. Consequently, the main goal of the lift, rather than allowing the whole species' population downstream to move upstream the dam (as it would in the case of anadromous or catadromous species), is to prevent fragmentation of potamodromous populations between different river segments [25]. For such species that carry out their life cycles downstream and upstream the dam [30], simple documentation of them passing upstream is sufficient [76], providing enough evidence that a considerable proportion of individuals used the fish lift, assuring a long-term sustainability of fish populations.

It should also be noted that the low ratios observed in the months outside the migratory season (for barbel, nase and trout) do not necessarily represent low lift efficacy, as they may reflect the absence of migratory stimuli and the consequent lack of motivation to overcome the obstacle. Trout was the species that theoretically performed the lowest, as shown by their lowest mean ratio (0.10) when compared to the other two species, as well to the corresponding mean value (0.35) in the literature for salmonids using fish lifts [26]. This is unlikely to reflect a lower performance of trout upon negotiating water velocities to enter the lift, as trouts are typically better swimmers and withstand higher velocities than cyprinids, but instead their natural low abundance in the present cyprinid-type river segment [77]. Future studies should try to associate efficacy to other indicators, namely efficiency and delay, to achieve a broader assessment of fish passage through a fishway [78]. In the particular case of fish lifts, it would be important to try to quantify the two components of attraction efficiency [17]: guidance (i.e., arrival at the entrance) in response to attraction currents, and entry (i.e., decision to enter). In this sense, biotelemetry techniques could be applied to monitor such fine-scale activity.

Finally, it should be pointed out that actions to improve the efficacy of upstream movements of potamodromous species at fish lifts may not always be the best practice. Fish lifts are unidirectional systems, transporting fish from downstream to upstream of dams, but do not operate on the reverse side (i.e., from upstream to downstream), therefore not allowing subsequent downstream migration. When this is coupled with the absence of suitable spawning and growth habitats upstream (even if they are present downstream), fish lifts may act as ecological traps, doing more harm than good to fish populations [79]. Although a unidirectional fishway, the Touvedo fish lift is not likely to be acting as an ecological trap, due to the existence of a network of good quality habitats upstream the Touvedo dam (see Study Area). Taken together, the Touvedo fish lift enables the upstream migration of a "considerable" number of adult potamodromous fish in the proper seasonal timing, which is a positive step towards the maintenance of populations above and below the dam, potentially contributing to their future sustainability. However, different fish species were found to be affected differently by the peak-flow magnitudes (nase preferentially migrating during periods of turbine shutdown, whereas barbel and trout making broader use of the lift when the powerhouse was operating), which points to the need of a proper peak-flow management during the species reproductive season. Future studies should consider determining to what extent fish can safely use the spillway gates or the turbines as a pathway in their descendent routes.

**Author Contributions:** Conceptualization, J.P., M.T.F. and J.M.S.; methodology, D.M., R.R., J.M.O., J.P. and J.M.S.; formal analysis, D.M. and J.M.S.; investigation, D.M. and J.M.S., writing—original draft preparation: D.M. and J.M.S.; writing—review and editing, R.R., J.M.O., J.P. and M.T.F.; supervision, M.T.F. and J.M.S. All authors have read and agreed to the published version of the manuscript.

**Funding:** Forest Research Centre (CEF) is a research unit funded by Fundação para a Ciência e a Tecnologia I.P. (FCT), Portugal (UID/AGR/00239/2019). This study was supported by EDP, S.A. (Energias de Portugal), who allowed the use of their facilities and equipment (video recording system). The funders had no role in study design, data collection and analysis, decision to publish, or preparation of the manuscript. Daniel Mameri is supported by a Ph.D. grant from the FLUVIO–River Restoration and Management programme funded by FCT (PD/BD/142885/2018). José Maria Santos is presently the recipient of a FCT researcher contract (IF/00020/2015).

**Acknowledgments:** The authors wish to thank Raul Arenas, Paulo Branco and André Fabião for their help during the electrofishing sampling campaigns. Thanks are also due Ulisses Cabral and António Leite Marinho for assistance, field and video logistics. The Institute for Nature Conservation and Forests (ICNF) provided the necessary fishing and handling permits. Finally, the authors want to thank the comments and suggestions from three anonymous reviewers, who greatly improved an early version of this manuscript.

**Conflicts of Interest:** The authors declare no conflict of interest.

## References


1. Dudgeon, D.; Arthington, A.H.; Gessner, M.O.; Kawabata, Z.I.; Knowler, D.J.; Lévêque, C.; Naiman, R.J.; Prieur-Richard, A.H.; Soto, D.; Stiassny, M.L.J.; et al. Freshwater biodiversity: Importance, threats, status and conservation challenges. *Biol. Rev.* **2005**, *81*, 163–182. [CrossRef] [PubMed]
2. Vörösmarty, C.J.; McIntyre, P.B.; Gessner, M.O.; Dudgeon, D.; Prusevich, A.; Green, P.; Glidden, S.; Bunn, S.E.; Sullivan, C.; Liermann, C.R.A.; et al. Global threats to human water security and river biodiversity. *Nature* **2010**, *467*, 555–561. [CrossRef] [PubMed]
3. Branco, P.; Segurado, P.; Santos, J.M.; Pinheiro, P.; Ferreira, M.T. Does longitudinal connectivity loss affect the distribution of freshwater fish? *Ecol. Eng.* **2012**, *48*, 70–78. [CrossRef]
4. Fullerton, A.H.; Burnett, K.M.; Steel, E.A.; Flitcroft, R.L.; Pess, G.R.; Feist, B.E.; Torgersen, C.E.; Miller, D.J.; Sanderson, B.L. Hydrological connectivity for riverine fish: Measurement challenges and research opportunities. *Freshw. Biol.* **2010**, *55*, 2215–2237. [CrossRef]
5. Maceda-Veiga, A. Towards the conservation of freshwater fish: Iberian Rivers as an example of threats and management practices. *Rev. Fish Biol. Fish.* **2012**, *23*, 1–22. [CrossRef]
6. Boavida, I.; Santos, J.M.; Ferreira, T.; Pinheiro, A. Barbel habitat alterations due to hydropeaking. *J. Hydro-Environ. Res.* **2015**, *9*, 237–247. [CrossRef]
7. Hayes, D.; Moreira, M.; Boavida, I.; Haslauer, M.; Unfer, G.; Zeiringer, B.; Greimel, F.; Auer, S.; Ferreira, T.; Schmutz, S. Life Stage-Specific Hydropeaking Flow Rules. *Sustainability* **2019**, *11*, 1547. [CrossRef]
8. Oliveira, J.M.; Ferreira, M.T.; Pinheiro, A.N.; Bochechas, J.H. A simple method for assessing minimum flows in regulated rivers: The case of sea lamprey reproduction. *Aquat. Conserv.* **2004**, *14*, 481–489. [CrossRef]
9. Santos, J.M.; Godinho, F.; Ferreira, M.T.; Cortes, R. The organisation of fish assemblages in the regulated Lima basin, Northern Portugal. *Limnologica* **2004**, *34*, 224–235. [CrossRef]
10. Bruder, A.; Tonolla, D.; Schweizer, S.P.; Vollenweider, S.; Langhans, S.D.; Wüest, A. A conceptual framework for hydropeaking mitigation. *Sci. Total Environ.* **2016**, *568*, 1204–1212. [CrossRef]
11. Capra, H.; Plichard, L.; Bergé, J.; Pella, H.; Ovidio, M.; McNeil, E.; Lamouroux, N. Fish habitat selection in a large hydropeaking river: Strong individual and temporal variations revealed by telemetry. *Sci. Total Environ.* **2017**, *578*, 109–120. [CrossRef] [PubMed]
12. Enders, E.C.; Watkinson, D.A.; Ghamry, H.; Mills, K.H.; Franzin, W.G. Fish age and size distributions and species composition in a large, hydropeaking Prairie River. *River Res. Appl.* **2017**, *33*, 1246–1256. [CrossRef]
13. Schmutz, S.; Bakken, T.H.; Friedrich, T.; Greimel, F.; Harby, A.; Jungwirth, M.; Melcher, A.; Unfer, G.; Zeiringer, B. Response of Fish Communities to Hydrological and Morphological Alterations in Hydropeaking Rivers of Austria. *River Res. Appl.* **2014**, *31*, 919–930. [CrossRef]
14. Baras, E.; Lucas, M. Impacts of man's modifications of river hydrology on the migration of freshwater fishes: A mechanistic perspective. *Ecohydrol. Hydrobiol.* **2001**, *1*, 291–304.
15. Benitez, J.P.; Matondo, B.N.; Dierckx, A.; Ovidio, M. An overview of potamodromous fish upstream movements in medium-sized rivers, by means of fish passes monitoring. *Aquat. Ecol.* **2015**, *49*, 481–497. [CrossRef]

16. Benitez, J.P.; Dierckx, A.; Matondo, B.N.; Rollin, X.; Ovidio, M. Movement behaviours of potamodromous fish within a large anthropised river after the reestablishment of the longitudinal connectivity. *Fish. Res.* **2018**, *207*, 140–149. [CrossRef]

17. Roscoe, D.W.; Hinch, S.G. Effectiveness monitoring of fish passage facilities: Historical trends, geographic patterns and future directions. *Fish Fish.* **2010**, *11*, 12–33. [CrossRef]

18. Clay, C.H. *Design of Fishways and Other Fish Facilities*, 2nd ed.; Lewis Publishers: Boca Raton, FL, USA, 1995; 256p.

19. Food and Agriculture Organization (FAO); Deutsche Vereinigung für Wasserwirtschaft, Abwasser und Abfall (DVWK). *Fish Passes—Design, Dimensions and Monitoring*; FAO: Rome, Italy, 2002; 119p.

20. Santos, J.M.; Silva, A.; Katopodis, C.; Pinheiro, P.; Pinheiro, A.; Bochechas, J.; Ferreira, M.T. Ecohydraulics of pool-type fishways: Getting past the barriers. *Ecol. Eng.* **2012**, *48*, 38–50. [CrossRef]

21. Larinier, M. Biological factors to be taken into account in the design of fishways, the concept of obstruction to upstream migration. *Bull. Fr. Pêche Piscic.* **2002**, *364*, 28–38. [CrossRef]

22. Allan, J.D.; Castillo, M.M. *Stream Ecology—Structure and Function of Running Waters*; Springer: Dordrecht, The Netherlands, 2007; 436p. [CrossRef]

23. Croze, O.; Bau, F.; Delmouly, L. Efficiency of a fish lift for returning Atlantic salmon at a large-scale hydroelectric complex in France. *Fish. Manag. Ecol.* **2008**, *15*, 467–476. [CrossRef]

24. Schletterer, M.; Reindl, R.; Thonhauser, S. Options for re-establishing river continuity, with an emphasis on the special solution "fish lift": Examples from Austria. *Rev. Eletrônica Gestão Tecnol. Ambient.* **2016**, *4*, 109. [CrossRef]

25. Travade, F.; Larinier, M. Fish locks and fish lifts. *Bull. Fr. Pêche Piscic.* **2002**, *364*, 102–118. [CrossRef]

26. Noonan, M.J.; Grant, J.W.A.; Jackson, C.D. A quantitative assessment of fish passage efficiency. *Fish Fish.* **2011**, *13*, 450–464. [CrossRef]

27. Pompeu, P.D.S.; Martinez, C.B. Efficiency and selectivity of a trap and truck fish passage system in Brazil. *Neotrop. Ichthyol.* **2007**, *5*, 169–176. [CrossRef]

28. Larinier, M.; Chanseau, M.; Bau, F.; Croze, O. The use of radio telemetry for optimizing fish pass design. In *Aquatic Telemetry: Advances and Applications, Proceedings of the Fifth Conference on Fish Telemetry held in Ustica, Italy, 9–13 June 2003*; FAO/COISPA: Rome, Italy, 2005.

29. Santos, J.M.; Rivaes, R.; Oliveira, J.; Ferreira, T. Improving yellow eel upstream movements with fish lifts. *J. Ecohydraulics* **2016**, *1*, 50–61. [CrossRef]

30. Santos, J.M.; Ferreira, M.T.; Godinho, F.N.; Bochechas, J. Performance of fish lift recently built at the Touvedo Dam on the Lima River, Portugal. *J. Appl. Ichthyol.* **2002**, *18*, 118–123. [CrossRef]

31. Morán-López, R.; Uceda Tolosa, O. Image techniques in turbid rivers: A ten-year assessment of cyprinid stocks composition and size. *Fish. Res.* **2017**, *195*, 186–193. [CrossRef]

32. Benitez, J.P.; Ovidio, M. The influence of environmental factors on the upstream movements of rheophilic cyprinids according to their position in a river basin. *Ecol. Freshw. Fish* **2017**, *27*, 660–671. [CrossRef]

33. Katopodis, C.; Williams, J.G. The development of fish passage research in a historical context. *Ecol. Eng.* **2012**, *48*, 8–18. [CrossRef]

34. Nieminen, E.; Hyytiäinen, K.; Lindroos, M. Economic and policy considerations regarding hydropower and migratory fish. *Fish Fish.* **2016**, *18*, 54–78. [CrossRef]

35. Boavida, I.; Jesus, J.B.; Pereira, V.; Santos, C.; Lopes, M.; Cortes, R.M.V. Fulfilling spawning flow requirements for potamodromous cyprinids in a restored river segment. *Sci. Total Environ.* **2018**, *635*, 567–575. [CrossRef] [PubMed]

36. Ferreira, M.T.; Sousa, L.; Santos, J.M.; Reino, L.; Oliveira, J.; Almeida, P.R.; Cortes, R.V. Regional and local environmental correlates of native Iberian fish fauna. *Ecol. Freshw. Fish* **2007**, *16*, 504–514. [CrossRef]

37. Lucas, M.C.; Mercer, T.; Peirson, G.; Frear, P.A. Seasonal movements of coarse fish in lowland rivers and their relevance to fisheries management. In *Management and Ecology of River Fisheries*; Cowx, I.G., Ed.; Fishing New Books: Oxford, UK, 2000; pp. 87–100.

38. Lucas, M.C.; Frear, P.A. Effects of a flow-gauging weir on the migratory behaviour of adult barbel, a riverine cyprinid. *J. Fish Biol.* **1997**, *50*, 382–396. [CrossRef]

39. Doadrio, I.; Perea, S.; Garzón-Heydt, P.; González, J.L. *Ictiofauna Continental Española. Bases Para Su Seguimiento*; D.G. Medio Natural y Política Forestal; MARM: Madrid, Spain, 2011; 616p.

40. Rodriguez-Ruiz, A.; Granado-Lorencio, C. Spawning period and migration of three species of cyprinids in a stream with Mediterranean regimen (SW Spain). *J. Fish Biol.* **1992**, *41*, 545–556. [CrossRef]

41. Santos, J.M.; Ferreira, M.T.; Godinho, F.N.; Bochechas, J. Efficacy of a nature-like bypass channel in a Portuguese lowland river. *J. Appl. Ichthyol.* **2005**, *21*, 381–388. [CrossRef]

42. Ovidio, M.; Baras, E.; Goffaux, D.; Giroux, F.; Philippart, J.C. Seasonal variations of activity pattern of brown (*Salmo trutta*) in a small stream, as determined by radio-telemetry. *Hydrobiologia* **2002**, *470*, 195–202. [CrossRef]

43. García-Vega, A.; Sanz-Ronda, F.J.; Fuentes-Pérez, J.F. Seasonal and daily upstream movements of brown trout *Salmo trutta* in an Iberian regulated river. *Knowl. Manag. Aquat. Ecosyst.* **2017**, *418*, 9. [CrossRef]

44. Santos, J.M.; Oliveira, J.M.; Rivaes, R.; Pizarro, R.A.; Ferreira, M.T.; Pádua, J.; Marin, C. *Plano de Ação para a Otimização do Ascensor de Peixes do Aproveitamento Hidroelétrico de Touvedo*; Relatório Final; Instituto Superior de Agronomia da Universidade de Lisboa e EDP Labelec: Lisboa, Portugal, 2014; 142p.

45. Jonsson, N. Influence of water flow, water temperature and light on fish migration in rivers. *Nordic J. Freshw. Res.* **1991**, *66*, 20–35.

46. Bunt, C.M.; Castro-Santos, T.; Haro, A. Performance of fish passage structures at upstream barriers to migration. *River Res. Appl.* **2012**, *28*, 457–478. [CrossRef]

47. Larinier, M. Fish passage experience at small-scale hydro-electric power plants in France. *Hydrobiologia* **2008**, *609*, 97–108. [CrossRef]

48. Laine, A.; Jokivirta, T.; Katopodis, C. Atlantic salmon, *Salmo salar* L., and sea trout, *Salmo trutta* L., passage in a regulated northern river—Fishway efficiency, fish entrance and environmental factors. *Fish. Manag. Ecol.* **2002**, *9*, 65–77. [CrossRef]

49. R Core Team. *R: A Language and Environment for Statistical Computing*; R Foundation for Statistical Computing: Vienna, Austria, 2018.

50. Venables, W.N.; Ripley, B.D. *Modern Applied Statistics with S*, 4th ed.; Springer: New York, NY, USA, 2002; 498p. [CrossRef]

51. Plichard, L.; Capra, H.; Mons, R.; Pella, H.; Lamouroux, N. Comparing electrofishing and snorkelling for characterizing fish assemblages over time and space. *Can. J. Fish. Aquat. Sci.* **2017**, *74*, 75–86. [CrossRef]

52. Santos, J.M.; Pinheiro, P.J.; Ferreira, M.T.; Bochechas, J. Monitoring fish passes using infrared beaming: A case study in an Iberian river. *J. Appl. Ichthyol.* **2008**, *24*, 26–30. [CrossRef]

53. De Leeuw, J.J.; Winter, H.V. Migration of rheophilic fish in the large lowland rivers Meuse and Rhine, the Netherlands. *Fish. Manag. Ecol.* **2008**, *15*, 409–415. [CrossRef]

54. Ovidio, M.; Capra, H.; Philippart, J.C. Field protocol for assessing small obstacles to migration of brown trout *Salmo trutta*, and European grayling *Thymallus thymallus*: A contribution to the management of free movement in rivers. *Fish. Manag. Ecol.* **2007**, *14*, 41–50. [CrossRef]

55. Granado-Lorencio, C. Fish species ecology in Spanish freshwater ecosystems. *Limnetica* **1992**, *8*, 255–261.

56. Flodmark, L.E.W.; Vollestad, L.A.; Forseth, T. Performance of juvenile brown trout exposed to fluctuating water level and temperature. *J. Fish Biol.* **2004**, *65*, 460–470. [CrossRef]

57. Alexandre, C.M.; Quintella, B.R.; Ferreira, A.F.; Romão, F.A.; Almeida, P.R. Swimming performance and ecomorphology of the Iberian barbel *Luciobarbus bocagei* (Steindachner, 1864) on permanent and temporary rivers. *Ecol. Freshw. Fish* **2014**, *23*, 244–258. [CrossRef]

58. Vilizzi, L.; Copp, G.H. An analysis of 0+ barbel (*Barbus barbus*) response to discharge fluctuations in a flume. *River Res. Appl.* **2005**, *21*, 421–438. [CrossRef]

59. Webb, J.; Hawkins, A.D. The movement and spawning behaviour of adult salmon in the Girnock Burn, a tributary of the Aberdeenshire Dee, 1986. In *Scotish Fisheries Research Report*; Department of Agriculture and Fisheries for Scotland: Edinburgh, Scotland, 1989; Volume 40, 41p.

60. Fraser, D.F.; Gilliam, J.F.; Akkara, J.T.; Albanese, B.W.; Snider, S.B. Night feeding by guppies under predator release: Effects on growth and daytime courtship. *Ecology* **2004**, *85*, 312–319. [CrossRef]

61. Lobón-Cerviá, J.; Rincón, P.A. Environmental determinants of recruitment and their influence on the population dynamics of stream-living brown trout *Salmo trutta*. *Oikos* **2004**, *105*, 641–646. [CrossRef]

62. Piecuch, J.; Lojkasek, B.; Lusk, S.; Marek, T. Spawning migration of brown trout, *Salmo trutta* in the Morávka reservoir. *Folia Zool.* **2007**, *56*, 201–212.

63. Santos, J.M.; Reino, L.; Porto, M.; Oliveira, J.; Pinheiro, P.; Almeida, P.R.; Cortes, R.; Ferreira, M.T. Complex size-dependent habitat associations in potamodromous fishspecies. *Aquat. Sci.* **2010**, *73*, 233–245. [CrossRef]

64. Romão, F.; Quintella, B.R.; Pereira, T.J.; Almeida, P.R. Swimming performance of two Iberian cyprinids: The Tagus nase *Pseudochondrostoma polylepis* (Steindachner, 1864) and the bordallo *Squalius carolitertii* (Doadrio, 1988). *J. Appl. Ichthyol.* **2012**, *28*, 26–30. [CrossRef]

65. Plaut, I. Critical swimming speed: Its ecological relevance. *Comp. Biochem. Physiol. Part A Mol. Integr. Physiol.* **2001**, *131*, 41–50. [CrossRef]

66. Silva, A.T.; Santos, J.M.; Ferreira, M.T.; Pinheiro, A.N.; Katopodis, C. Effects of water velocity and turbulence on the behaviour of Iberian barbel (*Luciobarbus bocagei*, Steindachner 1864) in an experimental pool-type fishway. *River Res. Appl.* **2011**, *27*, 360–373. [CrossRef]

67. Eyler, S.M.; Welsh, S.A.; Smith, D.R.; Rockey, M.M. Downstream Passage and Impact of Turbine Shutdowns on Survival of Silver American Eels at Five Hydroelectric Dams on the Shenandoah River. *Trans. Am. Fish. Soc.* **2016**, *145*, 964–976. [CrossRef]

68. Song, C.; Omalley, A.; Roy, S.G.; Barber, B.L.; Zydlewski, J.; Mo, W. Managing dams for energy and fish tradeoffs: What does a win-win solution take? *Sci. Total Environ.* **2019**, *669*, 833–843. [CrossRef]

69. Mateus, C.S.; Quintella, B.R.; Almeida, P.R. The critical swimming speed of Iberian barbel *Barbus bocagei* in relation to size and sex. *J. Fish Biol.* **2008**, *73*, 1783–1789. [CrossRef]

70. Tudorache, C.; Viaene, P.; Blust, R.; Vereecken, H.; De Boek, G. A comparison of swimming capacity and energy use in seven European freshwater fish species. *Ecol. Freshw. Fish* **2008**, *17*, 284–291. [CrossRef]

71. Amaral, S.D.; Branco, P.; da Silva, A.T.; Katopodis, C.; Viseu, T.; Ferreira, M.T.; Pinheiro, A.N.; Santos, J.M. Upstream passage of potamodromous cyprinids over small weirs: The influence of key-hydraulic parameters. *J. Ecohydraulics* **2016**, *1*, 79–89. [CrossRef]

72. Amaral, S.D.; Branco, P.; Romão, F.; Viseu, T.; Ferreira, M.T.; Pinheiro, A.N.; Santos, J.M. The effect of weir crest width and discharge on passage performance of a potamodromous cyprinid. *Mar. Freshw. Res.* **2018**, *69*, 1795–1804. [CrossRef]

73. Cooke, S.J.; Hinch, S.G. Improving the reliability of fishway attraction and passage efficiency estimates to inform fishway engineering, science, and practice. *Ecol. Eng.* **2013**, *58*, 123–132. [CrossRef]

74. Teixeira-de-Mello, F.; Kristensen, E.A.; Meerhoff, M.; González-Bergonzoni, I.; Baattrup-Pedersen, A.; Iglesias, C.; Kristensen, P.B.; Mazzeo, N.; Jeppesen, E. Monitoring fish communities in wadeable lowland streams: Comparing the efficiency of electrofishing methods at contrasting fish assemblages. *Environ. Monit. Assess.* **2013**, *186*, 1665–1677. [CrossRef] [PubMed]

75. Lucas, M.C.; Baras, E. Methods for studying spatial behaviour of freshwater fishes in the natural environment. *Fish Fish.* **2000**, *1*, 283–316. [CrossRef]

76. Eberstaller, J.; Hinterhofer, M.; Parasiewicz, P. The effectiveness of two nature-like bypass channels in an upland Austrian river. In *Fish Migration and Fish Bypasses*; Jungwirth, M., Schmutz, S., Weiss, S., Eds.; Fishing News Books: Oxford, UK, 1998; pp. 363–383.

77. INAG-Instituto da Água. *Desenvolvimento de um Indice de Qualidade para a Fauna Piscícola*; Ministério da Agricultura, Mar, Ambiente e Ordenamento do Território: Lisbon, Portugal, 2012; 17p.

78. Castro-Santos, T.; Cotel, A.; Webb, P.W. Fishway evaluations for better bioengineering—An integrative approach. *Am. Fish. Soc. Symp.* **2009**, *69*, 557–575.

79. Pelicice, F.M.; Agostinho, A.A. Fish-Passage Facilities as Ecological Traps in Large Neotropical Rivers. *Conserv. Biol.* **2008**, *22*, 180–188. [CrossRef]

 *sustainability*

*Article*

# Conceptual Approach for Positioning of Fish Guidance Structures Using CFD and Expert Knowledge

Linus Feigenwinter [1,*], David F. Vetsch [1], Stephan Kammerer [1], Carl Robert Kriewitz [2] and Robert M. Boes [1]

[1]   Laboratory of Hydraulics, Hydrology and Glaciology, ETH Zurich, 8093 Zurich, Switzerland; vetsch@vaw.baug.ethz.ch (D.F.V.); kammerer@vaw.baug.ethz.ch (S.K.); boes@vaw.baug.ethz.ch (R.M.B.)

[2]   BKW Energie AG, BKW Engineering Viktoriaplatz 2, 3013 Bern, Switzerland; robert.kriewitz@bkw.ch

*   Correspondence: linus.feigenwinter@gmx.ch

Received: 26 December 2018; Accepted: 14 March 2019; Published: 19 March 2019

**Abstract:** The longitudinal connectivity of many rivers is interrupted by man-made barriers preventing the up- and downstream migration of fishes. For example, dams, weirs, and hydropower plants (HPP) are insuperable obstructions for upstream migration if no special measures like fish passes are put into effect. While upstream fishways have been implemented successfully and are still being optimized, the focus of current research is more and more on effective fish protection and guiding devices for downstream migration. According to current knowledge fish guidance structures (FGS) have a high potential in supporting the downstream migration by leading fishes to a bypass as an alternative to turbine passage. This work presents a structured and straightforward approach for the evaluation of potential locations of FGS combining traditional dimensioning principles with computational fluid dynamics (CFD) and novel findings from etho-hydraulic research. The approach is based on three key aspects: fish fauna, structural conditions, and hydraulic conditions, and includes three assessment criteria, which are used in an iterative process to define potential FGS locations. The hydraulic conditions can be investigated by means of hydrodynamic 3D simulations and evaluated at cross sections of potential FGS positions. Considering fundamentals of fish biology and ethology allows for rating of the flow conditions and thus for a suitability assessment of various locations. The advantage of the proposed procedure is the possibility to assess FGS configurations without implementing the FGS in the numerical model, thus limiting the computational expense. Furthermore, the implementation of various operation conditions is straightforward. The conceptual approach is illustrated and discussed by means of a case study.

**Keywords:** fishway; downstream migration; fish guidance structure; numerical 3D model; OpenFOAM

---

## 1. Introduction

Man-made barriers like hydropower plants (HPP), weirs, and dams interrupt the longitudinal connectivity of rivers, which hinders migration of fishes and delimitates their natural habitats [1]. The negative impacts on fish communities and consequently on the whole ecosystem underline the need to restore river connectivity allowing for up- and downstream migration of fishes again [2]. The importance of healthy river ecosystems is manifested by international regulations and guidelines. For instance, the EU's Water Framework Directive states that impacted water bodies need to achieve "good ecological status" and national laws stipulate river restorations (e.g., Federal Act on the Protection of Waters in Switzerland). The upstream migration of fishes has been studied in detail during the last decades, leading to a considerable number of upstream fish passage designs [3,4]. They can be classified into nature-like structures (e.g., fish ramps), technical structures (e.g., vertical slot fishways), and special purpose structures (e.g., eel ladders) [5]. Devices improving a successful downstream migration have been investigated less and are a topic of current research [3,6–10]. The two main approaches for the reestablishment of downstream migration and guidance of fishes from the headwater into the tailwater of HPP are conveyance and bypass [8,11]. The former requires the application of fish-friendly turbines (e.g., minimum gap runner or Alden turbine) or a fish-friendly facility management. These measures can result in very high installation costs or considerable losses in energy production, respectively. The latter approach includes fish guidance structures (FGS) in terms of racks that block fish passage physically using narrow bar spacing or eliciting fish movement along the rack via generated flow fields as well as non-physical barriers like acoustic deterrents or strobe lights that cause fish deterrence. FGS have an especially high potential to support downstream migration [6–8,10–16]. Sensory systems are based on fish-specific behavior that shows high variation among different species [10–12].

FGS use the concept of guiding fishes along the rack or screen axis to a bypass. Usually they are positioned with a horizontal angle $\alpha < 45°$ to the approach flow to achieve the favorable ratio between the tangential velocity $v_t$ and normal velocity $v_n$ above one (Figure 1a), which is considered as a suitable criterion for guiding efficiency along the rack [8,13]. These so-called angled FGS can be installed either with horizontally or vertically oriented bars. The bar spacing of horizontal angled screens is usually between 10 and 20 mm so that fish with a body height greater than this cannot physically pass (Figure 1b). To date, horizontal FGS were mostly installed at HPP with design discharges below $100 \text{ m}^3/\text{s}$ (e.g., HPP Raghun, Germany with $Q = 88 \text{ m}^3/\text{s}$) to avoid normal velocities $v_n$ larger than the sustained fish swimming speed, which would press fish against the rack [12] (Section 2.1.1). The most common vertical screens are louvers and bar racks, which are frequently used in the north-eastern United States [14,15]. Louvers consist of vertical bars that are generally positioned perpendicular to the flow direction (Figure 1c), while the bars of bar racks are oriented normal to the rack plane (Figure 1d). The bar spacing of bar racks and louvers is usually larger than for horizontal screens so that a considerable share of the fish species can pass them. However, depending on fish size they may prevent physical passage as well [6,10,16].

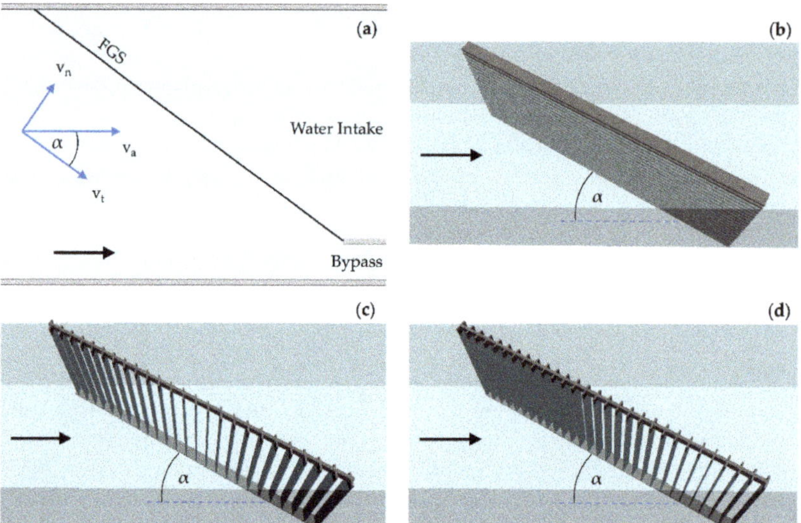


**Figure 1.** (**a**) Layout of a fish guidance structure (FGS) with horizontal angle $\alpha$ with respect to the approach flow velocity $v_a$ (velocity components $v_n$ and $v_t$ are normal and tangential to the rack, respectively); (**b**) example geometry of a horizontal FGS; (**c**) example geometry of a louver; (**d**) example geometry of a bar rack. The black arrows represent the flow direction.

Despite the relatively large bar spacing of vertical FGS, an avoidance effect is achieved by turbulent flow structures created by each bar and sensed by the fish when approaching the rack [10,16], while the guiding effect along the FGS comes from $v_t/v_n > 1$ [6,8]. The primary mechanisms for guiding fish movement at FGS are specified as follows:

1.  Of primary importance are the hydraulic cues caused by the rack bars, i.e., the turbulent zones and flow separation along the rack edges [16], triggering an avoidance reaction in fish. These turbulent structures cause only a few fish to swim through the FGS [10,17]. However, the avoidance reaction decreases with increasing bar spacing [12]. In fact, two fish species-specific reaction patterns could be observed in ethohydraulic experiments [10,17]. Some species, such as barbel, brown trout, and eel show investigative behavior and do not avoid rack contact. Instead, they hug the upstream rack surface and actively use it as a guidance structure but avoid passing. Grayling and spirlin, for example, behave differently as they avoid structures and only touch the screen in exceptional situations.

2.  The second mechanism is called sweeping flow, i.e., the parallel flow component $v_t$ actively drifting the fish towards the bypass. This mechanism is decisive for the functionality of FGS because otherwise the fishes are not directed towards the bypass. In case of exhaustion, the fish will be pressed against the bars of horizontal FGS and in case of FGS with vertical bars they can pass through the rack.

In addition to the classical bar rack, there exist further developments that are not arranged orthogonally to the rack axis, such as those with straight (Modified Bar Rack, MBR [10,17]) or curved bar cross-section shapes (Curved Bar Rack [18–20]). First investigations of this FGS type show a lower flow deflection and more symmetric flow field in the tailwater as well as lower hydraulic losses compared to conventional FGS.

Furthermore, screens can also be positioned with an inclination ("inclined racks") or even horizontal to the riverbed plane. However, this type of FGS is dependent on a relatively shallow water depth, as the angle of inclination should be <10° [8]. For larger hydropower plants, this screen type is therefore usually unsuitable.

The proper positioning of FGS requires good knowledge of the hydraulic conditions prevailing at possible locations and a well-founded understanding of behavior patterns of the predominant fish communities [3,21]. Due to today's computing power, computational fluid dynamics (CFD) has evolved as a powerful tool for investigating the flow under varying conditions and therefore has a high potential for fishway designs, for both up- and downstream migration. Various researchers applied CFD to investigate upstream fish passages, for example to assess the hydraulic conditions in pool-type fishways [22–27] or to optimize the attraction flow of fishway entrances [28,29]. However, downstream migration numerical simulations have been rarely used. A few authors implemented CFD models of turbine flows to assess fish mortality or injury rates [30,31]. Raynal et al. [32] investigated the hydraulic impacts of horizontal inclined bar racks in a rectangular channel by means of model- and real-scale CFD simulations. Mulligan et al. [33,34] performed numerical simulations and acoustic doppler velocimeter (ADV) measurements for surface partial-depth guiding walls in a laboratory flume to assess the guiding efficiency for surface-oriented fish species. A similar study was performed by Lundström et al. [35] who investigated the guiding potential of a surface guide wall at the Sikfors HPP in Sweden assessing the surface-orientated downstream migration of fish salmon smolts.

In contrast to most of the studies applying CFD to assess fishways, which focus on general facility optimization or are related to site-specific fish passage investigations, this paper presents a conceptual approach for the positioning of FGS using CFD, fish biology, and expert knowledge to support the planning of measures for fish downstream migration. The conceptual approach is illustrated and discussed by means of a case study of the HPP Brügg in Switzerland. Firstly, we will introduce the different aspects of the suggested approach and present a possible way to link the hydraulics with species-specific fish behavior patterns to rate the suitability of potential FGS configurations. Next, the application of the conceptual approach is introduced step by step for the HPP Brügg. Finally, we discuss the advantages and drawbacks of the conceptual approach, draw the main conclusions, and give recommendations for future research.

## 2. Methods

### 2.1. Conceptual Approach

The successful downstream guidance of fishes into the tailwater of HPP is a complex challenge, involving knowledge from different fields of research and concerns of various stakeholders [3]. In order to consider all essential aspects, the conceptual approach shown in Figure 2 is suggested to identify potential FGS positions. The approach is based on the three key aspects of fish fauna, structural conditions of HPP, and hydraulic and hydrologic conditions, and includes three assessment criteria, which are used in an iterative process to define potential FGS locations as described hereafter.

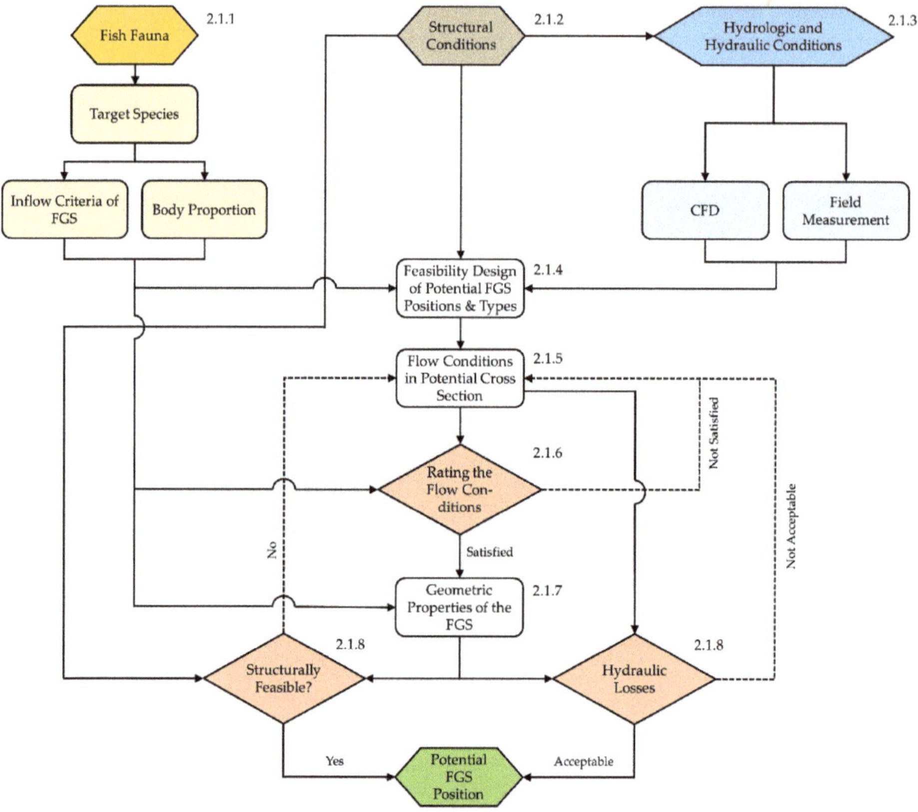


**Figure 2.** Conceptual approach for the positioning of FGS using computational fluid dynamics (CFD) and expert knowledge.

### 2.1.1. Fish Fauna

Considering a river reach for restoration, awareness, and understanding of existing fish communities and their role in the ecosystem of the river is of major importance. This mainly includes knowledge of required habitats, migration behavior, and reproduction cycles. Based on the findings one or several target species may be identified. Subsequently, the fish biological fundamentals related to the defined target species must be scrutinized involving the migration period, body proportion, and species-specific swimming speed capabilities to be compared to the inflow conditions of the FGS.

Fish body proportions can be characterized by the relative body width $w_{fish,rel}$ and height $h_{fish,rel}$ as well as the proportion index $P$ [12,36]. The relative body widths and heights are defined as the ratio between maximum width $w_{fish}$ and height $h_{fish}$, respectively, and the total length $L$ of the fishes, whereas the proportion index $P$ is known as the ratio between $h_{fish}$ and $w_{fish}$. For fishes with $P > 1$ the critical length $L_{crit}$ for the physical rack passage in relation to the bar spacing $s_{bar}$ can be determined as:

$$L_{crit} = \frac{s_{bar}}{w_{fish,rel}} \tag{1}$$

This criterion is strictly valid for vertically oriented bars. According to the present state of knowledge, fish change their natural body alignment only conditionally and rarely pass the barrier in a lateral position [37]. Therefore, for horizontal bars a higher protective effect can be expected for fishes having $P > 1$. As no general fish behavior patterns concerning the rack passage are known yet, it is recommended to use the same principles for horizontal bars as for vertical ones [12].

Various models have been developed for the design of incident flow velocities and horizontal angles of angled FGS, such as the approaches of Pavlov [36] and Bates and Vinsonhaler [14]. However, the method of O'Keeffe & Turnpenny [9] indicates the most favorable conditions for effective fish protection. They state that FGS must be designed in such a way that the normal speed $v_n$ perpendicular to the rack axis does not exceed the sustained swimming speed $v_{sust}$ of the fish:

$$v_n \leq v_{sust} \tag{2}$$

The sustained swimming speed refers to the swimming activity that can be maintained by the fish over several hours (often $t = 200$ min is defined as the minimum characteristic duration) without fatigue [38,39]. Many empirical general and species-specific univariate and multivariate models were developed to estimate $v_{sust}$. The works of Wolter and Arlinghaus [40], Peake [41], and Ebel [12] provide an overview of the relevant literature. Ebel [12] analyzed 785 studies from 21 countries including 80 fish species. Based on these data, he developed different multivariate models to quantify the sustained swimming speed $v_{sust}$ of European fish species and provide design recommendations for FGS and bypass systems at HPP. For a first assessment of $v_{sust}$, the general multivariate model of Ebel can be applied, (derived from 225 datasets on 22 European fish species). In this model, $L$ represents the body length [m], $T$ the water temperature [°C], and $t$ the swimming duration [s]:

$$\log(v_{sust}) = 0.5130 + 0.7941 \cdot \log(L) - 0.0906 \cdot \log(t) + 0.2921 \cdot \log(T) \tag{3}$$

For a more precise definition of $v_{sust}$, Ebel proposes empirical multivariate approaches for rheophilic (4) and non-rheophilic species (5):

$$\log(v_{sust}) = 0.5460 + 0.7937 \cdot \log(L) - 0.0902 \cdot \log(t) + 0.2813 \cdot \log(T) \tag{4}$$

$$\log(v_{sust}) = 0.3674 + 0.7692 \cdot \log(L) - 0.0982 \cdot \log(t) + 0.3649 \cdot \log(T) \tag{5}$$

It should be noted that these approaches are not applicable to genera with special swimming styles such as lampreys (*Petromyzontidae*), sturgeons (*Acipenseridae*), or eels (*Anguilidae*) where species-specific models must be used. The swimming ability of lampreys can be assessed with the approach of Beamish [42], where $p$ represents the body weight [g]:

$$v_{sust} = 0.7671 + 0.1309 \cdot \log(p) - 0.2632 \cdot \log(t) + 0.0077 \cdot T \tag{6}$$

For sturgeons the model of Peake et al. [43] applies:

$$v_{sust} = \frac{3.1782 + 2.26 \cdot L + 0.0547 \cdot T - \log(t)}{4.55 + 0.0536 \cdot T - 1.85 \cdot L} \tag{7}$$

and for eels the approach of Ebel [12] is suggested:

$$\log(v_{sust}) = 0.425 + 0.567 \cdot \log(L) - 0.133 \cdot \log(t), \text{ for } T > 10\,^{\circ}\text{C} \tag{8}$$

It must be mentioned that these models represent fish in optimal conditions, which is not the case when fish migrate.

With regard to permissible flow velocities, regulators of the US [8] and UK [9] provide some recommendations for FGS that are listed in Table 1. Both reports include many case studies, but most of their recommendations are not explicitly related to the behavior patterns of the target species. In the regulations for the UK [9], it is mentioned that $v_n$ should be smaller than the 90th percentile of the maximum sustained swimming speed, which is reasonably consistent with the recommendations of Ebel [12]. However, an overview of the sustained swimming speed $v_{sust}$ is only given for some fish species.

**Table 1.** Recommendations concerning permissible flow velocities of the regulators of the US [8] and UK [9] ($v_n$ = normal velocity, $v_a$ = approach flow velocity, $v_{sust}$ = sustained swimming speed).

| Literature | Physically Impassable | Physically Passable |
|:---:|:---|:---|
| [8] | $v_n$ < 0.12 m/s for salmonid fry<br>$v_n$ < 0.24 m/s for salmonid fingerling | $v_a$ = 0.3–0.6 m/s for small and weak swimming fish<br>$v_a$ = 0.8–1.5 m/s for larger and strong swimming fish |
| [9] | $v_n$ < 90th percentile of $v_{sust}$ | $v_a$ = 0.3–1.0 m/s<br>$v_n$ < 90th percentile of $v_{sust}$ |

### 2.1.2. Structural Conditions

The second pillar of the suggested approach represents the HPP and the associated structural conditions. It must be determined what type of HPP or migration obstacle is present, and its geometric dimensions must be specified. This includes the river width, the width and length of the headrace channel, the elevation difference to be overcome, and the prevailing flow depths.

### 2.1.3. Hydrologic and Hydraulic Conditions

The hydrologic and hydraulic conditions in the inflow area of the HPP also have to be assessed. In a first step, the flow duration curve of the investigated river stretch should be evaluated as to typical scenarios relevant for downstream migration of the target species, thereby accounting for their migration periods. From this, typical discharges result that should underlie the detailed hydraulic investigations. For this purpose, numerical modelling and field measurements can be used to determine the flow fields in detail for the relevant scenarios. Best knowledge of the prevailing hydraulic conditions can be achieved when both methods are combined, and field data can be used to calibrate and validate the numerical model. Considering FGS in real-scale numerical 3D-simulations of HPP is usually restricted to a smaller section, because the required grid resolution to resolve the whole FGS would lead to very large computing time making the conceptual approach inexpedient. Therefore, the hydraulic conditions of the present state without FGS are used as a reference, while the effects of FGS on the flow fields and on rack-near fish behavior known from detailed etho-hydraulic modelling [6,7,10,13,17,19,20,44–47] are taken into account separately in the assessment phase, see Section 2.3.

### 2.1.4. Feasibility Design of Potential FGS Positions and Types

Based on the numerical results, the HPP geometry, and an initial estimate of the fish biological conditions, advantageous regions in the headrace channel for the installation of FGS can be determined. First considerations about the FGS type should be undertaken as well. Overviews of the application range of different FGS types are given in References [8,9,12].

### 2.1.5. Flow Conditions in Cross Section of Potential Positions

In a next step, a possible FGS position is defined for each advantageous region. The hydraulic conditions obtained from the numerical model are then evaluated in the cross-section of the defined FGS position. For FGS with horizontally angled flow, the tangential velocities along the rack and the normal velocities perpendicular to the rack axis are of interest. In the case of non-angled racks, the subsequent evaluation is based on the approach flow velocity. For the definition of the relevant hydraulic parameters for other FGS types, References [8,9,12] can be used. It is important to select the FGS type with regard to the target species. For instance, investigations concerning the downstream migration of barbels showed that bar racks support a successful passage, while louvers are less suitable as a guiding device [13,17].

### 2.1.6. Rating the Flow Conditions with Respect to Target Species

The determined flow conditions, and thus the suitability of the rack positions to support the downstream migration of fish, can then be assessed using Equation (2) in combination with the decisive sustained swimming speeds $v_{sust}$ of the target species (Equations (3)–(8)). To achieve a good guiding efficiency along the rack, the ratio between the tangential velocity $v_t$ and normal velocity $v_n$ has to be above one [8].

This evaluation is considered as the primary assessment criterion for a potential FGS position. If the conditions are not fulfilled, the position of the FGS should be adjusted and the assessment must be carried out again.

### 2.1.7. Geometric Properties of FGS

After identifying a suitable position, the geometric characteristics of the FGS can be determined based on the fish biology, the site-specific conditions, and the chosen type of FGS. In the case of physically impassable barriers, the bar spacing must be defined in relation to the body proportions of the target species. To this end, Equation (1) can be rearranged to provide a criterion for the bar spacing $s_{bar}$, which requires the definition of an appropriate value for $L_{crit}$. It is recommended to choose the body length of the target fish species during its first migration stage in the life cycle as $L_{crit}$. For vertical bars that are passable for fish, the negative correlation between the bar spacing and the protective effect must be taken into account [12]. The design of the rack geometry should also consider the rack stability, low susceptibility for rack vibration, and the installation of the rack cleaning system.

### 2.1.8. Structural Feasibility and Hydraulic Losses

Finally, the structural feasibility of the FGS realization and hydraulic losses caused by the FGS must be examined. Particular attention should be paid to the installation of a rack cleaning system and whether a certain configuration allows satisfactory connection with both bottom-near and surface-near bypass entrances. Basic dimensions for the bypass design and the rack cleaning system can be found in References [8,9,12]. Another important criterion to consider is the installation cost of the FGS.

For evaluation purposes, hydraulic head loss can be estimated with conventional formulas, like the approaches of Raynal et al. [13], Albayrak et al. [44], and Beck et al. [19] for vertical angled racks or the methods of Albayrak et al. [45] for horizontal angled racks. For inclined FGS the method of Raynal et al. [46] can be applied. The hydraulic head loss of FGS are particularly dependent on the flow velocity, the bar shape, and the bar spacing [13,44]. For HPP, these losses should be minimized to keep the deficits in electricity production as low as possible. If empirical equations are applied to estimate the hydraulic head loss, it must be noted that these formulae were developed based on laboratory tests or empirically from data sets, mainly for homogeneous inflow. Accordingly, for highly inhomogeneous approaches, flow condition deviations must be expected.

It is crucial that these two assessment criteria are appraised in close cooperation with the HPP operator. If a requirement is not met, the position of the FGS can be adjusted and the whole assessment must be carried out again starting with the evaluation of the hydraulic conditions in the adjusted cross-section. If no satisfactory FGS position is found even after repeated iteration through the concept, the investigated region should be excluded from the feasibility design. If this procedure is carried out for each region identified in the feasibility design, one or several potential FGS positions may be defined.

### 2.2. Numerical Modelling

The hydraulic conditions in the inflow area of HPP can be investigated by means of hydrodynamic 3D-simulations. To this end, the opensource CFD software OpenFOAM [48] constitutes a useful tool, which was used in the present study. For simulations of free-surface gravity flows the Eularian solver interFOAM is most suitable [49]. The interFOAM solver uses the Reynolds-averaged Navier-Stokes equations combined with a turbulence model acting as closure relation to the Reynolds-Stresses as governing flow equations [50]. The equations are solved with a finite volume method in combination with the PIMPLE algorithm for pressure-velocity coupling [50,51]. The interFOAM solver identifies the interface between water and air based on the volume of fluid (VOF) method, which is based on the water fraction parameter $\alpha_w$ having values between 0 (air) and 1 (water). The interFOAM solver uses an artificial compression term in the transport equation of $\alpha_w$ to define a fluid interface ($\alpha_w = 0.5$) and to avoid the use of interface reconstruction schemes. Detailed information about the VOF method and its implementation in the interFOAM solver can be found in References [26,51,52].

Another advantage of OpenFOAM besides its free availability is the use of an unstructured computational grid. The additional use of polyhedral cells (in addition to hexahedral cells) allows even complex geometries to be accurately mapped.

### 2.3. Influences of FGS on the Flow Field

The presented approach considers the hydraulic conditions without FGS. Hence, it is not possible to directly simulate the hydraulic effects of FGS on the flow field in its vicinity. Therefore, expert knowledge of these effects on the fish guidance efficiency should enter in this phase of the conceptual approach.

Raynal et al. [13,32] investigated the influences of angled bar racks on the flow field in their vicinity by means of laboratory investigations and numerical simulations in a rectangular channel. Based on systematic ethohydraulic model studies, Kriewitz [10] and Albayrak et al. [17] modified the classical angled bar racks by varying the bar angle $\beta$ independent of $\alpha$. Likewise, Beck et al. [19,20] developed curved bar racks and also tested these in laboratory flumes for both hydraulic and fish guidance performance. For all tested combinations of the horizontal angle $\alpha$, the bar spacing, and the bar shape in the mentioned studies, a flow deflection and acceleration along the rack axis could be observed. The ratio of tangential to normal velocity increased continuously along the rack axis, mainly induced by increasing tangential velocities, as the normal velocities increased only slightly along the rack. These effects of the bar racks on the upstream flow field mainly increase with decreasing horizontal angle $\alpha$, while the bar spacing and shape have minor influences on the hydraulic cues in the upstream near-rack zone.

In contrast to vertical angled bar racks, angled FGS with horizontal bars, particularly hydrodynamic bar shapes, which are the preferred types to minimize head losses, hardly affect the flow field, neither in the near-rack upstream nor downstream zones [47], except at flow depths close to the bed and to the surface if bottom and top overlays, respectively, are used.

Based on these findings it can be concluded that the evaluation of potential FGS positions following the suggested approach would not change fundamentally if the FGS could be included in the numerical models, as FGS mainly increase the tangential velocity along the rack axis.

## 3. Results: Application of Conceptual Approach to Case Study Hydropower Plant Brügg

HPP Brügg is located on the Aare river, a tributary of the Rhine river, about 2.2 km downstream of Lake Biel in Switzerland (Figure 3). It is a bay type HPP equipped with two double regulated Kaplan turbines having frontal approach flow. The HPP is placed next to the weir Port, which has five weir bays each equipped with a double hook gate. The design flow rate of the HPP is 220 m$^3$/s. At higher flow rates, the additional discharge is spilled over the weir.

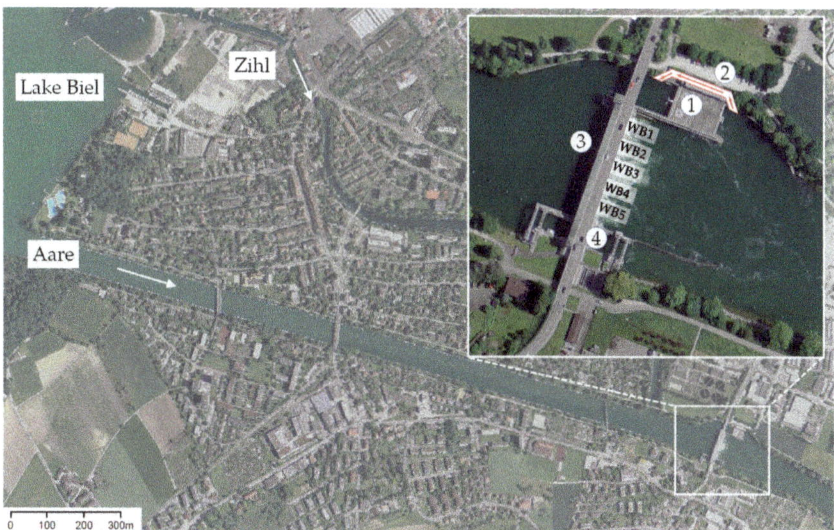


**Figure 3.** Situation of hydropower plant (HPP) Brügg ①, upstream fishway ②, weir Port with weir bays WB1–WB5 ③, and ship lock ④. The white arrows indicate the flow direction (photos: map.geo.admin.ch).

The facility in Port/Brügg features an upstream fishway in the form of a vertical slot pass. However, this fish pass was not considered for the investigations of possible FGS positions, as it has to be rebuilt and the new exit will possibly be located upstream of the headrace channel.

Downstream migration of fish is currently only possible by passing through the Kaplan turbines, during weir operation, and for a limited time through the adjoining ship lock.

The revised Federal Act on the Protection of Waters in Switzerland requires restoration of the longitudinal connectivity of Swiss rivers [53]. Thus, obligations to improve the fish passage at the Port/Brügg facility are of a "very high" priority. Based on this prioritization, remediation measures are to be implemented by 2020. Hereafter, the results from applying the conceptual approach described in Section 2 to HPP Brügg are presented.

### 3.1. Fish Fauna

Through efforts to re-establish a self-sustaining salmon (*Salmo salar*) population in the Rhine river basin, this fish species is also expected in the Aare system in the long term. Therefore, salmon and barbel (*Barbus barbus*) have been defined as target species for the design of fishways along the Aare river. Salmon smolts in the life cycle state of first downstream migration have a body length between 10 and 20 cm considered as $L_{crit}$ [12,54]. The decisive body length of the barbel is estimated to be in the same range [55,56]. The relative body width $w_{fish,rel}$ is 0.10 and 0.11 for salmon and barbel, respectively. The proportion indices of salmon and barbel are $P = 1.80$ and $P = 1.55$, respectively [12]. The sustained swimming speeds $v_{sust}$ were assessed with Equation (3) for a swimming duration of $t = 200$ min and are between 0.36 and 0.85 m/s, depending on the water temperature ($T = 5$–$15\,°C$).

### 3.2. Structural Conditions

Upstream of the HPP the river width is about 80 m. The headrace channel has a length of 46 m, a width of 24 m, and its entry width from the inlet pier to the left river bank is about 60 m. The water depth in the headrace channel is more than 6 m. The usable head of the HPP is 0.8–3.0 m, which leads to an annual electricity production of 25 GWh. The road bridge over the facility has a bridge pier placed in the middle of the headrace channel with a width of 1 m and a length of 3.5 m.

### 3.3. Hydrologic and Hydraulic Conditions

The five weir gates of HPP Brügg remain closed up to the design flow rate of 220 $m^3$/s and the entire water volume is turbined. These conditions represent the most demanding situation for downstream migrating fish. Due to the closed weir, only the route via the powerhouse is possible and with the turbines running near full capacity, the flow velocities in the headrace channel are highest. HPP Brügg has an average annual runoff of 270 $m^3$/s. The smallest average monthly discharge occurs in January and is 190–200 $m^3$/s. Thus, the migration periods of fish play a subordinate role, as the relevant operational condition occurs throughout the year. Moreover, the goal of installing fish guidance structures at HPP Brügg is to ensure safe downstream migration for every native species all year long, which requires to consider the described operation condition to be decisive.

Accordingly, the hydraulic conditions were simulated for an operating condition with turbine operation at closed weir bays. The discharge was 195 $m^3$/s and thus slightly below the design discharge. We did not simulate the whole design flow rate as no water level measurements to validate the numerical results were available for this operating condition. The hydraulic conditions in the inflow area of the HPP Brügg were investigated by means of hydrodynamic 3D-simulations using the opensource CFD software OpenFOAM (release v1712), as introduced in Section 2.2. The k-ε-RNG turbulence model was applied for the simulations performed herein.

For the sake of efficiency, a cascade of three simulations was carried out, whereby the model perimeter was reduced in size while the spatial resolution was increased. The model geometry of the first simulation (S1) includes the HPP Brügg, the weir Port, as well as 930 m of the Aare river upstream and 710 m downstream (Figure 4a). In the second simulation (S2), the HPP and weir were mapped (Figure 4b) and in the third simulation (S3), only the headrace channel and turbine admission flow area were included (Figure 4c). The initial resolutions of the computational grids are listed in Figure 4a–c. The computational grid of the S3 model, consisting of about 6.1 million cells, is shown in Figure 4d.

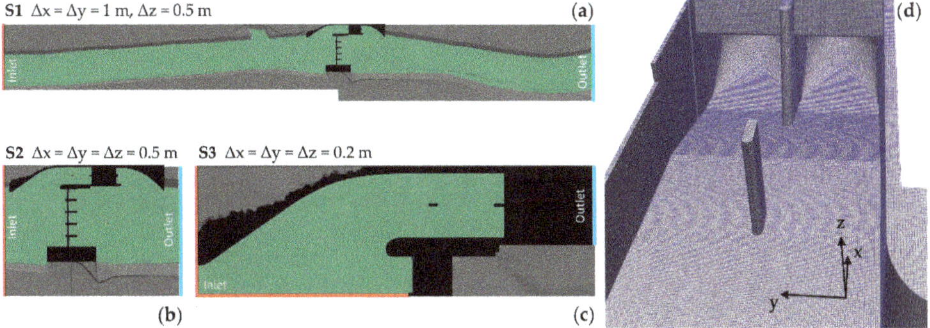


**Figure 4.** (a–c) Model perimeters of the simulation sequence with the respective grid spacing; (d) Computational grid of the third simulation and orientation (coordinate system).

For simulation S1, different boundary conditions for water and air at the inflow and outflow borders were set based on the known water surface elevations (Table A1 in Appendix A). As initial conditions, different water surface elevations and flow velocities were applied for the head and tail waters. For simulations S2 and S3, the steady-state solution of the previous calculation was used as the initial condition and the boundary values $\alpha_w$, $u$, $p$, $k$, and $\varepsilon$ were mapped accordingly.

In order to verify convergence of the solution to a steady state, the mass balances of water and the water surface elevation in the head and tail water were recorded over time. The solution was considered stationary when the mass balance was fulfilled, and constant water surface elevations were reached (Figure A1 in Appendix A).

The numerical models were validated using water level measurements in the head and tail waters of the HPP. The difference of the simulated and measured water surface elevations amounts to only a few centimeters and is thus in good agreement.

For simulation S3, the hydraulic conditions at steady state are shown in Figure 5. These results were used in the further application of the conceptual approach.

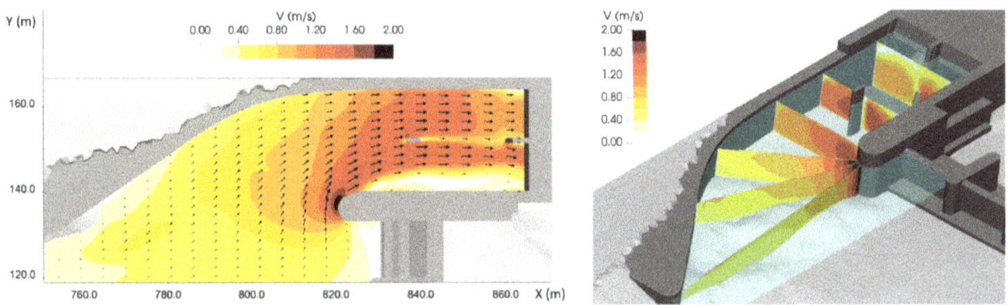


**Figure 5.** Flow velocity contours in the headrace channel at steady state for simulation S3: (**left**) flow velocity at horizontal plane at water depth of 3 m; (**right**) flow velocity at different vertical cross sections. The light blue transparent plane represents the water surface at water fraction $\alpha_w = 0.5$.

### 3.4. Feasibility Design of Potential FGS Positions and Types

Prior to this study, two different studies concerning the installation of FGS at HPP Brügg had been carried out focusing on structural feasibility [57,58]. These studies spatially narrowed the feasibility of FGS implementation to the headrace channel. Due to the high flow depth of more than 6 m, inclined FGS are not an option, while angled FGS are considered to have the greatest potential of supporting fish downstream migration. Based on these findings and the simulated hydraulic conditions, two different regions for the installation of FGS in the headrace channel were defined (Figure 6).

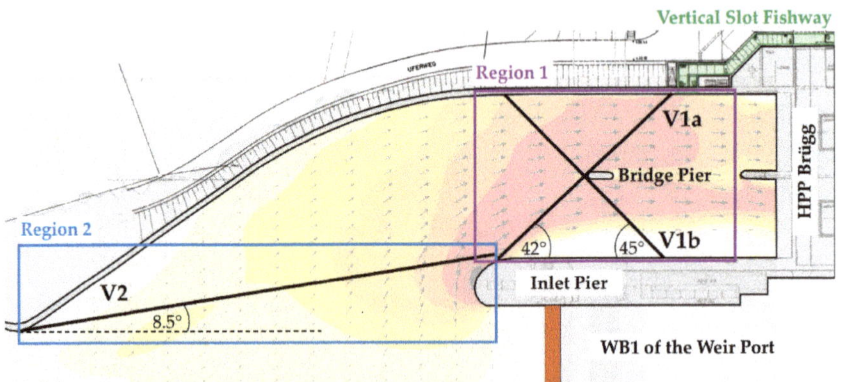


**Figure 6.** Potential configurations of FGS in the headrace channel of HPP Brügg, where horizontal angles to main approach flow direction are indicated. The existing vertical slot fishway for upstream migration is shown in green. The black arrows indicate the flow direction and the arrow lengths represent the velocity magnitude.

### 3.5. Flow Conditions in Cross Sections of Potential Positions

Possible FGS configurations were defined for each region determined in Section 3.4. Two configurations were selected for region 1 (V1a and V1b), while there is only one configuration in region 2 (V2). From a qualitative point of view, configuration V1b is unfavorable, as it is placed almost perpendicular to the flow vectors. However, this configuration was still considered as it allows for a very short bypass inside the inlet pier of the HPP down to the tailwater. The downstream end of configuration V2 is positioned on the left side of the inlet pier since a bypass through weir bay 1 (WB1) is not possible. The FGS of all configurations are positioned horizontally angled to the main flow direction, to favor the ratio of tangential to normal velocities being above one. For these three positions, the hydraulic conditions obtained from the numerical model were evaluated in the corresponding cross-section (Figure A2 in Appendix A).

### 3.6. Rating the Flow Conditions with Respect to Target Species

The hydraulic conditions were rated with respect to the target species. To this end, Equation (2) was applied and the ratio of tangential to normal velocities was determined (Figure 7). Ethohydraulic experiments with various fish species have shown that vertical bar racks with horizontally angled approach flow guide fish to the bypass under laboratory conditions [10,17]. Thus, a rather high value of sustained swimming speed of $v_{sust}$ = 0.7 m/s, but within the range of 0.36 to 0.85 m/s determined in Section 3.1, was applied to rate the hydraulics.

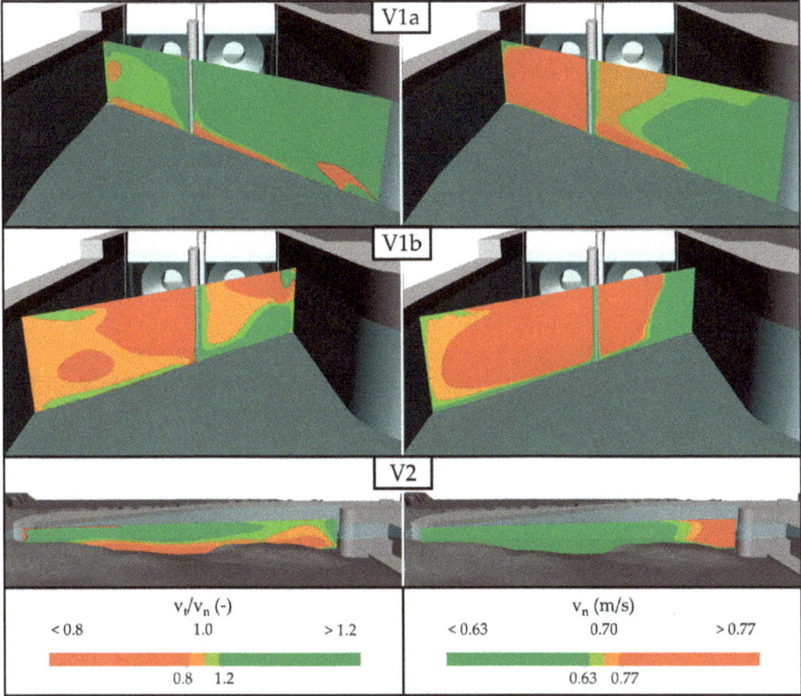



**Figure 7.** Rating the flow conditions with respect to the target species: (**left**) ratio of tangential ($v_t$) to normal velocities ($v_n$); if the ratio is above unity, good fish guiding efficiency can be expected; (**right**) normal velocities compared to $v_{sust}$ = 0.7 m/s. The light blue transparent planes indicate the water surface at water fraction $\alpha_w$ = 0.5.

The results show that a good guiding efficiency can be expected for configurations V1a and V2 since the tangential velocities of these configurations are mostly higher than the normal velocities. At configuration V1b the ratio between the tangential and normal velocities is largely below one, likely resulting in a lower guiding efficiency than for the other configurations. The rating of the flow conditions indicates that the normal velocities of configurations V1a and V1b exceed $v_{sust}$, especially for the left half of the rack, which incorporates the risk of fish being jammed at the rack. The same applies to configuration V2 at the inlet pier of the HPP, where high normal velocities occur due to flow separation (see also Figure 5).

Consequently, none of the investigated configurations meets the assessment criteria concerning the hydraulic conditions. Since no other FGS configuration is possible in region 2, this area has not been considered further. Configuration V1b was initially considered because it would allow a favorable bypass connection. Since the rating of the hydraulic conditions is very negative and there would hardly be a better evaluation even with an adjusted position, this configuration was not further investigated. Configuration V1a only fails with regard to the normal velocities at the left half of the rack. To reduce the normal velocities in front of the left turbine inlet, a milder horizontal angle is required. Thus, an adjusted configuration of V1a, named V1a*, was investigated where the horizontal angle $\alpha$ of the left half of the rack was reduced from 42° to 30°. For this adjusted configuration, the hydraulic conditions were again evaluated in the corresponding cross-section and rated with the same assessment criteria as described above (Figure 8). The adjusted configuration shows potential for a good fish guiding efficiency and the normal velocities are mostly below the limit value of $v_{sust}$. Consequently, this configuration meets the assessment criteria concerning the hydraulic conditions.

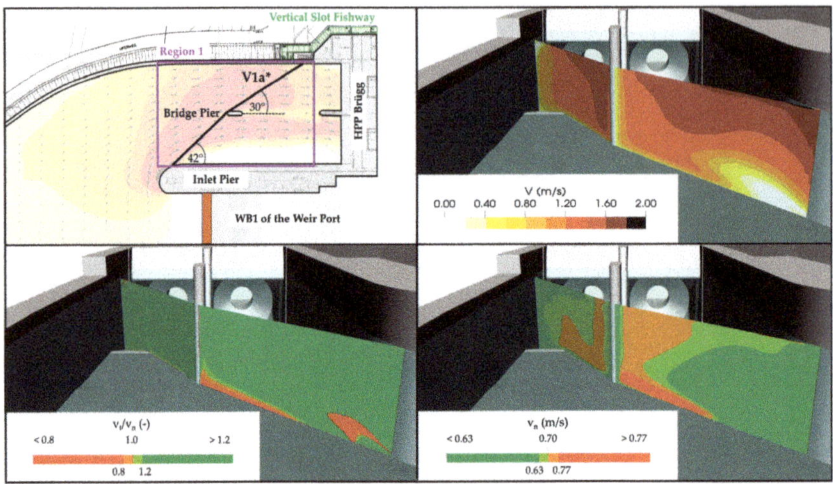


**Figure 8.** Assessment of the adjusted configuration V1a*: (**upper left**) adjusted FGS position; (**upper right**) absolute flow velocities in the corresponding cross-section; (**lower left**) ratio of tangential ($v_t$) to normal velocities ($v_n$); (**lower right**) normal velocities compared to $v_{sust}$ = 0.7 m/s. The light blue transparent planes represent the water surface at water fraction $\alpha_w$ = 0.5.

### 3.7. Geometric Properties of FGS

As described in Section 3.4 angled FGS are considered to have the greatest potential supporting the fish downstream migration. Configuration V1a* requires a rack length of approximately 42 m and a height of more than 6 m. According to Equation (1) a bar spacing $s_{bar}$ between 1.0 and 2.2 cm is required for FGS that are physically impassable. Since the proportion indices of salmon and barbel are above one, a high protective effect is to be expected with horizontal bars. For FGS with vertical bars (louvers and bar racks) there is no general rule about bar spacing. Bar spacings between a few centimeters up to more than 30 cm are reported in the literature [8–10]. For similar situations to HPP Brügg, a bar spacing in the order of 5–10 cm can be found in the same literature.

### 3.8. Structural Feasibility and Hydraulic Losses

Whether an FGS such as in configuration V1a* can be installed, it must be examined in detail in a further step, which is beyond the scope of this study. The bypass might be aligned similar to that of the existing fishway and its discharge can potentially be partly used as attraction flow for upstream migration. Special attention must be paid to the bridge pier to investigate how it affects the design and operation of the rack cleaning system.

Recent findings indicate lower hydraulic head losses for horizontal racks compared to vertically oriented bars, especially for hydrodynamic bar shapes [45,47]. Therefore the hydraulic head losses for configuration V1a* with horizontal bars were estimated based on the approach of Albayrak et al. [45], assuming rounded bar shape with a bar spacing of 1.5 cm, a bar thickness of 0.8 cm, and a bar depth of 8 cm. From the numerical simulations a mean approach flow velocity of 1.25 m/s was determined and the mean horizontal angle between the velocity vectors and the screen axis is about 30°. For such conditions the estimated hydraulic losses amount to 2–3 cm. To compare with the hydraulic head losses of configuration V1a* with vertical bars, we applied the approach of Beck et al. [19] for a curved bar rack with rounded bar tips. The calculation was performed for a bar spacing of 8 cm, a bar thickness of 0.8 cm, and a bar depth of 10 cm. The approach flow velocity and the mean angle between the velocity vectors were the same as for horizontal bars. With these assumptions, the resulting hydraulic head losses amount to 4–5 cm. In contrast to classical and modified angled bar racks with rather asymmetric and inhomogeneous flow fields downstream of the rack, curved bar racks (i) cause significantly reduced head losses and (ii) align the flow, ensuring a symmetric turbine admission flow [20]. This is particularly important for configuration V1a* as the rack is immediately upstream of the turbines. Note that in these examples, no bottom and top overlays were assumed.

However, these results must be considered with caution as the flow conditions in the headrace channel are highly inhomogeneous. It is the task of the HPP operator to define what hydraulic head losses due to FGS are acceptable. Thus, it cannot be definitively stated here whether the location V1a* represents a suitable FGS position concerning the two assessment criteria of structural feasibility and hydraulic losses.

## 4. Discussion

Silva et al. [3] developed a multidisciplinary approach providing a general framework to consider the interests of all stakeholders in the development and implementation of fishways. The presented approach in this work may be considered as a specification of this framework to address downstream fish migration by means of FGS involving essential aspects for a successful layout. With the proposed concept, we intend to provide a tool that combines traditional dimensioning principles of fish descent aids as reported in References [8–10,12] with computational fluid dynamics in order to evaluate suitable FGS configuration in a structured and straightforward way.

The proposed procedure uses the hydraulic conditions of the present state without FGS as a reference since the geometric representation of the FGS in the numerical models would require very small cell sizes (some millimeters) that would lead to a very high computing effort impeding an efficient layout procedure. On the one hand, omitting the FGS in the numerical simulations has the advantage that all possible FGS locations can be examined using the results of the same simulation, which reduces the computing effort considerably. On the other hand, the influences of the FGS on the surrounding flow field are not directly considered with this approach and have to be accounted for by expert judgement separately. As to the effect of FGS on the upstream flow field sensed by approaching fish, distinction should be made between FGS with vertical and horizontal bars. Based on findings on the flow impact of various kinds of vertical angled FGS in their vicinity [13,19,20,32], the suggested approach gives rather conservative results in terms of fish guidance efficiency, as the rack-parallel sweeping flow component tends to be intensified by the racks ("guidance effect"). Regarding angled FGS having horizontal bars without overlays, the flow field is quasi unaffected [47] so that the approach proposed herein is fully applicable. If bottom and/or top overlays are applied, the guidance effect for bottom- and surface-oriented fish is intensified, so that guidance efficiencies are again likely underestimated by the present approach. The overall assessment would thus not change considerably if FGS were included in the numerical models. However, further research on this topic involving numerical simulations and laboratory experiments for different FGS types is recommended to reduce uncertainties in the understanding of the impact caused by FGS on the upstream flow field.

The connection between hydraulics and the related behavioral responses of fishes is summarized in only one variable, which is $v_{sust}$. This may be criticized as a too restrictive simplification of the complex behavioral pattern of fish species. Indeed, many authors claim that there are gaps in knowledge regarding species-specific behavior of fish, which must be approached in future research [1,3,59–61]. For example, the velocity increase along the rack is another important criterion, which can also be assessed easily based on the CFD results. The United States Bureau of Reclamation (USBR) [8] states that louvers should be operated with bypass-to-approach flow velocity ratios between 1.1 and 1.5. The ratio for vertical and horizontal FGS is in a similar, slightly lower range, but is still the subject of ongoing research. Nevertheless, the presented approach is structured in such a way that it is possible and desirable to consider future findings when rating the hydraulic conditions.

Fish downstream migration devices must incorporate four features [3]. Firstly, they must shield fish from the zones of potential injury, e.g., turbine passage. Secondly, they have to guide the fish to an alternative migration corridor. Thirdly, fishes must find their way into the bypass and last, but not least, fishes should successfully migrate through the bypass into the tailwater without damage. Thus, after finding a suitable configuration of an FGS by means of the presented approach, a proper design of the bypass is a crucial prerequisite for supporting a successful downstream migration. For the implementation of a bypass in the numerical model, flow obstruction by the FGS has to be taken into account. As discussed above, depending on the type of angled FGS, a part of the approach flow is diverted parallel to the FGS. This is reasonable from a hydrodynamic point of view, because flow obstruction first occurs at the most upstream part of the FGS causing a small increase of the water surface elevation, i.e., the tangential velocity will increase being in favor of fish guidance efficiency. As a first attempt, a similar behavior might be simulated with the numerical model by using a baffle at the location of the FGS and specifying a hydraulic head loss estimated by empirical approaches as given in the literature. However, the capability of baffles to reproduce the flow field near FGS must be validated by further investigations, and further research on how to consider FGS in a simplified manner is needed. Once this has been achieved, the presented concept can be extended by carrying out additional CFD simulations for the identified potential configurations, in which the bypass and the FGS are resolved and directly taken into account.

## 5. Conclusions

The successful downstream guidance of fishes into the tailwater of HPP by fish guidance structures (FGS) is a complex challenge. In order to consider all essential aspects, a structured and straightforward conceptual approach for the evaluation of potential FGS positions by means of CFD was presented. The applicability of the concept was illustrated and discussed based on the case study of HPP Brügg in Switzerland. The use of numerical models instead of field measurements to investigate the flow situation allows for assessing the hydraulic conditions in any desired section and for different scenarios and operating conditions. Furthermore, numerical results can be used for additional investigations, such as the evaluation and optimization of the turbine admission flow. In the present conceptual approach, potential FGS positions can be assessed without implementing the FGS in the numerical model to facilitate efficient application. However, this states also the main limitation because the hydrodynamic influence of FGS on the local flow field cannot be directly simulated but has to be assessed indirectly by considering recent findings form experimental etho-hydraulic and detailed numerical investigations ("expert judgement"). Future research has to develop new approaches to include FGS in a simplified manner while reproducing the main flow features near FGS at reasonable computational expense. Such a contribution would make the present approach more powerful. Furthermore, interactions between hydraulics and related species-specific behavioral responses of fishes are the subject of current research and related new findings can easily be included in the concept.

Applying the approach to HPP Brügg showed that the conceptual approach serves as an optimization tool for potential FGS configurations. A suitable configuration concerning the hydraulic conditions could be determined for which the normal velocities exceed the sustained swimming speed $v_{sust}$ of the target species only locally and which shows a ratio of tangential to normal velocity above one. Further steps in the design process, such as checking structural stability of the FGS and technical aspects of the rack cleaning system, are not part of the case study herein but are important in real-world applications.

**Author Contributions:** The numerical modelling and writing of the manuscript was performed by L.F. under the supervision of D.F.V., S.K. and R.M.B., C.R.K. provided the technical data of the hydropower plant Brügg and supported the feasibility assessment of the different alternatives. S.K. provided the boundary conditions and processed the bathymetry data of the Aare river as well as the model geometry of the hydropower plant Brügg. All authors were involved in editing the manuscript.

**Funding:** This research received no external funding.

**Conflicts of Interest:** The authors declare no conflict of interest.

## Appendix A

**Table A1.** Boundary conditions of the first OpenFOAM model (S1) for all hydraulic variables. An extended definition of their numerical implementation can be found in [62].

| Boundary | $\alpha_w$ | $u$ | $p$ | $k$ | $\varepsilon$ | *nut* |
|---|---|---|---|---|---|---|
| inlet water | fixedValue | flowrate-InletVelocity | fixedFlux-Pressure | fixedValue | fixedValue | calculated |
| inlet air | zeroGradient | noSlip | fixedFlux-Pressure | kqRWall-Function | epsilonWall-Function | nutkWall-Function |
| outlet water | fixedValue | inletOutlet | fixedValue | inletOutlet | inletOutlet | calculated |
| outlet air | inletOutlet | fixedValue | zeroGradient | inletOutlet | inletOutlet | calculated |
| atmosphere | inletOutlet | pressureInlet-OutletVelocity | totalPressure | inletOutlet | inletOutlet | calculated |
| walls | zeroGradient | noSlip | fixedFlux-Pressure | kqRWall-Function | epsilonWall-Function | nutkWall-Function |

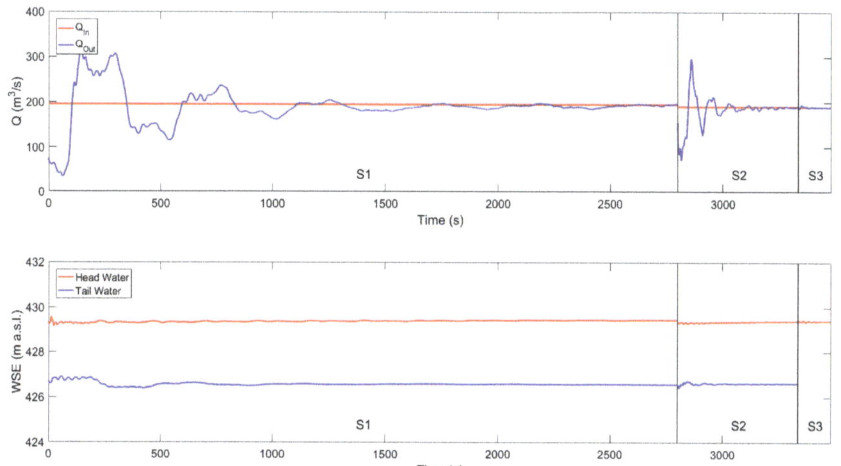

**Figure A1.** Assessment of the convergence to a steady state for the three simulations (S1), (S2) and (S3): (**top**) mass balance of water over time; (**bottom**) water surface elevation over time in the head and tailwater of the HPP. For the simulation (S3), the tail water was not included in the model.

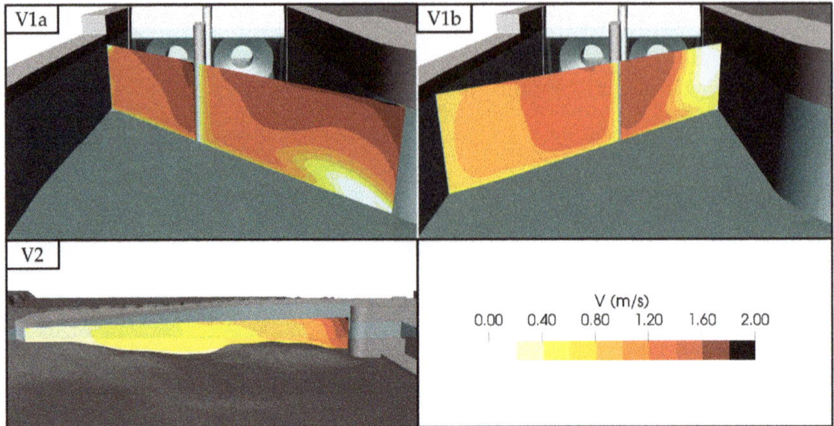

**Figure A2.** Hydraulic conditions in the cross-sections of the defined FGS positions. The light blue transparent planes indicate the water surface at water fraction $\alpha_w = 0.5$.

## References


1. Katopodis, C.; Aadland, L.P. Effective dam removal and river channel restoration approaches. *Int. J. River Basin Manag.* **2006**, *4*, 153–168. [CrossRef]

2. Lynch, A.J.; Myers, B.J.E.; Chu, C.; Eby, L.A.; Falke, J.A.; Kovach, R.P.; Krabbenhoft, T.J.; Kwak, T.J.; Lyons, J.; Paukert, C.P.; et al. Climate Change Effects on North American Inland Fish Populations and Assemblages. *Fisheries* **2016**, *41*, 346–361. [CrossRef]

3. Silva, A.T.; Lucas, M.C.; Castro-Santos, T.; Katopodis, C.; Baumgartner, L.J.; Thiem, J.D.; Aarestrup, K.; Pompeu, P.S.; O'Brien, G.C.; Braun, D.C.; et al. The future of fish passage science, engineering, and practice. *Fish Fish.* **2017**, *19*, 340–362. [CrossRef]

4. Clay, C.H. *Design of Fishways and Other Fish Facilities*, 2nd ed.; Lewis Publishers: Boca Raton, FL, USA, 1994.

5. FAO/DVWK. *Fish Passes—Design, Dimensions and Monitoring*; FAO: Rome, Italy, 2002.

6. Boes, R.; Albayrak, I. Fish guidance structures: New head loss formula and fish guidance efficiencies. In Proceedings of the 37th IAHR World Congress, Kuala Lumpur, Malaysia, 13–18 August 2017.

7. Szabo-Meszaros, M.; Navaratnam, C.U.; Aberle, J.; Silva, A.T.; Forseth, T.; Calles, O.; Fjeldstad, H.P.; Alfredsen, K. Experimental hydraulics on fish-friendly trash-racks: An ecological approach. *Ecol. Eng.* **2018**, *113*, 11–20. [CrossRef]

8. U.S. Department of the Interior. *Fish Protection at Water Diversions—A Guide for Planning and Designing Fish Exclusion Facilities*; Water Resources Technical Publication, Bureau of Reclamation: Denver, CO, USA, 2006.

9. O'Keeffe, N.; Turnpenny, W.H. *Screening for Intake and Outfalls: A Best Practice Guide*; Environment Agency: Bristol, UK, 2005.

10. Kriewitz, C.R. Leitrechen an Fischabstiegsanlagen: Hydraulik und fischbiologische Effizienz ('Guiding racks at fish descent systems: Hydraulics and fish biological efficiency'). In *VAW-Mitteilungen 230*; Boes, R., Ed.; Laboratory of Hydraulics, Hydrology and Glaciology (VAW), ETH Zürich: Zürich, Switzerland, 2015. (In German)

11. Larinier, M.; Travade, F. Downstream migration: Problems and facilities. *Bull. Fr. Peche Piscic.* **2002**, *364*, 181–207. [CrossRef]

12. Ebel, G. *Fischschutz und Fischabstieg an Wasserkraftanlagen: Handbuch Rechen- und Bypasssysteme: Ingenieurbiologische Grundlagen, Modellierung und Prognose, Bemessung und Gestaltung ('Fish Protection and Fish Downstream Migration at Hydropower Plants: Handbook on Rack and Bypass Systems: Fundamentals of Engineering Biology, Modelling and Forecasting, Dimensioning and Design')*; Büro für Gewässerökologie und Fischereibiologie Dr. Ebel: Halle (Saale), Germany, 2013. (In German)

13. Raynal, S.; Chatellier, L.; Courret, D.; Larinier, M.; David, L. An experimental study on fish-friendly trashracks—Part 2. Angled trashracks. *J. Hydraul. Res.* **2013**, *51*, 67–75. [CrossRef]

14.  Bates, D.W.; Vinsonhaler, R. Use of Louvers for Guiding Fish. *Trans. Am. Fish. Soc.* **1957**, *86*, 38–57. [CrossRef]
15.  Amaral, S.V.; Winchell, F.C.; McMahon, B.J.; Dixon, D.A. Evaluation of angled bar racks and louvers for guiding silver phase American eels. *Am. Fish. Soc.* **2003**, *33*, 367–376.
16.  Electric Power Research Institute (EPRI); Dominion Millstone Laboratories (DML). *Evaluation of Angled Bar Racks and Louvers for Guiding Fish at Water Intakes*; Report No. 1005193; EPRI: Palo Alto, CA, USA, 2001.
17.  Albayrak, I.; Boes, R.M.; Kriewitz-Byun, C.R.; Peter, A.; Tullis, B. Fish guidance structures: Hydraulic performance and fish guidance efficiencies. *J. Ecohydraulics* **2019**. in review.
18.  Beck, C.; Albayrak, I.; Boes, R. Improved hydraulic performance of fish guidance structures with innovative bar design. In Proceedings of the 12th International Symposium on Ecohydraulics (ISE 2018), Tokyo, Japan, 19–24 August 2018.
19.  Beck, C.; Albayrak, I.; Meister, J.; Boes, R.M. Hydraulic performance of fish guidance structures with curved bars—Part 1: Head loss assessment. *J. Hydraul. Res.* **2019**. in review.
20.  Beck, C.; Albayrak, I.; Boes, R.M. Hydraulic performance of fish guidance structures with curved bars—Part 2: Flow fields. *J. Hydraul. Res.* **2019**. in review.
21.  Wilkes, M.; Baumgartner, L.; Boys, C.; Silva, L.G.M.; O'Connor, J.; Jones, M.; Stuart, I.; Habit, E.; Link, O.; Webb, J.A. Fish-Net: Probabilistic models for fishway planning, design and monitoring to support environmentally sustainable hydropower. *Fish Fish.* **2018**, *19*, 677–697. [CrossRef]
22.  Khan, L.A. A three-dimensional computational fluid dynamics (CFD) model analysis of free surface hydrodynamics and fish passage energetics in a vertical-slot fishway. *N. Am. J. Fish. Manag.* **2006**, *26*, 255–267. [CrossRef]
23.  An, R.D.; Li, J.; Liang, R.F.; Tuo, Y.C. Three-dimensional simulation and experimental study for optimising a vertical slot fishway. *J. Hydro-Environ. Res.* **2016**, *12*, 119–129. [CrossRef]
24.  Quaranta, E.; Katopodis, C.; Revelli, R.; Comoglio, C. Turbulent flow field comparison and related suitability for fish passage of a standard and a simplified low-gradient vertical slot fishway. *River Res. Appl.* **2017**, *33*, 1295–1305. [CrossRef]
25.  Plymesser, K.; Cahoon, J. Pressure gradients in a steeppass fishway using a computational fluid dynamics model. *Ecol. Eng.* **2017**, *108*, 277–283. [CrossRef]
26.  Fuentes-Perez, J.F.; Silva, A.T.; Tuhtan, J.A.; Garcia-Vega, A.; Carbonell-Baeza, R.; Musall, M.; Kruusmaa, M. 3D modelling of non-uniform and turbulent flow in vertical slot fishways. *Environ. Model. Softw.* **2018**, *99*, 156–169. [CrossRef]
27.  Quaresma, A.L.; Romao, F.; Branco, P.; Ferreira, M.T.; Pinheiro, A.N. Multi slot versus single slot pool-type fishways: A modelling approach to compare hydrodynamics. *Ecol. Eng.* **2018**, *122*, 197–206. [CrossRef]
28.  Andersson, A.G.; Lindberg, D.E.; Lindmark, E.M.; Leonardsson, K.; Andreasson, P.; Lundqvist, H.; Lundstrom, T.S. A Study of the Location of the Entrance of a Fishway in a Regulated River with CFD and ADCP. *Mod. Simul. Eng.* **2012**, *2012*, 327929. [CrossRef]
29.  Gisen, D.C.; Weichert, R.B.; Nestler, J.M. Optimizing attraction flow for upstream fish passage at a hydropower dam employing 3D Detached-Eddy Simulation. *Ecol. Eng.* **2017**, *100*, 344–353. [CrossRef]
30.  Richmond, M.C.; Serkowski, J.A.; Ebner, L.L.; Sick, M.; Brown, R.S.; Carlson, T.J. Quantifying barotrauma risk to juvenile fish during hydro-turbine passage. *Fish. Res.* **2014**, *154*, 152–164. [CrossRef]
31.  Zangiabadi, E.; Masters, I.; Williams, A.J.; Croft, T.N.; Malki, R.; Edmunds, M.; Mason-Jones, A.; Horsfall, I. Computational prediction of pressure change in the vicinity of tidal stream turbines and the consequences for fish survival rate. *Renew. Energy* **2017**, *101*, 1141–1156. [CrossRef]
32.  Raynal, S.; Chatellier, L.; David, L.; Courret, D.; Larinier, M. Numerical Simulations of Fish-Friendly Angled Trashracks at Model and Real Scale. In Proceedings of the 35th IAHR World Congress, Vols I and Ii, Chengdu, China, 8–13 September 2013; pp. 2557–2566.
33.  Mulligan, K.B.; Towler, B.; Haro, A.; Ahlfeld, D.P. A computational fluid dynamics modeling study of guide walls for downstream fish passage. *Ecol. Eng.* **2017**, *99*, 324–332. [CrossRef]
34.  Mulligan, K.B.; Towler, B.; Haro, A.; Ahlfeld, D.P. Downstream fish passage guide walls: A hydraulic scale model analysis. *Ecol. Eng.* **2018**, *115*, 122–138. [CrossRef]
35.  Lundström, T.S.; Brynjell-Rahkola, M.; Ljung, A.L.; Hellstrom, J.G.I.; Green, T.M. Evaluation of Guiding Device for Downstream Fish Migration with in-Field Particle Tracking Velocimetry and CFD. *J. Appl. Fluid Mech.* **2015**, *8*, 579–589. [CrossRef]

36. Pavlov, D.S. *Structures Assisting the Migrations of Non-Salmonid Fish: USSR*; FAO Fisheries Technical Paper 308, 97 S; Food and Agriculture Organisation of the United Nations: Rome, Italy, 1989.

37. Stewart, J.; Ferrell, D.J. Escape panels to reduce by-catch in the New South Wales demersal trap fishery. *Mar. Freshw. Res.* **2002**, *53*, 1179–1188. [CrossRef]

38. Webb, P.W. *Hydrodynamics and Energetics of Fish Propulsion*; Department of the Environment Fisheries and Marine Service: Ottawa, ON, Canada, 1975.

39. Beamish, F.W.H. Swimming Capacity. In *Fish Physiology 7*; Hoar, W.S., Randall, D.J., Eds.; Academic Press: New York, NY, USA, 1978; pp. 101–187.

40. Wolter, C.; Arlinghaus, R. Navigation impacts on freshwater fish assemblages: The ecological relevance of swimming performance. *Rev. Fish Biol. Fish.* **2003**, *13*, 63–89. [CrossRef]

41. Peake, S. *Swimming Performance and Behavior of Fish Species to Newfoundland and Labrador: A Literature Review for the Purpose of Establishing Design Water Velocity Criteria for Fishways and Culverts*; Canadian Manuscript Report of Fisheries and Aquatic Sciences 1488-53872843; Department of Fisheries and Oceans: Ottawa, ON, Canada, 2008.

42. Beamish, F.W.H. Swimming Performance of Adult Sea Lamprey, Petromyzon-Marinus, in Relation to Weight and Temperature. *Trans. Am. Fish. Soc.* **1974**, *103*, 355–358. [CrossRef]

43. Peake, S.; Beamish, F.W.H.; McKinley, R.S.; Scruton, D.A.; Katopodis, C. Relating swimming performance of lake sturgeon, Acipenser fulvescens, to fishway design. *Can. J. Fish. Aquat. Sci.* **1997**, *54*, 1361–1366. [CrossRef]

44. Albayrak, I.; Kriewitz, C.R.; Hager, W.H.; Boes, R.M. An experimental investigation on louvres and angled bar racks. *J. Hydraul. Res.* **2018**, *56*, 59–75. [CrossRef]

45. Albayrak, I.; Maager, F.; Boes, R.M. An experimental investigation on fish guidance structures with horizontal bars. *J. Hydraul. Res.* **2019**. accepted.

46. Raynal, S.; Courret, D.; Chatellier, L.; Larinier, M.; David, L. An experimental study on fish-friendly trashracks—Part 1. Inclined trashracks. *J. Hydraul. Res.* **2013**, *51*, 56–66. [CrossRef]

47. Meister, J.; Fuchs, H.; Albayrak, I.; Boes, R.M. Horizontal bar rack bypass systems for fish downstream migration: State of knowledge, limitations, and gaps. In Proceedings of the 12th International Symposium on Ecohydraulics (ISE 2018), Tokyo, Japan, 19–24 August 2018.

48. Greenshields, C.J. OpenFOAM: The Open Source CFD Toolbox—Programmer's Guide; OpenFOAM Foundation Ltd. 2015. Available online: http://foam.sourceforge.net/docs/Guides-a4/ProgrammersGuide.pdf (accessed on 19 March 2019).

49. Ubbink, O. *Numerical Prediction of Two Fluid Systems with Sharp Interfaces*; Department of Mechanical Engineering Imperial College of Science, Technology & Medicine, University of London: London, UK, 1997.

50. Holzmann, T. *Mathematics, Numerics, Derivations and Openfoam®*, 4th ed.; Holzmann CFD: Leoben, Austria, 2017.

51. Hirt, C.W.; Nichols, B.D. Volume of Fluid (Vof) Method for the Dynamics of Free Boundaries. *J. Comput. Phys.* **1981**, *39*, 201–225. [CrossRef]

52. Berberovic, E.; van Hinsberg, N.P.; Jakirlic, S.; Roisman, I.V.; Tropea, C. Drop impact onto a liquid layer of finite thickness: Dynamics of the cavity evolution. *Phys. Rev. E* **2009**, *79*. [CrossRef]

53. Federal Act on the Protection of Waters of Switzerland (Gewässerschutzgesetz). SR 814.20 of 24 January 1991. 1 January 2017. Available online: https://www.admin.ch/opc/en/classified-compilation/19910022/201701010000/814.20.pdf (accessed on 19 March 2019).

54. Morais, P.; Daverat, F. *An Introduction to Fish Migration*; Taylor & Francis Group, LLC: Boca Raton, FL, USA; London, UK, 2016.

55. Hunt, P.C.; Jones, J.W. Population Study of Barbus-Barbus (L) in River Severn, England. 2. Movements. *J. Fish. Biol.* **1974**, *6*, 269–278. [CrossRef]

56. Baras, E.; Lambert, H.; Philippart, J.C. A Comprehensive Assessment of the Failure of Barbus-Barbus Spawning Migrations through a Fish Pass in the Canalized River Meuse (Belgium). *Aquat. Living Resour.* **1994**, *7*, 181–189. [CrossRef]

57. Höttges, A.; Sausen, S. Einsatz von Fischleitrechen am Aare-Kraftwerk Brügg ('Use of Fish Guiding Screens at the Aare Power Plant in Brügg'). Master' Thesis, Laboratory of Hydraulics, Hydrology and Glaciology (VAW), ETH Zürich, Zürich, Switzerland, 2016, unpublished. (In German)

58. BKW. *Vorstudien—Sanierung Fischgängigkeit am WKW Brügg ('Prestudies—Remediation of fish migration at the HPP Brügg')*; Bielersee Kraftwerke: Biel, Switzerland, 2016. (In German)

59. Katopodis, C. Developing a toolkit for fish passage, ecological flow management and fish habitat works. *J. Hydraul. Res.* **2005**, *43*, 451–467. [CrossRef]

60. Romao, F.; Quaresma, A.L.; Branco, P.; Santos, J.M.; Amaral, S.; Ferreira, M.T.; Katopodis, C.; Pinheiro, A.N. Passage performance of two cyprinids with different ecological traits in a fishway with distinct vertical slot configurations. *Ecol. Eng.* **2017**, *105*, 180–188. [CrossRef]

61. Wilkes, M.A.; Webb, J.A.; Pompeu, P.S.; Silva, L.G.M.; Vowles, A.S.; Baker, C.F.; Franklin, P.; Link, O.; Habit, E.; Kemp, P.S. Not just a migration problem: Metapopulations, habitat shifts, and gene flow are also important for fishway science and management. *River Res. Appl.* **2018**. [CrossRef]

62. OpenFoam. Standard Boundary Conditions. Available online: https://www.openfoam.com/documentation/user-guide/standard-boundaryconditions.php (accessed on 2 February 2019).

 *sustainability*

*Article*

# Repulsive Effect of Stroboscopic Light Barriers on Native Salmonid (*Salmo trutta*) and Cyprinid (*Pseudochondrostoma duriense* and *Luciobarbus bocagei*) Species of Iberia

Joaquim Jesus [1,2,*], Amílcar Teixeira [3], Silvestre Natário [2] and Rui Cortes [1]

[1]   CITAB, Centre for the Research and Tecnhology of Agro-Environmental and Biological Sciences, University of Trás-os-Montes e Alto Douro, Quinta dos Prados, 5000-911 Vila-Real, Portugal; rcortes@utad.pt

[2]   OriginAL Solutions, Estrada do Cando, Casa da Fraga, 5400-010 Chaves, Portugal; snatario@gmail.com

[3]   Centro de Investigação de Montanha (CIMO), Instituto Politécnico de Bragança, Campus de Santa Apolónia, 5300-253 Bragança, Portugal; amilt@ipb.pt

*   Correspondence: jjesus@utad.pt; Tel.: +351-963310059

Received: 30 December 2018; Accepted: 27 February 2019; Published: 4 March 2019

**Abstract:** A repulsive effect, that some induced primary stimuli, like sound and light, is known to be provoked in fish behavior. In the present study, two strobe light frequencies, 350 flashes/minute and 600 flashes/minute, were tested in laboratorial conditions, using three native freshwater fish species of northern Portugal: Brown trout (*Salmo trutta*), Northern straight-mouth nase (*Pseudochondrostoma duriense*) and Iberian barbel (*Luciobarbus bocagei*). The results showed a differential repulsive behavior of the fish species to light stimulus, and particularly to a frequency of 600 flashes/minute. *S. trutta* presented the most repulsive behavior, whereas the *L. bocagei* showed less repulsion to the light stimulus. No relevant differences were found between pre-test and post-assessments, confirming a rapid recovery of natural fish behavior after the deterrent effect. The results highlighted the potential of behavioral barriers, particularly in salmonid streams, based on strobe light stimulus.

**Keywords:** underwater light; behavioral barriers; brown trout; endemic cyprinids; deterrent effect

---

## 1. Introduction

Mediterranean freshwater ecosystems and particularly native fish are severely threatened by human activities, such as river regulation, responsible for dramatic habitat modifications [1], leading to the reduction of habitats (e.g., breeding, feeding, and shelter) and increased biotic interactions (e.g., competition and predation) with the non-indigenous species [2]. In Iberia, a large number of small and large dams are responsible for the disruption of river connectivity, which can directly affect fish movements. The migratory reproductive routes of diadromous and inclusively potamodromous species to the spawning habitats are inhibited, with consequences for their survival [3,4], as well as the interruption of the migrations downstream of the cyprinids have great importance in the life cycle of these fish, especially in the trophic and refuge migrations [5–7]. Other movements of Iberian cyprinids, often in the downstream direction, may result also from severe droughts, where fish may look for refuge in pools with favorable morphological and physicochemical conditions, namely well-developed canopies [8]. The existence of certain barriers becomes a serious handicap for their survival. According to the Red Book of the Portuguese vertebrates [9], 69% of native freshwater fish species are extremely vulnerable, justifying the development of in-situ conservation measures.

Native salmonid (*S. trutta*) and cyprinid (*L. bocagei*, *P. duriense*) species of northern Iberia have higher displacements along the river, namely during the potamodromous reproductive seasons, and their interest for conservation (e.g., threatened populations) and exploitation (e.g., angling activity) purposes is relevant for different management plans. Furthermore these species show a vulnerability in River Douro catchment: Trout is restricted to the upper part of a few tributaries [10], negatively impacting the upstream and downstream movements [11], whereas the other two cyprinids are strongly impacted by regulation, coupled with the natural wide variability of flow, common in the Iberian Peninsula [12,13]. Moreover, Iberian rheophilic cyprinids have exigent habitat requirements [12], and frequently make displacements especially in spawning activity, during which they have to find well-oxygenated gravel-bed spawning areas but also to complete their life cycle [14].

Baras and Lucas [3] state that in regulated rivers, spawning grounds (breeding migrations upstream) and feeding and sheltering sites (trophic migrations downstream) can be at great distances, forcing these species to migrate further and further away. The authors concluding that potamodromy may correspond to an adaptive response for these species.

However, these movements are highly conditioned by the presence of obstacles, and high fish mortalities has been registered [15,16]. Abrupt changes in pressure, cavitation, shear forces, turbulence, and mechanical shock are some of the effects experienced by fish in the adduction to the hydroelectric turbine [17–19]. For these reasons, deterrent systems have been used in different management applications and physical and nonphysical barriers developed to prevent fish from spreading or to guide fishes away from sources of mortality [4,20,21].

Non-physical barriers can use behavioral and/or physiological stimuli to control fish movements, since fish may exhibit attraction or repulsion behavior, caused by various environmental stimuli including sound [15,22,23], light [20,21,24,25], electric [26,27], chemical [28,29], and mixed [30,31]). The efficiency of these non-physical barriers depends on the fish species, the environmental conditions, and the potential habituation to a particular stimulus [32,33].

Light has been used over the centuries, both as a repulsive/attractive stimulus, for fish, and other aquatic animals [4]. Luminous stimuli are particularly important to fish because they use vision for food, reproduction or to avoid predation, and responses to light are therefore crucial for their survival [34]. The use of strobe light on behavioral barriers for fish has demonstrated repulsive efficiency with some species, and at various tested frequencies [29,35–37]. However, even if there are studies concerning the movements of Iberian rheophilic cyprinids, they are limited to the influence of environmental factors on longitudinal displacements in rivers (e.g., Lucas & Baras [7]; Ovidio et al. [38]; Benitez & Ovidio [39]), and not to laboratory conditions to determine deterrent effects that can be transposed to natural conditions, in order to mitigate the effect of obstacles. Only the particular case of hydropeaking has been recently subject to indoor flume tests in Iberian barbel [40]. It is, therefore, an unprecedented study of great relevance, since the applied knowledge resulting from these experiments can contribute to the protection and safeguarding of these species through the deterrent effects of these stimuli on fish, avoiding their access, for example, to electric production turbines (upstream dam), pumping systems (downstream dam), and adductor systems, thereby reducing their mortality rate caused by the physical impacts typical of those structures [15]. Different light barriers had been used to protect native fish populations by re-routing them to the proper passages in hydroelectric power plants [41], and to avoid or, at least, slow down the spread of invasive non-indigenous species [4,42–45].

The main objective of the present study was to evaluate the repulsive behavioral response, in laboratory conditions in three native freshwater species of Northern Portugal: Brown trout *(S. trutta)*, a salmonid species, and two endemic cyprinids, Northern straight-mouth *(P. duriense)*, and Iberian barbel *(L. bocagei)*, exposed to two stroboscopic light frequencies, either in day and night periods. We hyphotetyze for different reactions considering the specific vision and trophic factors, characteristics of each taxa.

## 2. Materials and Methods

### 2.1. Study Site

The study took place in the fish farm facilities of governmental services, the Portuguese Conservation of Nature and Forest Institute (Posto Aquícola de Castrelos) in northeastern Portugal, which took place between May 2011 to December 2014. The laboratorial recirculating tanks were used to develop the experiments, supplied with a good water quality from a headstream, the River Baceiro. During experiments. The water temperature ranged from 14 to 16 °C and a high dissolved oxygen concentration (> 9.0 mg·L$^{-1}$) was maintained in the laboratorial tanks.

### 2.2. Target Species

Three native species of the Douro basin, *S. trutta*, *P. duriense*, and *L. bocagei* were used in the tests with the following biometric data: [Total length TL: mean ± SD (cm); Mass M: mean ± SD (g))]: (1) *S. trutta*-TL: 16.1 ± 2.3 cm; M: 59.2 ± 23.9 g; (2) *P. duriense*-TL: 12.4 ± 3.2 cm; M: 21.7 ± 11.6 g; and (3) *L. bocagei*-TL: 12.9 ± 2.5 cm; M: 32.8 ± 23.1 g. The fish were captured by electrofishing (Hans Grassl ELTII, DC 300/600 V) in the River Sabor (Douro basin) and transported in appropriated conditions of low temperature (T < 16 °C) and oxygenation (DO > 9.0 mg $O_2$·L$^{-1}$). Before the experimental trials, fish spent a quarantine week, after which they were distributed separately by different maintenance tanks, under similar water quality conditions with the test tank until the beginning of each trial (2 to 3 weeks). After the tests, the captured fish were in good health and were released in the same area of the river where they had been previously captured.

### 2.3. Test Stimuli

The trials were carried out under laboratory conditions and discriminated by each species, considering a group of 20 individuals replaced at each trial. To evaluate the reaction of fish it was induced a strobe stimuli: 350 flashes/minute, with opening of 30 ms and 984 lx for the daytime period and 739 lx (1 lx = 1 lm/m$^2$) for the night period and 600 flashes/min with aperture of 30 ms and 1226 lx for the daytime period and 739 lx for the night period. Lux values were recorded underwater (0.3 m distance) with Lux Meter RS-PRO Model 180-7133 with an accuracy of 4%.

### 2.4. Experimental Design

Experimental rectangular test tanks, with dimensions of 1.5 × 1.0 × 0.5 m and a capacity of 750 L were used, with oxygenation system and temperature control. The walls of the tanks, with the exception of the strobe lamp window, were coated with black matte paper and by avoiding visual influences from outside. The placement of a central black span (Figure 1) 0.70 × 0.36 × 0.45 m, served to promote the circular movement of fish. To create the adequate water flow conditions in the tanks (30 L·min$^{-1}$), a distributed pumping system was fitted and adjusted to the available internal space in order to stimulate the rheophilic behavior of these species. Each aquarium was equipped with a strobe light placed in the middle of one of the lateral corridors, where the fish counting line was defined (Figure 1). The test tanks were equipped with two video cameras (Model Sony 600TVLines, underwater camera), involving an infra-red system for night recording of fish movement. One of the video cameras (video 1—Figure 1) was placed just above the counting line, thus allowing a recording of the passage of the fish. The second video camera (video 2—Figure 1) was placed close to the test tank, allowing a more comprehensive video recording of the entire tank (Figure 1).

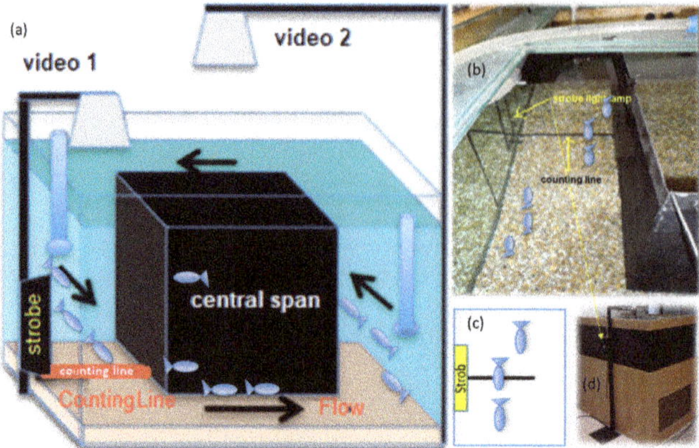



**Figure 1.** Design and conceptual model of the laboratorial test tank: (**a**) Functional diagram; (**b,c**) Counting line; (**d**) Test tank.

The same tests were performed for each target species, using the two mentioned frequencies during both nocturnal and diurnal periods. Each fish group (20 individuals) was placed in the aquarium two hours before starting each test for proper acclimation. The trials lasted for 130 min, corresponding to 60 min of pre-test (strobe light off), 60 min of Test (strobe light on), and 10 min of Post assessment (strobe light off). For each species and frequency tested (350/600), 3 replications were performed, both in the daytime and at night, totaling 36 trials for the three species (720 fish). The efficacy of each frequency was checked by video analysis and by direct observation of fish passages at the counting line, by comparing the means of the number of passages of the pre-test (strobe light off) to the means of the number of passages of the treatment test (strobe light on). Fish were counted each minute by freezing the video images, resulting in counts accumulated of 5 min. Data were organized in 12 periods of 5 min, completing the hour of pre-test and treatment tests. For post assessment tests only 2 periods of 5 min were considered. Through the comparative analysis between the counts obtained in the pre-test, and test, the behavioral response of each species was analyzed.

The relative efficiency corresponds to the fraction (percentage) that was affected repulsively by each the strobe light stimulus, and was calculated according to the following formula:

$$\text{Relative efficiency } (\%) = \left(1 - \left(\frac{\text{no. fishes counting line} - \text{tests (strobe on)}}{\text{no. fishes counting line} - \text{pre-tests (strobe of)}}\right)\right) \times 100$$

*2.5. Statistical Analysis*

The statistical analyses were initially performed using the Shapiro-Wilk procedure to test if data had a Gaussian distribution and by the Bartlett test to verify the homogeneity of variance. Since normality assumptions were not verified, no parametric tests were applied.

In the first test, a 2-way permutational multivariate analysis of variance (PERMANOVA) (type-III) was conducted, which was applied to the overall data (involving simultaneously day and night periods), considering the following factors: Stroboscopic light effect (2 levels: off and light stimuli), light frequencies (2 levels: 350 flashes/min and 600 flashes/min), and fish species (3 levels: *S. trutta*, *P. duriense*, and *L. bocagei*) all fixed factors used to assess the effect of each variable on the deterrent behavior response. The analysis included a random factor with different sets of fishes tested (6 levels). It used a balanced cross-design of the fixed factors. Additional PERMANOVA pairwise comparisons were used to statistically analyze the reaction of fish species. These computations are based on a similarity matrix using Euclidean distances.

In the second approach a 3-way PERMANOVA (type-III) was used for each fish species, taking into account the following factors: Stroboscopic light effect (2 levels: off and light stimuli), light frequencies (2 levels: 350 flashes/min and 600 flashes/min), and day period (2 levels: day and night), as fixed factors, in order to verify the individual fish species responses.

All PERMANOVAs were performed with 999 permutations on the basis of Euclidean distances (Anderson 2001). PERMANOVAs were computed with PRIMER 7 & PERMANOVA+ (Primer-E, UK) for Windows.

## 3. Results

The global results of 2-way PERMANOVA showed a significant effect of the stroboscopic light ($p < 0.01$) and of the fish species ($p < 0.01$), but not of frequency type ($p > 0.05$). Only the interaction between species and the considered assemblages was significant ($p < 0.05$). The pair-wise tests for fish species allowed significant differences ($p < 0.05$) between *S. trutta* and *L. bocagei* and between *P. duriense* and *L. bocagei* to be determined, but not between *S. trutta* and *P. duriense*.

Based on the 3-way PERMANOVA tests developed for each fish species, it was possible to highlight the following results: (1) *S. trutta*- the light effect (Pseudo F = 22.89, $p < 0.001$) and light frequencies (Pseudo F = 3.46, $p < 0.05$) effects were significant; (2) *P. duriense*- only light effects (Pseudo F = 6.84, $p < 0.01$) were significant; (3) *L. bocagei*- no significant differences were detected.

*S. trutta* appeared to be more sensitive to light stimulus comparatively with both endemic cyprinid species (Table 1). The relative efficiency was indeed more effective for *S. trutta* during the day, with 88% (600 flashes/min), 77% (350 flashes/min). The relative efficiencies between both cyprinid species were also distinct. The *L. bocagei* exhibited the lower deterrent effect relatively to the stroboscopic light effect, since the relative efficiencies were on average, and during day and night periods, only near 28%, and 25%, respectively, while for *P. duriense* these values were always above 40%, reaching 64.5% for 350 flashes/min/night (Table 1). Extract of video that shows the behavior of the fishes (trout in 600 flashes/min/day and nase 350 flashes/min/night) with the beginning of strobe stimulus can be seen in the "Supplementary Material".

**Table 1.** Relative efficiency (%) of the various stimuli tested in *S. trutta*, *P.duriense* and *L. bocagei*.

| Specie (dates) | Trial | | Counting Line [1] (n°. of fishes) | | Relative Efficiency (%) |
|---|---|---|---|---|---|
| | | | Pre-test | Testing | |
| *S. trutta* (19/5/2011–8/8/2011) | 350 flashes/min | Day | 2354 | 535 | 77.27 |
| | | night | 1606 | 819 | 49.00 |
| | 600 flashes/min | day | 1558 | 189 | 87.87 [2] |
| | | night | 1328 | 437 | 67.09 [3] |
| *P. duriense* (13/6/2011–8/7/2011) | 350 flashes/min | day | 2329 | 1376 | 40.92 |
| | | night | 2273 | 374 | 64.50 |
| | 600 flashes/min | day | 3156 | 1298 | 58.87 |
| | | night | 1740 | 975 | 43.97 |
| *L. bocagei* (16/6/2011–10/7/2011) | 350 flashes/min | day | 2538 | 1822 | 28.21 |
| | | night | 2466 | 2176 | 11.76 |
| | 600 flashes/min | day | 2428 | 1771 | 27.06 |
| | | night | 2055 | 1529 | 25.60 |

[1] Average of counted fish in the 3 repetitions, in each trial; [2] Maximum day; [3] Maximum night.

Figure 2 shows the time sequence of the fish counts (mean values − counting line) in each of the three trail tests for *S. trutta* and *P.duriense*: Pre-test treatment -and post-assessment tests. It should be noted that, after the light stimuli of the treatment tests (strobe light on), the fish species behavior during the post-assessment tests were similar to the behavior observed in the pre-test periods. The *L. bocagei* were not considered for this graphical presentation because it did not show significant repulsive behavior (relative efficiency values always lower than 30%).

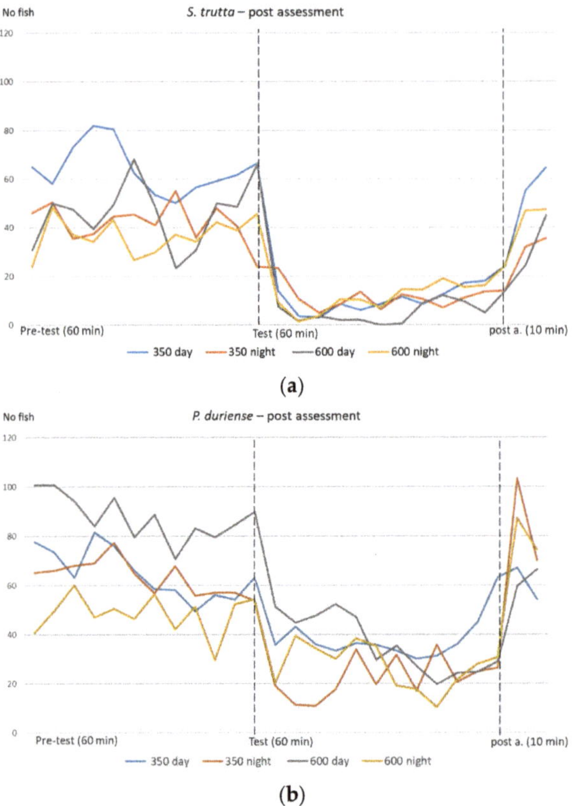


**Figure 2.** Fish count (mean values — counting line)) in the tests: 350 flashes/min/day (350 day), 350 flashes/min/night (350 night), 600 flashes/min/day (600 day), 600 flashes/min/night (600 night) with the species: *S. trutta* (**a**) *and P.duriense* (**b**): pre-test (60 min), test (60 min) and post-assessment (10 min).

## 4. Discussion

Non-physical barriers to guide fish movement can be a useful technique, particularly for the protection of freshwater native species. We studied brown trout to show their behavior [46,47]. However, for endemic cyprinids and inclusively for brown trout populations of Iberia, these techniques have not been applied to promote the success of fish downstream migration, through their behavior. Of course, the use of these techniques requires adequate research to determine the specific behavioral responses of each endemic species, since the reaction and response effects to different kind of stimuli remain unknown.

In fact, strobe lights have been successful in altering the behavior of fish and are the most widely underwater light system used for fish deterrent purposes [21,48,49], particularly for salmonid species [24,50–52]. Patrick et al. [53], in experiments with several species of freshwater and estuarine fish, concluded that stroboscopic light has a greater repulsive effect than continuous light. Sager et al. [32] reports that the repulsive efficacy of stroboscopic light in fish depends greatly on the frequency used (flashes/minute), and Coutant [51] adds that most strobe lights tested are set at frequencies of 300 flashes/minute or higher. The literature refers to the use of light stimuli in fish with strobe light at various frequencies: 60 flashes/minute [31,54], 86 flashes/minute [36]; 300 flashes/minute [35,37], and considering the evaluation of both the day-time period and the nocturnal period [29,35,55].

As verified in the results of the present study, *S. trutta* presented in general and particularly during the daytime a greater repulsive behavior than the cyprinid species, namely related to *L. bocagei*. The difference in repulsive behavior of stroboscopic light, between salmonid and cyprinids is not surprising, since Amaral et al. [55] also observed a superior repulsive behavior of salmonid Chinook salmon in relation to the repulsive behavior of the cyprinid *Ptychocheilus oregonensis* in the Yakima River (USA). Hansen et al. [56] verified that Chinook salmon, which exhibited a greater repulsive behavior to various light stimuli in the diurnal period and suggests, in its study, that this result may be related to vision and trophic factors. Nevertheless, our results also display clear behavioral differences between fish species, even inside the same family.

Salmonids are predominantly visual predators [57], feeding on small fish, crustaceans, and insects [58] during the day-time [59,60] and crepuscular periods [61], due to the higher ability to detect preys, which may be justified by the higher visual sensitivity during these periods, also observed in the present study for *S. trutta* (87.87%). Relatively to cyprinid species, it was observed for *P. duriense*, an absolute maximum value of the repulsive behavior (64.5%) and *L. bocagei* (28.21%). This result may also be related to the visual capacity, which is variable according to the food strategy, activity cycle (diurnal or nocturnal), water depth, and transparency [62,63]. *P. duriense*, includes a diet composed of plant material, organic debris, but also by small invertebrates [48]. It is a species that in trophic terms lays between *S. trutta* and *L. bocagei*, exhibiting visual capacity for predation, but the adapted mouth (lower corneal lip) allows to feed an algae and invertebrates, which may justify the results of the present study. *L. bocagei* presented a reduced sensitivity to the luminous stimulus for the two tested frequencies. The low sensitivity may be related to a higher dependence on, less illuminated benthic habitats and, consequently, presents sensory systems well developed like the olfactory and acoustic system [54].

Our results indicate that a repulsive effect of the light stimulus was maintained at a high rate throughout the 1-h of each treatment test, however there seems to be a slight habituation at the end of the test period, in both species, in the 350 flashes/min/day (Figure 2). The post-assessment results seem to reinforce the observation of a low resident effect, because immediately after the end of the light stimulus the fish return to its position similar to the pre-test situation (Figure 2). Mesquita et al. [54] found a habituation behavior of a cyprinid species *Danio rerio*, after 22.5 min, although it is important to note that they used a low frequency stroboscopic light (60 flashes/minute). However Hamel et al. [20] verified in the tests performed in-situ with strobes lights (360 flashes/minute and 450 flashes/minute) in Lake Oahe, that rainbow smel (*Osmerus mordax*) showed no habituation phenomena within the 4 h period of the trials.

The present work contributes to important knowledge of the behavior of native fish populations in the Iberian region, namely related with the objective of its conservation. The results, in general, indicate the high potential of stroboscopic light as a repulsive behavioral stimulus of fish, especially for the species *S. trutta* and *P. duriense*. The specific behavioral differences of each species found in the present study, if deepened, can allow the development of selective behavioral systems, of great utility in the safeguarding of these native species. Other trials, with longer periods, should then be carried out with these species, and with invasive species to add information to their habituation and avoidance patterns. Furthermore, this deterrent technology must be tested in natural conditions, using this effect alone or integrated with other behavioral barriers (e.g., bubble curtains, acoustic), in order to safeguard threatened fish species. The development of behavioral barriers specifically adapted to this endemic species of northern Iberia seems to be strategic, particularly in regulated rivers.

**Supplementary Materials:** The following are available online at http://www.mdpi.com/2071-1050/11/5/1332/s1.

**Author Contributions:** Conceptualization J.J., A.T. and R.C.; methodology J.J., S.N., and A.T.; formal analysis J.J., A.T., and R.C.; investigation, J.J., S.N., and A.T.; resources, J.J. and A.T.; writing—review and editing, J.J., A.T., and R.C.; Supervision, J.J. and A.T.

**Funding:** The project n°13737: Original Solutions—ENI and CITAB-UTAD was funded project: ANI/QREN/FEDER. European Investment Funds by FEDER/COMPETE/POCI—Operational Competitiveness and Internationalization Programme, under Project POCI-01-0145- FEDER-006958 and National Funds by FCT—Portuguese Foundation for Science and Technology, under the project UID/AGR/04033/2013.

**Acknowledgments:** The present study was technically supported by Instituto da Conservação da Natureza e Florestas (ICNF—Departamento Norte), namely the facilities at the fish farm "Posto Aquicola de Castrelos", for the execution of the experimental design.

**Conflicts of Interest:** The authors declare no conflict of interest.

## References

1. Hermoso, V.; Clavero, M. Threatening processes and conservation management of endemic freshwater fish in the Mediterranean basin: A review. *Mar. Freshw. Res.* **2011**, *62*, 244–254. [CrossRef]

2. Dudgeon, D.; Arthington, A.H.; Gessner, M.O.; Kawabata, Z.-I.; Knowler, D.J.; Lévêque, C.; Naiman, R.J.; Prieur-Richard, A.-H.; Soto, D.; Stiassny, M.L. Freshwater biodiversity: Importance, threats, status and conservation challenges. *Biol. Rev.* **2006**, *81*, 163–182. [CrossRef] [PubMed]

3. Baras, E.; Lucas, M.C. Impacts of man's modifications of river hydrology on the migration of freshwater fishes: A mechanistic perspective. *Int. J. Ecohydrol. Hydrobiol.* **2001**, *1*, 291–304.

4. Noatch, M.R.; Suski, C.D. Non-physical barriers to deter fish movements. *Environ. Rev.* **2012**, *20*, 71–82. [CrossRef]

5. Pitcher, T.J. Population Dynamics and Schooling Behaviour in the Minnow. Ph.D. Thesis, University of Oxford, Oxford, UK, 1971.

6. Fredrich, F. Preliminary studies on daily migration of chub (Leuciscus cephalus) in the Spree River. In *Underwater Biotelemetry, Proceedings of the First Conference and Workshop on Fish Telemetry in Europe, Liège, Belgium, 4–6 April 1995*; University of Liège: Liège, Belgium, 1995.

7. Lucas, M.; Baras, E. *Migration of Freshwater Fishes*; John Wiley & Sons: New York, NY, USA, 2008.

8. Pires, D.F.; Pires, A.M.; Collares-Pereira, M.J.; Magalhães, M.F. Variation in fish assemblages across dry-season pools in a Mediterranean stream: Effects of pool morphology, physicochemical factors and spatial context. *Ecol. Freshw. Fish* **2010**, *19*, 74–86. [CrossRef]

9. Cabral, M.J.; Almeida, J.; Almeida, P.R.; Dellinger, T.; Ferrand de Almeida, N.; Oliveira, M.E.; Palmeirim, J.M.; Queirós, A.I.; Rogado, L.; Santos-Reis, M. *Livro Vermelho dos Vertebrados de Portugal*; Instituto da Conservação da Natureza: Lisboa, Portugal, 2005.

10. Teixeira, A.; Cortes, R.M. PIT telemetry as a method to study the habitat requirements of fish populations: Application to native and stocked trout movements. In *Developments in Fish Telemetry*; Springer: Berlin/Heidelberg, Germany, 2007; pp. 171–185. [CrossRef]

11. Gosset, C.; Rives, J.; Labonne, J. Effect of habitat fragmentation on spawning migration of brown trout (*Salmo trutta* L.). *Ecol. Freshw. Fish* **2006**, *15*, 247–254. [CrossRef]

12. Santos, J.M.; Reino, L.; Porto, M.; Oliveira, J.; Pinheiro, P.; Almeida, P.R.; Cortes, R.; Ferreira, M.T. Complex size-dependent habitat associations in potamodromous fish species. *Aquat. Sci.* **2011**, *73*, 233–245. [CrossRef]

13. Clavero, M. Shifting baselines and the conservation of non-native species. *Conserv. Biol.* **2014**, *28*, 1434–1436. [CrossRef]

14. Ovidio, M.; Philippart, J.-C. Movement patterns and spawning activity of individual nase *Chondrostoma nasus* (L.) in flow-regulated and weir-fragmented rivers. *J. Appl. Ichthyol.* **2008**, *24*, 256–262. [CrossRef]

15. Abernethy, C.S.; Amidan, B.G.; Cada, G.F. *Simulated Passage through a Modified Kaplan Turbine Pressure Regime: A Supplement to "Laboratory Studies of the Effects of Pressure and Dissolved Gas Supersaturation on Turbine-Passed Fish"*; Pacific Northwest National Lab. (PNNL): Richland, WA, USA, 2002.

16. Rytwinski, T.; Algera, D.A.; Taylor, J.J.; Smokorowski, K.E.; Bennett, J.R.; Harrison, P.M.; Cooke, S.J. What are the consequences of fish entrainment and impingement associated with hydroelectric dams on fish productivity? A systematic review protocol. *Environ. Evid.* **2017**, *6*, 8. [CrossRef]

17. Čada, G.F.; Coutant, C.C.; Whitney, R.R. *Development of Biological Criteria for the Design of Advanced Hydropower Turbines*; DOE/ID-10578; US Department of Energy, Idaho Operations Office: Idaho Falls, ID, USA, 1997.

18. Čada, G.F. The development of advanced hydroelectric turbines to improve fish passage survival. *Fisheries* **2001**, *26*, 14–23. [CrossRef]

19. Hydro, B.C. *Developing Measures for the Aquatic Habitat Attribute in BC Hydro's 2005 Integrated Electricity Plan*; Ecofish Research Ltd.: Courtenay, BC, Canada, 2005.

20. Hamel, M.J.; Brown, M.L.; Chipps, S.R. Behavioral responses of rainbow smelt to in situ strobe lights. *N. Am. J. Fish. Manag.* **2008**, *28*, 394–401. [CrossRef]

21. Kim, J.; Mandrak, N.E. Effects of strobe lights on the behaviour of freshwater fishes. *Environ. Biol. Fishes* **2017**, *100*, 1427–1434. [CrossRef]

22. Sand, O.; ENGER, P.; Karlsen, H.E.; Knudsen, F.R. To Intense Infrasound In Juvenile Salmonids. In Proceedings of the American Fisheries Society Symposium, Phoenix, AZ, USA, 20–21 August 2001; Volume 26, pp. 183–193.

23. Maes, J.; Turnpenny, A.W.H.; Lambert, D.R.; Nedwell, J.R.; Parmentier, A.; Ollevier, F. Field evaluation of a sound system to reduce estuarine fish intake rates at a power plant cooling water inlet. *J. Fish Biol.* **2004**, *64*, 938–946. [CrossRef]

24. Maiolie, M.A.; Harryman, B.; Ament, B. Response of free-ranging kokanee to strobe lights. In Proceedings of the Behavioral Technologies for Fish Guidance: American Fisheries Society Symposium, Baltimore, MD, USA, 20–23 May 2001; p. 27.

25. Königson, S.; Fjälling, A.; Lunneryd, S.-G. Reactions in individual fish to strobe light. Field and aquarium experiments performed on whitefish (*Coregonus lavaretus*). *Hydrobiologia* **2002**, *483*, 39–44. [CrossRef]

26. Sparks, R.E.; Barkley, T.L.; Creque, S.M.; Dettmers, J.M.; Stainbrook, K.M. Evaluation of an electric fish dispersal barrier in the Chicago Sanitary and Ship Canal. In *American Fisheries Society Symposium*; American Fisheries Society: Bethesda, MD, USA, 2010; Volume 74.

27. Clarkson, R.W. Effectiveness of electrical fish barriers associated with the Central Arizona Project. *N. Am. J. Fish. Manag.* **2004**, *24*, 94–105. [CrossRef]

28. Kates, D.; Dennis, C.; Noatch, M.R.; Suski, C.D. Responses of native and invasive fishes to carbon dioxide: Potential for a nonphysical barrier to fish dispersal. *Can. J. Fish. Aquat. Sci.* **2012**, *69*, 1748–1759. [CrossRef]

29. Da Silva, L.G.M. Estudo de Sistemas para Repulsão de Peixes Como Alternativas de Mitigação de Impacto Ambiental em Usinas Hidrelétricas e Canais para Abastecimento de Água. Ph.D. Thesis, UFMG, Belo Horizonte, Brzail, 2010.

30. Ruebush, B.C. In-situ Tests of Sound-Bubble-Strobe Light Barrier Technologies to Prevent the Range Expansions of Asian Carp. 2011. Available online: http://hdl.handle.net/2142/26112 (accessed on 24 February 2019).

31. Stewart, H.A.; Wolter, M.H.; Wahl, D.H. Laboratory investigations on the use of strobe lights and bubble curtains to deter dam escapes of age-0 Muskellunge. *N. Am. J. Fish. Manag.* **2014**, *34*, 571–575. [CrossRef]

32. Sager, D.R.; Hocutt, C.H.; Stauffer, J.R., Jr. Estuarine fish responses to strobe light, bubble curtains and strobe light/bubble-curtain combinations as influenced by water flow rate and flash frequencies. *Fish. Res.* **1987**, *5*, 383–399. [CrossRef]

33. Johnson, R.L.; Simmons, M.A.; McKinstry, C.A.; Simmons, C.S.; Cook, C.B.; Brown, R.S.; Tano, D.K.; Thorsten, S.L.; Faber, D.M.; Lecaire, R. *Strobe Light Deterrent Efficacy Test and Fish Behavior Determination at Grand Coulee Dam Third Powerplant Forebay*; Pacific Northwest National Lab. (PNNL): Richland, WA, USA, 2005.

34. Lythgoe, J.N. *Ecology of Vision*; Clarendon Press: Oxford, UK, 1979.

35. Puckett, K.J.; Anderson, J.J. *Behavioral Responses of Juvenile Salmonids to Strobe and Mercury Lights*; Fisheries Research Institute: Seattle, WA, USA, 1987.

36. Richards, N.S.; Chipps, S.R.; Brown, M.L. Stress response and avoidance behavior of fishes as influenced by high-frequency strobe lights. *N. Am. J. Fish. Manag.* **2007**, *27*, 1310–1315. [CrossRef]

37. Silva, L.G.M.; Martinez, C.B.; Formagio, P. Uso de luz estroboscópica para repulsão de peixes de áreas de risco em usinas hidrelétricas. *SIMPÓSIO Bras. SOBRE PEQUENAS E MÉDIAS CENTRAIS HIDRELÉTRICAS* **2006**, *5*, 3–6.

38. Ovidio, M.; Parkinson, D.; Philippart, J.-C.; Baras, E. Multiyear homing and fidelity to residence areas by individual barbel (*Barbus barbus*). *Belg. J. Zool.* **2007**, *137*, 183–190.

39. Benitez, J.-P.; Ovidio, M. The influence of environmental factors on the upstream movements of rheophilic cyprinids according to their position in a river basin. *Ecol. Freshw. Fish* **2018**, *27*, 660–671. [CrossRef]

40. Costa, M.J.; Boavida, I.; Almeida, V.; Cooke, S.J.; Pinheiro, A.N. Do artificial velocity refuges mitigate the physiological and behavioural consequences of hydropeaking on a freshwater Iberian cyprinid? *Ecohydrology* **2018**, *11*, e1983. [CrossRef]

41. Coutant, C.C.; Whitney, R.R. Fish behavior in relation to passage through hydropower turbines: A review. *Trans. Am. Fish. Soc.* **2000**, *129*, 351–380. [CrossRef]

42. Taylor, R.M.; Pegg, M.A.; Chick, J.H. Response of bighead carp to a bioacoustic behavioural fish guidance system. *Fish. Manag. Ecol.* **2005**, *12*, 283–286. [CrossRef]

43. Vetter, B.J.; Cupp, A.R.; Fredricks, K.T.; Gaikowski, M.P.; Mensinger, A.F. Acoustical deterrence of silver carp (Hypophthalmichthys molitrix). *Biol. Invasions* **2015**, *17*, 3383–3392. [CrossRef]

44. Schilt, C.R. Developing fish passage and protection at hydropower dams. *Appl. Anim. Behav. Sci.* **2007**, *104*, 295–325. [CrossRef]

45. Perry, R.W.; Romine, J.G.; Adams, N.S.; Blake, A.R.; Burau, J.R.; Johnston, S.V.; Liedtke, T.L. Using a non-physical behavioural barrier to alter migration routing of juvenile chinook salmon in the sacramento–san joaquin river delta. *River Res. Appl.* **2014**, *30*, 192–203. [CrossRef]

46. Nedwell, J.R.; Turnpenny, A.W.; Lovell, J.M.; Edwards, B. An investigation into the effects of underwater piling noise on salmonids. *J. Acoust. Soc. Am.* **2006**, *120*, 2550–2554. [CrossRef] [PubMed]

47. Vowles, A.S.; Kemp, P.S. Effects of light on the behaviour of brown trout (Salmo trutta) encountering accelerating flow: Application to downstream fish passage. *Ecol. Eng.* **2012**, *47*, 247–253. [CrossRef]

48. Popper, A.N.; Carlson, T.J. Application of sound and other stimuli to control fish behavior. *Trans. Am. Fish. Soc.* **1998**, *127*, 673–707. [CrossRef]

49. Bullen, C.R.; Carlson, T.J. Non-physical fish barrier systems: Their development and potential applications to marine ranching. *Rev. Fish Biol. Fish.* **2003**, *13*, 201–212. [CrossRef]

50. Nemeth, R.S.; Anderson, J.J. Response of juvenile coho and chinook salmon to strobe and mercury vapor lights. *N. Am. J. Fish. Manag.* **1992**, *12*, 684–692. [CrossRef]

51. Coutant, C.C. Integrated, multi-sensory, behavioral guidance systems for fish diversion. In *Behavioral Technologies for Fish Guidance: American Fisheries Society Symposium*; American Fisheries Society: Bethesda, MD, USA, 2001; p. 105.

52. Taft, E.P.; Dixon, D.A.; Sullivan, C.W. Electric Power Research Institute's (EPRI) research on behavioral technologies. In *Behavioral Technologies for Fish Guidance: American Fisheries Society Symposium*; American Fisheries Society: Bethesda, MD, USA, 2001; p. 115.

53. Patrick, P.H.; Christie, A.E.; Sager, D.; Hocutt, C.; Stauffer, J., Jr. Responses of fish to a strobe light/air-bubble barrier. *Fish. Res.* **1985**, *3*, 157–172. [CrossRef]

54. De Oliveira Mesquita, F.; Godinho, H.P.; de Azevedo, P.G.; Young, R.J. A preliminary study into the effectiveness of stroboscopic light as an aversive stimulus for fish. *Appl. Anim. Behav. Sci.* **2008**, *111*, 402–407. [CrossRef]

55. Amaral, S.V.; Winchell, F.C.; Pearsons, T.N. *Behavioral Technologies for Fish Guidance: American Fisheries Society Symposium*; American Fisheries Society: Bethesda, MD, USA, 2001; pp. 125–144.

56. Hansen, M.J.; Cocherell, D.E.; Cooke, S.J.; Patrick, P.H.; Sills, M.; Fangue, N.A. Behavioural guidance of Chinook salmon smolts: The variable effects of LED spectral wavelength and strobing frequency. *Conserv. Physiol.* **2018**, *6*, coy032. [CrossRef] [PubMed]

57. Ruetz III, C.R.; Hurford, A.L.; Vondracek, B. Interspecific interactions between brown trout and slimy sculpin in stream enclosures. *Trans. Am. Fish. Soc.* **2003**, *132*, 611–618. [CrossRef]

58. Kottelat, M.; Freyhof, J. *Handbook of European Freshwater Fishes*; Publications Kottelat: Cornol, Switzerland, 2007.

59. McIntosh, A.R.; Townsend, C.R. Contrasting predation risks presented by introduced brown trout and native common river galaxias in New Zealand streams. *Can. J. Fish. Aquat. Sci.* **1995**, *52*, 1821–1833. [CrossRef]

60. Fraser, N.H.C.; Metcalfe, N.B. The costs of becoming nocturnal: Feeding efficiency in relation to light intensity in juvenile Atlantic salmon. *Funct. Ecol.* **1997**, *11*, 385–391. [CrossRef]

61. Bachman, R.A.; Reynolds, W.W.; Casterlin, M.E. Diel locomotor activity patterns of wild brown trout (*Salmo trutta* L.) in an electronic shuttlebox. *Hydrobiologia* **1979**, *66*, 45–47. [CrossRef]

62. Muntz, W.R. Visual adaptations to different light environments in Amazonian fishes. *Rev. Can. Biol. Exp.* **1982**, *41*, 35–46. [PubMed]
63. Pankhurst, N.W. The relationship of ocular morphology to feeding modes and activity periods in shallow marine teleosts from New Zealand. *Environ. Biol. Fishes* **1989**, *26*, 201–211. [CrossRef]

MDPI

St. Alban-Anlage 66

4052 Basel

Switzerland

Tel. +41 61 683 77 34

Fax +41 61 302 89 18

www.mdpi.com

*Sustainability* Editorial Office

E-mail: sustainability@mdpi.com

www.mdpi.com/journal/sustainability